Hyper Bio Assembler for 3D Cellular Systems

Tatsuo Arai • Masayuki Yamato • Fumihito Arai
Editors

Hyper Bio Assembler for 3D Cellular Systems

Springer

Editors
Tatsuo Arai
Department of Systems Innovation
Osaka University
Toyonaka
Japan

Fumihito Arai
Department of Micro-Nano Systems
Engineering
Nagoya University
Nagoya
Japan

Masayuki Yamato
Institute of Advanced Biomedical
Engineering and Science
Tokyo Women's Medical University
Tokyo
Japan

ISBN 978-4-431-55296-3 ISBN 978-4-431-55297-0 (eBook)
DOI 10.1007/978-4-431-55297-0

Library of Congress Control Number: 2015937971

Springer Tokyo Heidelberg New York Dordrecht London

Printed on acid-free paper

Springer Japan KK is part of Springer Science+Business Media (www.springer.com)

Preface

Hyper Bio Assembler for 3D Cellular Innovation (Bio Assembler) is a 5-year research project that started in July 2011 with the support of a Grant-in-Aid for Scientific Research on Innovative Areas from the Japanese Ministry of Education, Culture, Sports, Science and Technology. The aim of the Bio Assembler is to build 3D cellular systems that can function in vitro. This is an entirely new research field that has not been previously explored. The Bio Assembler will elucidate the principles of ultra-high-speed measurement, the manipulation techniques, and function expression of cells and tissues.

One of the features of the research project is the close collaboration between robotics experts, who specialize in micro-nanoscale measurement and manipulation, and biological and medical researchers, who construct stem cells and multicellular systems and aim for the medical application of these systems. Another feature is that these experts and researchers work together to create an innovative methodology (the Bio Assembler) that will be able to measure and sort target cells at high speed and construct 3D cellular systems in vitro. By understanding the various characteristics of 3D cellular systems and establishing a technology for constructing them, we hope to make a breakthrough in tissue engineering/biology and establish a new technical area of high-speed micro-nano robotics. Our final goal is to combine the work done in these three research areas in order to create artificial 3D cellular systems with morphologies and functions that can be used in tissue engineering.

We have so far explored this field and obtained new knowledge for creating an artificial 3D cellular system that can be used in regenerative medicine, and established innovative measurement-control technologies based on micro-nano robotics. We now have summarized the results of the project and published our achievements in the innovative field of the Bio Assembler as a book.

The structure of the book reflects the structure of the new research field consisting of three research areas. Part I, "Cell Sort Engineering", describes the research in the area of measurement and control of cell characteristics, which focuses on measuring the physical properties of cells taken from living organisms at high speed, and sorting the target cells that are useful in constructing cellular systems. Part II, "3D Cellular System Design", describes the research in the area of "3D cellular system assembly", which aims at shaping and assembling 3D cellular systems with complex morphologies. Part III, "Sociocytology", describes the research in the area of analysis and evaluation of 3D cellular systems, which analyzes and evaluates the growth, differentiation-inducement, morphogenesis controls, and transplantation responses of the constructed 3D cellular systems, and conducts functional elucidation as well as comparison and verification in vivo, looking for ways to apply our outcomes to regenerative medicine.

Part I Cell Sort Engineering

Understanding the various properties of cells is crucial. Moreover, ultra-high-speed sorting of active and uncommon cells is important in the construction of three-dimensional cellular systems. This part of our research focuses on methodologies for measuring and sorting cells at high speed. Recently, fluorescent imaging using an optical microscope is being widely used to study the properties of cells. However, cellular mechanical properties are not well investigated. In this part, measurement methods of cellular mechanical properties using high-speed micro-nano robot technologies are shown. In the case of floating cells or objects such as spheroids, microfluidic chips can be used for continuous and sequential measurement. Chapter 1 describes a mechanical characterization method of floating cells which uses an on-chip robot with a microfluidic chip. The mechanical properties of floating cells are measured one by one continuously on the chip. Chapter 2 describes dimensionless evaluation of the deformability of floating cells using a microfluidic chip. Through the dimensional analysis performed on the microfluidic chip, three dimensionless parameters determining stiffness-based deformability are derived, and a new index is introduced based on these parameters. Chapter 3 describes the real-time image processing for active sensing of cells flowing in the microfluidic channel using a high-speed vision system. On the other hand, for the continuous and sequential measurement of the mechanical properties of adherent cells, the atomic force microscope is a more suitable and powerful system. Chapter 4 describes high-throughput measurements of single-cell rheology using atomic force microscopy. As cells have a variety of different properties, sorting techniques are needed for the construction of three-dimensional cellular systems. Many methodologies for sorting of specific cells have been proposed. Chapter 5 describes the application of dielectrophoresis for the discrimination of cells with specific antigens expressed on a cellular membrane. This method is suitable for the high-speed sorting of floating cells. As an example of an application of methodologies for measuring and sorting cells at high speed, Chapter 6 analyzes the physical characteristics of hematopoietic cells using microfluidics-based devices, which ensure efficient detection of rare cell populations in circulating blood. New methodologies for measuring and sorting target cells at high speed are quite important for the Hyper Bio Assembler for 3D Cellular Systems. The methods introduced here will contribute to a better understanding of the various properties of cells and of the synthesis of complex biological tissues in the future.

Part II 3D Cellular System Design

High-speed robotic construction is crucial to achieve three-dimensional cellular systems in vitro. Part II of our research, "3D Cellular System Design", looks into ways of achieving the construction of 3D cellular systems that function as tissue, and proposes innovative construction technologies for 3D cellular systems based on micro-nano robotics, microfluidics, and MEMS, starting from the construction of simple 3D cellular systems and ultimately aiming to build tissue with a high oxygen requirement. Significant progress has been made by working together with front-line researchers toward this goal.

This part covers each step of the construction methodologies, from 1D and 2D structures, up to large-scale, complicated 3D tissues. In Chapter 8, we discuss typical expansion methods from 1D to 2D, which provide high-throughput cell assembly featuring heterogeneous hydrogels produced by using microfluidic devices. Single cells are encapsulated at high speed in hydrogel materials, which function as unit structures in constructing further large tissues. Expansion methods from 2D to 3D are shown in Chaps. 9, 10, and 14. Chapter 9 describes an on-chip fabrication technique based on the manipulation and self-assembly of 3D cellular structures. Chapter 10 describes fabrication of 3D cellular tissue utilizing MEMS technologies based on the folding method and a micro magnetic plate. In Chapter 14, we propose a device for rapid transfer of living cell sheets, named "cell scooper". The cell part assembly and the introduction of vascular-like structures are the most crucial technologies for building large tissues. Chapter 7 introduces cell manipulation and cellular part assembly using micro hand systems and other robotic devices. Chapter 12 proposes an approach for engineering 3D tissue with inner tubular structures, based on electrochemical cell detachment. This approach can be applied to fabricate vascularized bone, liver, and other tissues, by integrating parenchymal cells and other biomaterials. Furthermore, in Chapter 11, we show photo fabrication techniques that enable the construction of tissues using hydrogel. Quantitative evaluation of cell–hydrogel adhesion by advanced optical techniques is shown in Chapter 13.

For the "Hyper Bio Assembler for 3D Cellular Systems", the technology of constructing large tissue-like structures with target cells is very important. The new methodologies introduced in this part will contribute to the realization of 3D cellular systems in vitro, and provide a better understanding of artificial tissues.

Part III Sociocytology

In life sciences, advanced technologies including rapid DNA sequencing, flow cytometry, confocal laser scanning microscopy, and transgenic animals have made remarkable progresses. We expect the "Hyper Bio Assembler" project to provide deep insights into the underlying mechanisms of living beings, as well as diseases, and the basis for preventive, diagnostic, and therapeutic strategies, including regenerative medicine.

In Part III of our study, we brought together several papers describing exciting topics for research in the life sciences from the point of view of sociocytology. Although tissues, organs, and all the multicellular living beings are composed of various types of cells and the extracellular matrix, in traditional, conventional biology based on reductionism, purified and isolated cells were subjected to investigation, and soluble factors have been believed to be the major factor to play a role in regulating cell functions. Recently the importance of interactions between various types of cells and solid extramatrix have been given much attention. We would like to call this novel approach, "sociocytology", as we believe that by introducing micro nanorobotics tools into this new field, the revolutionary understandings of cells and biology will be realized. In Chap. 15, the surrounding microenvironments modulating the growth and morphology of 3D tissues on hydrogels with different mechanical stiffness are discussed. Chapter 16 introduces calcium phosphate-based scaffold materials used for regenerating hard tissues in bone defects. A microfabricated culture device for assembling the spheroids of chondrocytes in a 3D cellular construct is also described in this chapter. Chapter 17 presents an in vivo imaging method, which sheds light on the organogenesis of transplanted liver buds, derived from induced pluripotent stem (iPS) cells. Chapter 21 describes the in vitro analysis of the formation and mineralization of bone tissue, derived from bone marrow stromal cells (BMSCs). Chapter 22 describes cell sheet-based bio-assembler technologies used for creating multicellular, functional 3D tissues. With the aid of advanced robotics and biomechanics, these technologies should solve problems that former technologies were not able to solve. A bionic simulator-on-a-chip using organ explants of embryonic chick heart is presented in Chap. 18. In particular, mechanoregulation of cell and tissue functions and properties are quantitatively described here. Also we present "old" unanswered questions that can be re-examined with these increasingly powerful methods. In Chap. 19, an in vitro 3D imaging system for the tempo-spatial dynamics of cellular mechanics which uses fluorescent microscopy and digital image analysis is presented. Chapter 20 describes the in vitro imaging of cellular migration and invasion on 3D matrices for evaluating tumor invasion.

Through shedding light on sociocytology, major discoveries may be made in the future. We believe the Hyper Bio Assembler can provide the ideal tools for the rapid dissemination and discussion of all aspects of sociocytology, including the interactions among various cell types and between cells and the extracellular matrix.

Tatsuo Arai
Masayuki Yamato
Fumihito Arai

Contents

Part I
Cell Sort Engineering

Chapter 1
Cell Mechanical Characterization Based on On-Chip Robotics

Fumihito Arai and Shinya Sakuma

Abstract In this chapter, we propose "on-chip robotics" that is based on robot technologies and lab-on-a-chip devices to achieve high-speed, high-accuracy, and quantitative cell manipulation and analysis on a chip. We presents a high-throughput cell mechanical characterization method using a robot integrated microfluidic chip (robochip). The robochip contains a magnetically driven on-chip probe, a force sensor, and microchannels. The displacement reduction mechanism is adopted to the on-chip probe for high resolution positioning of the tip of the probe, and we achieved noncontact actuation of the on-chip probe with 0.18 μm in repetitive positioning. We show the automated mechanical characterization of oocyte. The throughput of the measurement is approximately 15–20 s per oocyte, and we conclud that high-throughput cellular mechanical characterization is achieved. We also show measurements of the viscoelastic properties of oocyte using the robochip. The proposed approach based on a robochip is a promising technology for contributing to the analysis of the physical and biological properties of cells.

Keywords On-Chip robotics · Mechanical characterization · Robot integrated microfluidic chip · Displacement reduction mechanism · Oocyte · Microfluidic chip · Probe · Force sensor · Microchannel · MEMS · Lab-on-a-chip

1.1 Introduction

Recently, along with the development of micro/nanotechnology, many researchers have focused on the measurement of the mechanical characteristics of not only living tissue but also single cells. Conventional bio-characterization processes tend to be carried out using probe devices that act as end effectors, attached to mechanical micromanipulators [1–5]. A mechanical micromanipulator based on robot technology is capable of providing precise operation, a high output force, and multiple degrees of freedom (multi-DOF). In addition, these probes offer functions such as applying mechanical stimuli to a cell, and measuring the cellular mechanical or electrical

F. Arai (✉) · S. Sakuma
Department of Micro-Nano Systems Engineering, Nagoya University, Chikusa-ku, Japan
e-mail: arai@mech.nagoya-u.ac.jp

T. Arai et al. (eds.), *Hyper Bio Assembler for 3D Cellular Systems*,
DOI 10.1007/978-4-431-55297-0_1

responses. Thus, the mechanical micromanipulators approach is widely used for medical and life science applications. Although multi-DOF manipulators can manipulate a target cell through the application of complex procedures because they offer high accuracy, a high power output, and flexibility in the manipulation, the manipulation requires a high degree of skill on the part of the operator because it requires that the manipulators be controlled in multiple degrees of freedom. Moreover, the manipulators are placed outside the cell culture environment, where they are exposed to air due to the relatively huge size of the manipulator. This leads to cell contamination issues. In addition, there are many potential disturbances such as unexpected fluid flows due to open spaces and this encumbers precise motion. As a result, the success rate, repeatability, and productivity of the manipulations tend to depend on the operator's skill. Therefore, there is a need for a high-throughput sensing technique that can consistently measure many cells, because it is normal to find variability among cells even under the same conditions. On the other hand, lab-on-a-chip devices based on a microfluidic chip have a closed environment that helps to prevent cell contamination as well as provide a stable environment for robot actuation [6–12]. Moreover, the cost of microfluidic chips is generally low enough for them to be disposable because they are fabricated using micro electro mechanical systems (MEMS) processes that support mass production. These microfluidic chips provide a stable laminar flow environment because of the low Reynolds number, and this contributes to the high-throughput cell manipulation and analysis. The manipulation of cells in the confined space of a microfluidic chip, therefore, offers great advantages in the field of bio-characterization.

Against this background, we are proposing "on-chip robotics" that is based on a micromechanical manipulator and lab-on-a-chip devices [13, 14]. Figure 1.1 shows an outline of the on-chip robotics fields. Robot technology capable of high-precision, high-output operation is applied to a microfluidic chip that offers a high level of environmental control. To realize on-chip robotics, the following three requirements must be satisfied.

i. Primary structured environment:
   Design and development of the components that are directly required for cell manipulation and analysis.
   e.g., microchannel, on-chip robot, on-chip sensor

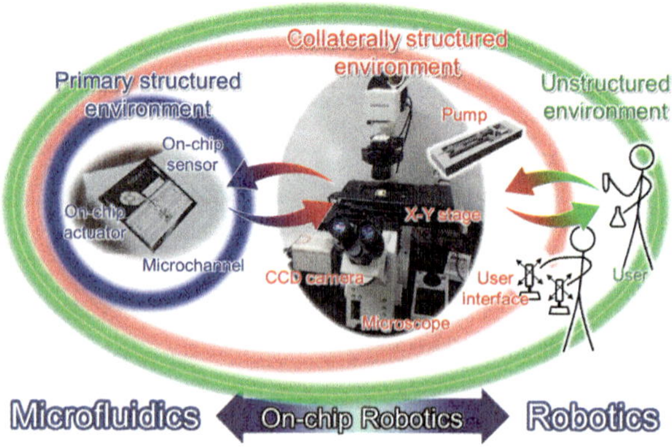

Fig. 1.1 Outline of on-chip robotics

ii. Collaterally structured environment:
Design and development of the components that are indirectly required to achieve the primary structured environment.
e.g., microscope, vision sensor, chip mount stage, flow control pump, on-chip robot/sensor control module, user interface
iii. Unstructured environment:
Design and development of components that are resilient to uncertainties.
e.g., usability, fail-safe, foolproof

Especially, in order to achieve high-speed, high-accuracy, and quantitative cell manipulation and analysis, it is necessary to build a primary, collaterally structured environment that can eliminate, as far as possible, uncertainty in its operation and analysis. In this chapter, we explain a cell mechanical characterization system that is based on the robot integrated microfluidic chip (robochip). Robochip provides a primary structured environment in a cell mechanical characterization system that offers a collaterally structured environment.

1.2 On-Chip Mechanical Characterization of Cell

1.2.1 On-Chip Mechanical Characterization of Cell

As mentioned in the first section, recent bioscience research has focused on the response and quality of cells. In particular, the mechanical stimulation of cells is important for analyzing their mechanical parameters. In general, the deformation of a single cell is required to evaluate its mechanical characteristics such as its spring coefficient and viscosity coefficient. The manipulation methods used in these studies have two important features: the force and resolution of the manipulation. Previously, several methods have been proposed for microscale cell manipulation, such as atomic force microscopy (AFM) [15], optical tweezers [16, 17], and magnetic actuators [18–20]. Regarding the AFM method, the manipulation force is in the nanonewton order, and the manipulation resolution is in the angstrom order. The AFM method can precisely measure the shape of a cell and its mechanical parameters. However, it is difficult to apply to floating cells because the method relies on a scanning technique. Moreover, high-throughput sensing is difficult because conventional systems are not suitable for the continuous measurement of floating cells in a microfluidic chip. On the other hand, optical tweezers can be used to measure floating cells by indirectly manipulating them by means of noncontact actuation using microtools. In this case, the manipulating force is in the piconewton order, and the manipulation resolution is in the nanometer order. This method can manipulate cells with a high resolution in a confined space. However, a force in the order of nanonewtons to millinewtons is generally required for the manipulation needed for cell deformation. Magnetic actuators are also used in the flowing environment of a microfluidic chip for the analysis of floating cells. Magnetic fields can be a suitable force source for an on-chip robot because of their noncontact

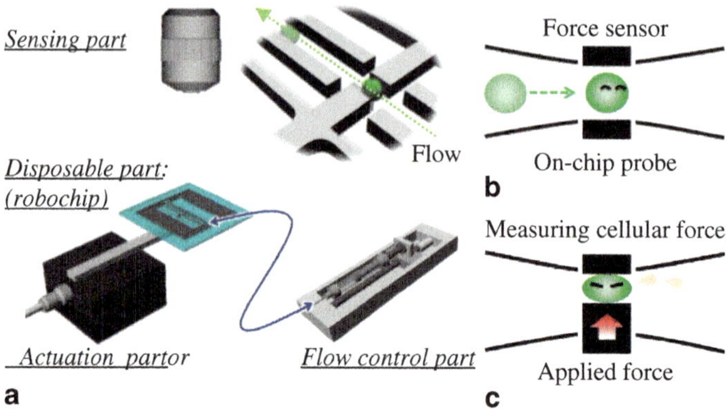

Fig. 1.2 Cell mechanical characterization system: **a** outline of the system, **b** transportation of target cell, and (**b**) cell

drive, low invasiveness with respect to the cell, and minimal production cost. Thus, considerable research has been carried out on magnetic actuators such as microstirrers, pumps, valves, and cell manipulators. Previously, we proposed the use of a magnetically driven microtool (MMT) to automate cell manipulation. MMTs were fabricated from magnetic materials and were actuated in a microfluidic chip by the magnetic force produced by an electromagnetic coil or a permanent magnet placed outside the microfluidic chip. This approach enables the manipulation of cells in a closed microfluidic chip with a force in the order of millinewtons. In a previous study, Hagiwara et al. successfully achieved MMT positioning with a 1-μm order resolution [20]. A wide range of applications for cell manipulation can be achieved using the drive method in a microfluidic chip. However, the positioning accuracy of MMTs is difficult to improve because MMTs are driven by a magnet on a motorized stage, and the magnetic actuation method consists of a dead band area where MMTs cannot be moved despite the displacement of the external magnetic field. Therefore, it is necessary to improve the positioning accuracy when measuring the cellular force in a microfluidic chip.

Herein, we present an on-chip cell mechanical characterization system based on the robochip [13, 14]. Figure 1.2 shows the fundamental concept of the on-chip cellular force measurement system. The system consists of four parts: a sensor, an actuator, a disposable robochip, and a flow controller. To avoid complicated cell manipulation in multiple degrees of freedom, the target cells are transported to the manipulation point and are controlled by the flow induced by an external syringe pump as shown in Fig. 1.2b. The robochip contains an on-chip probe and a force sensor in the wall of a microchannel at the manipulation point. The on-chip probe is described in detail in the following section. In short, the on-chip probe is driven by a permanent magnet, which is actuated by the magnetic force of another permanent magnet placed on the motorized stage. Using this configuration, we can easily obtain a high output force in the order of millinewtons. The transported target cell is deformed by the on-chip probe, and the cellular reaction force is determined by

Table 1.1 Classification of actuation methods of magnetically driven actuators

Classification	Characteristics
A. Control method of an external magnetic field	
1. Fixed electromagnet	Magnetic field is controlled by electromagnets
	ON/OFF control by switching the magnetic pole
	Continuous control by controlling the distribution of the magnetic field
2. Fixed permanent magnet	Magnetic fields are applied by permanent magnets.
	Continuous fixed actuators
3. Moving electromagnet	Magnetic fields are controlled by electromagnets.
	Controls the position of the electromagnet by actuators (piezoelectric, AC/DC motor, etc.) and controls the magnetic pole and magnetic field
4. Moving permanent magnet	Magnetic fields are applied by permanent magnets.
	Controls the position of the permanent magnet by actuators (piezoelectric, electromagnet, AC/DC motor, etc.)
B. Magnetic characteristics of actuator	
1. Variable reactance (VR) type	Uses materials with a high magnetic permeability
2. Permanent magnet (PM) type	Uses a permanent magnet
C. Degree of freedom	
1. 1-DOF x, (y, z)—lateral	Continuous, ON/OFF
2. 1-DOF rotation	Continuous, step
3. 2-DOF xy–lateral, lateral + rotation	Continuous, ON/OFF, step
4. Multi-DOF	Continuous, ON/OFF, step

measuring the displacement of the on-chip force sensor using a CCD camera attached to the microscope. Because the robochip is disposable, our concept can be used for biomedical applications.

1.2.2 Displacement Reduction Mechanism

In this section, we describe the displacement reduction mechanism that is used to achieve highly accurate manipulation [13]. The magnetically driven microtools are classified in Table 1.1, according to the type of external magnetic field, actuator material, and number of degrees of freedom. To obtain a high output force with a high resolution during the continuous positioning of the actuator, a permanent magnet microtool (B-2) is actuated by an external magnetic field produced by a permanent magnet placed on a motorized stage (A-4). Although the magnetically driven approach can produce a high output force, the positioning accuracy is difficult to improve because of the so-called "dead band." Figure 1.3 shows an idealized image of

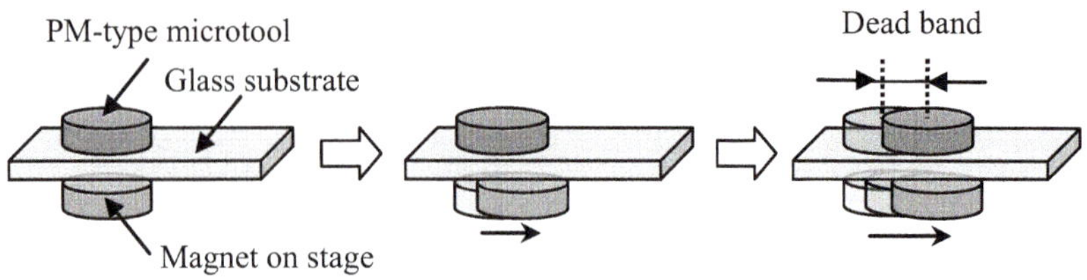

Fig. 1.3 Dead band when microtool is actuated

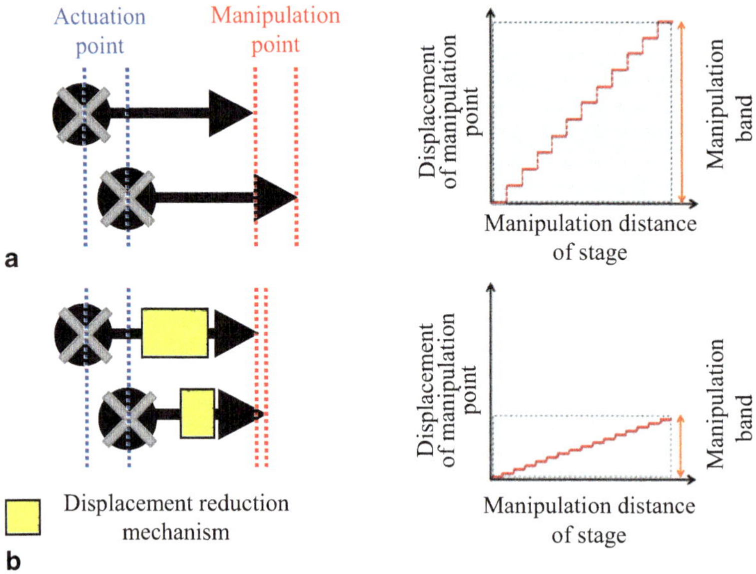

Fig. 1.4 Basic concept for high accuracy positioning: **a** without and **b** with displacement reduction mechanism

the dead band, where the microtools do not move despite a non-negligible displacement of the motorized stage. In this case, the displacement of the actuation point of the microtool is equal to that of its manipulation point, as shown in Fig. 1.4a. Since, there is a band area at the actuation point, the movement manifests itself as a stick-slip motion, and thus, the dead band limits the positioning accuracy. Figure 1.4b illustrates the basic concept of the displacement reduction mechanism. By using the proposed mechanism, the displacement of the actuation point including the dead band is reduced, and a high positioning resolution at the manipulation point is achieved. We have noted that the manipulation distance is large enough to deform a cell by designing an appropriate reduction rate between the displacement of the actuation point and the manipulation point, even though it is reduced by the mechanism.

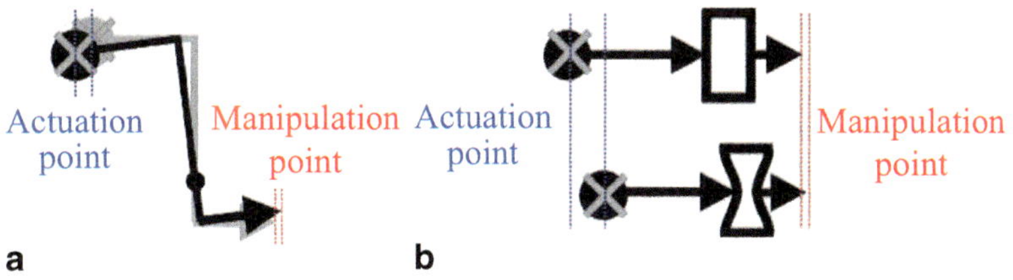

Fig. 1.5 Types of the displacement-reduction mechanisms: **a** cantilever type and **b** frame type

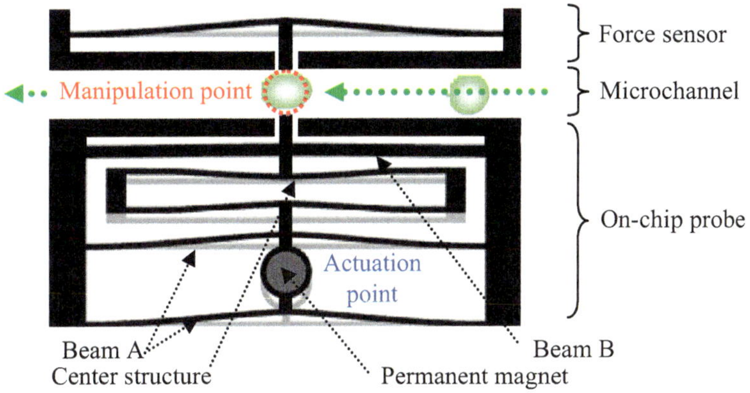

Fig. 1.6 Concept of cellular force measurement using an on-chip probe with a displacement-reduction mechanism

In order to reduce the displacement in a microfluidic chip, two general types of mechanism are considered: the cantilever type and the frame type, as shown in Fig. 1.5. The cantilever type can produce a large force through the application of leverage. However, it is difficult to achieve an ideal microscale hinge structure and robust manipulation in a flow environment because the fluctuation of the flow is also amplified by the leverage. On the other hand, the frame type is configured by combining simple structures such as beams, and can be actuated robustly in a flow environment by employing a parallel plate structure. In addition, only single degree of freedom lateral motion of the actuator is required to deform the cell. Therefore, the frame type is adopted with a parallel plate structure as the displacement reduction mechanism because magnetic actuation can produce sufficient power to deform the beam structure. Figure 1.6 shows the concept of cellular force measurement using a robochip that has an on-chip probe with a displacement reduction mechanism and a force sensor. The on-chip probe consists of beams (A, B), a center structure, and a permanent magnet to provide sufficient driving force to actuate the on-chip probe. The displacement of the actuation point including the dead band is reduced by combining the beams, and the target cell is deformed by the high-resolution on-

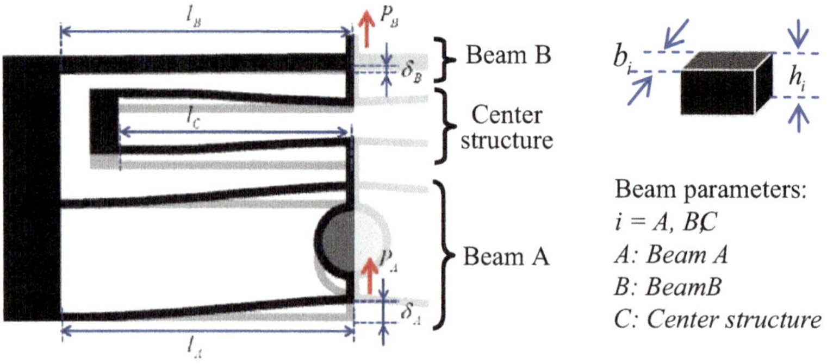

Fig. 1.7 Analytical model of on-chip probe

chip probe at the manipulation point. The features of robochip are summarized as follows.

1. The probe is actuated in a microfluidic chip with a high positioning resolution and a high output force.
2. Robust actuation from a disturbance is possible by employing a parallel plate structure.
3. The cellular force is measured using a force sensor placed opposite the on-chip probe.
4. The robochip is disposable and produced at low cost. Because the chip part is disposable, this concept is suitable for biomedical applications.

1.2.3 Analysis of On-Chip Probe with Displacement Reduction Mechanism

An analytical model of the on-chip probe is shown in Fig. 1.7. In this model, k_i, l_i, b_i, h_i, and E are the spring coefficients, the length, the width, the height, and Young's module of the beams, respectively. Subscript i indicates beams A, B, C. The spring coefficient of the beams is given by Eq. 1.1.

$$k_A = \frac{Eh_A b_A^{\ 3}}{12},\ k_B = \frac{Eh_B b_B^{\ 3}}{12},\ k_C = \frac{Eh_C b_C^{\ 3}}{12} \tag{1.1}$$

When the magnet is actuated by an external magnetic force, the displacements of beam A δ_A and beam B δ_B are expressed by Eqs. 1.2 and 1.3, respectively.

$$\delta_A = \frac{k_B + k_C}{(k_A + k_C)(k_B + k_C) - k_c^{\ 2}} P_A \tag{1.2}$$

$$\delta_B = \frac{k_C}{(k_A + k_C)(k_B + k_C) - k_c^{\ 2}} P_A \tag{1.3}$$

In this case, the manipulation force and the reduction ratio (δ_B/δ_A) are shown in Eqs. 1.4 and 1.5, respectively.

$$P_B = \frac{k_A}{k_B}\frac{1}{1+k_B/k_C}P_A \tag{1.4}$$

$$R_{Disp} = \frac{\delta_B}{\delta_A} = \frac{1}{1+k_B/k_C} \tag{1.5}$$

Equation 1.5 indicates that the reduction ratio is determined only by the ratio of stiffness between beam B and the center structure. If beam B and the center structure are fabricated by using the same material and have the same thickness, the reduction ratio can be determined from the shape of the structure. Consequently, a desired reduction ratio can be established by designing the shape of the structure. As a first step toward achieving cellular force measurement, an on-chip probe is designed for an oocyte with a diameter of approximately 150 μm. In this case, the reduction rate is required to be less than 1/5. In this chapter, the variables for the fabricated structure are as follows: l_A=8.5 mm, l_B=8.5 mm, l_C=7.3 mm, b_A=50 μm, b_B=58 μm, b_C=25 μm, h_A=h_B=h_C=200 μm. In this case, the reduction rate is approximately 1/7.5 (0.13).

1.2.4 Fabrication Process of Robochip

The separation of the on-chip probe and force sensor from the substrate surface is necessary to reduce the friction force between them. Therefore, a layer fabrication process is employed to realize the required gap. This fabrication process is based on a simple two-step procedure. First, the holder and device layers are fabricated in exactly the same way. The layers are then assembled to create a multi-layer structure. The device layer is composed of an on-chip probe, a force sensor, and a microchannel, and the hold layer is composed of a spacer pattern and a cage pattern, which holds the device layer in place. Figure 1.8 shows the process flow for the fabrication of the probe.

<Process flow of hold layer>

a. The first SU-8 (Nippon Kayaku Co., Ltd.) layer is patterned as a spacer layer after a Cr/Au is patterned on a glass substrate. The Cr/Au pattern is used to provide the alignment marks for the multi-step exposure of the holder layer. The thickness of the spacer layer defines the gap between the device structure and the substrate surface. In this case, the gap is designed to be 10 μm to avoid contact between the Si structure and the substrate surface because of the magnetic force and deflection caused by its own weight.
b. The second SU-8 layer is fabricated as a cage structure with a thickness of 220 μm. This structure holds the device layer.

<Process flow of device layer>

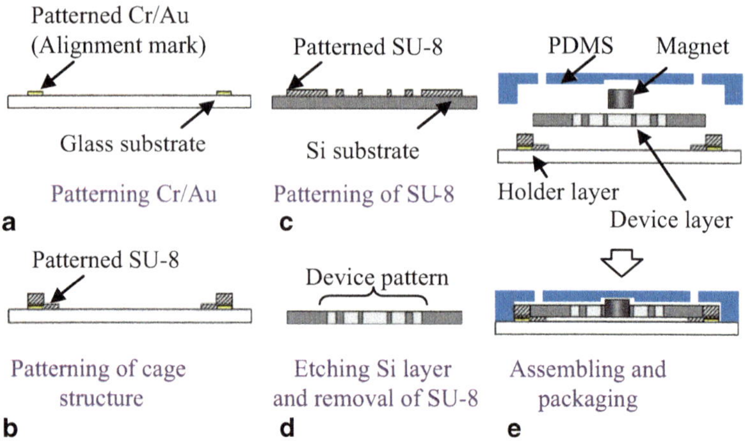

Fig. 1.8 Fabrication process of robochip. **a** Patterning Cr/Au, **b** Patterning of cage structure, **c** Patterning of SU-8, **d** Etching Si layer and removal of SU-8, **e** Assembling and packaging

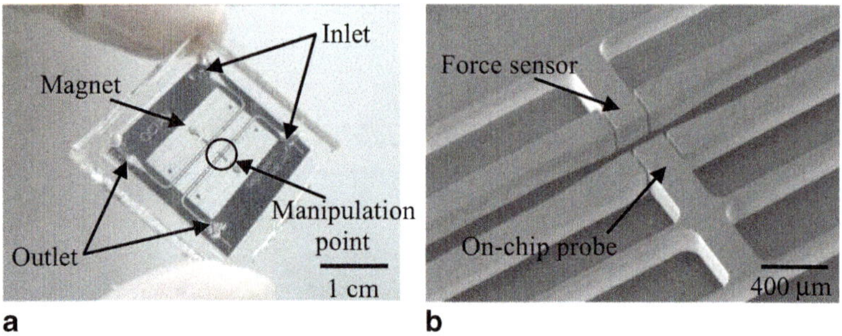

Fig. 1.9 Images of fabricated robochip. **a** Photograph of fabricated probe and **b** SEM image of manipulation point

c. An SU-8 layer is patterned on the surface of the Si substrate (thickness 200 μm). This pattern consists of an on-chip probe, a force sensor, and a microchannel for transporting the cells.
d. The Si substrate is etched by using a deep reactive ion etching (DRIE) technique, and the SU-8 is removed by O_2 plasma ashing after the DRIE process.
<Assembly and packaging>
e. The fabricated device layer is assembled onto the holder layer, and a permanent magnet (diameter: 1 mm, height: 0.5 mm, density of magnetic flux: 140 mT) is assembled onto the actuation point. Finally, a polydimethylsiloxane (PDMS) cover is bonded onto the holder layer.

A photograph of the fabricated probe and a SEM image of the manipulation point of the robochip are shown in Fig. 1.9.

1.3 Mechanical Characterization of Oocyte

1.3.1 System Configuration

Figure 1.10 is an overall view of the experimental setup. This system is composed of a microscope with a CCD camera, a motorized stage, and a robochip. A permanent magnet (diameter: 1 mm, height: 1 mm, density of the magnetic flux: 176 mT) is placed on the motorized stage to actuate the on-chip probe. The density of the magnetic flux on the glass substrate is 32 mT. The permanent magnet, which is assembled on the on-chip probe, is actuated by the permanent magnet placed on the motorized stage. The photographs in Fig. 1.11 illustrate the actuation of the on-chip probe. From Fig. 1.11, we can see that the permanent magnet at the actuation point is actuated by the external magnetic force, and that any large displacement of the actuation point is reduced by the center structure. As a result, fine displacement is achieved at the probe tip.

The repetitive positioning accuracy is evaluated using this system. The amplitudes of the displacement of the manipulation and actuation points are measured

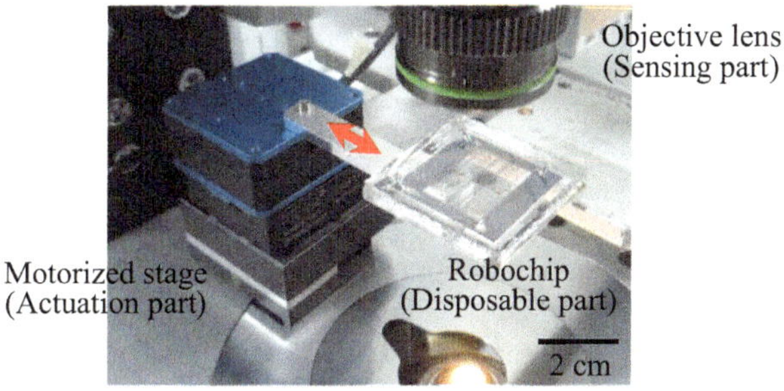

Fig. 1.10 Overall view of experimental setup

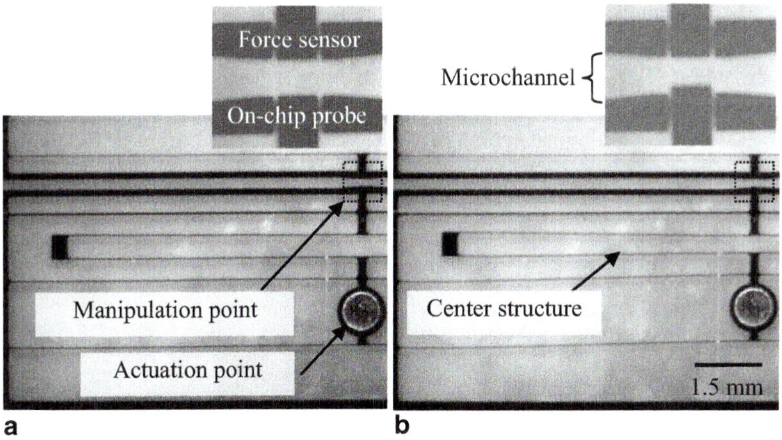

Fig. 1.11 Demonstration of reduction mechanism: **a** before actuation and **b** after actuation

Fig. 1.12 Repetitive positioning accuracy of probe

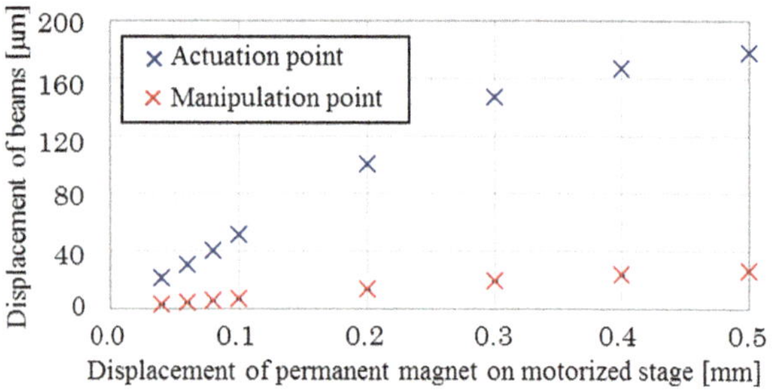

Table 1.2 Results of repetitive accuracy experiments

Displacement of permanent magnet on motorized stage [mm]	Actuation point displacement [μm]		Manipulation point displacement [μm]		Reduction rate
	AVE	STD	AVE	STD	
0.02	No movement		No movement		-
0.04	20.8	1.96	2.68	0.18	0.13
0.06	30.0	1.19	3.89	0.14	0.13
0.08	39.8	1.09	5.23	0.12	0.13
0.10	50.8	1.14	6.56	0.10	0.13
0.20	100.3	1.32	13.24	0.10	0.13
0.30	147.0	1.47	19.03	0.10	0.13

using the pixels of the images captured with the CCD camera (1 pix ≅ 0.08 μm) as a function of the amplitude of the displacement of the motorized stage. The drive frequency of the motorized stage is set to 0.25 Hz. Figure 1.12 shows the results obtained for the repetitive positioning accuracy. The horizontal axis represents the displacement of the permanent magnet on the motorized stage, and the vertical axis shows the displacement of the actuation and manipulation points. The plots indicate the average (10 times) of the displacements, and the error bars represent the standard deviation. Table 1.2 lists the value of each result. These results indicate that the on-chip probe has a dead band of approximately 20 μm. The dead band is reduced by employing a magnet, which is placed at the actuation point of the parallel plate structure and separated from the substrate surface. The dead band of the actuation point is, therefore, up to 10 times better than in the case of the hybrid-type MMT previously developed by our group. Moreover, these results indicate that the standard deviation of the manipulation point's displacement is less than 0.18 μm. In addition, the relationship between the manipulation distance of the motorized stage and the displacement of the manipulation point exhibit fairly good linearity when the manipulation distance is less than 0.3 mm and the reduction ratio is approximately 0.13. When the distance of the motorized stage is 0.4–0.5 mm, the standard deviation is 0.10–0.12 μm, and the repetitive positioning accuracy is again in the order of nanometers. The displacement of the beams is no longer linear, and the reduction ratio is 0.14–0.15. The nonlinearity within this range is caused by the large deformation of the beams, and the reduction ratio increases because the stiffness of the beams is relatively high. These results indicate that the positioning accuracy is also 10 times better than that of a horizontal polar drive (HPD) MMT that utilizes ultrasonic vibration and the HPD method to improve the positioning accuracy. In addition to the positioning accuracy, the standard deviation of the probe tip remains under 0.18 μm, and the positioning accuracy is high enough for cellular force measurements because the displacement measured by the CCD camera could be controlled, even though the displacement is nonlinear. As a result, we can conclude that the displacement of the actuation point and the dead band is reduced by the displacement reduction mechanism, and that nanometer-order noncontact actuation of the on-chip probe tip is achieved.

1.3.2 Oocyte Preparation

Bovine oocytes with cumulus cells were obtained from the Livestock Improvement Association of Japan, Inc. (Tokyo, Japan) and then transported to the laboratory within 24 h in cryogenic vials with a culture medium. The oocytes with cumulus cells were then transferred to a culture dish containing 1 % hyaluronidase (Sigma) in Medium 199 (Gibco), and the surrounding cumulus cells were removed enzymatically by pipetting. The oocytes were individually selected and then immersed in the prepared Medium 199 in a culture dish.

1.3.3 Automated Cell Measurement

Automated cell mechanical characterization was performed by using visual feedback control [14]. The system is illustrated in Fig. 1.13. A mechanical characterization system lets cells flow through the microchannel of the robochip and then individually positions them at the manipulation point. The cells at the manipulation point are deformed to some extent by the movement of the on-chip probe when measuring the force applied. The system utilizes a desktop computer that interfaces with a syringe pump and a motorized linear stage. Visual feedback control is available through a high-speed camera attached to the microscope. The position of the target cell, force sensor, and on-chip probe are measured by simple and fast image processing algorithms to provide a fast response. The deformation of the cells is measured by measuring the distance between a force sensor and on-chip probe, relative to the original diameter of the cell. The deformation information is then used for PID control of the motorized stage. With image processing being the bottleneck (about 300 Hz), PID control operates at 500 Hz to enable accurate positioning and sensing. Automated continuous cell measurement is possible using this system. Figure 1.14 shows typical results of continuous cell measurement. Oocytes are transported through the microchannel of the robochip and are then individually

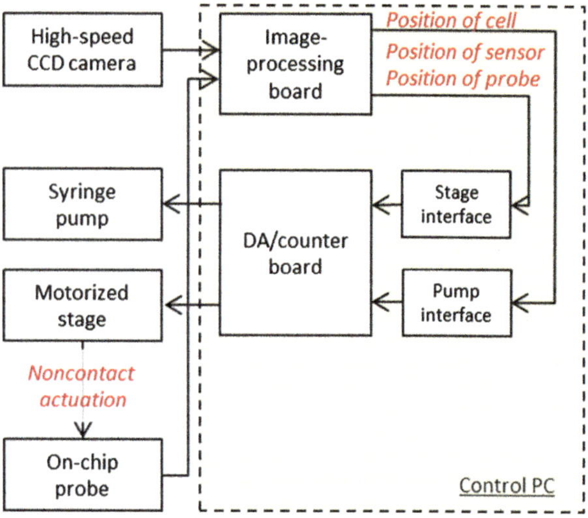

Fig. 1.13 System for automated cell mechanical characterization

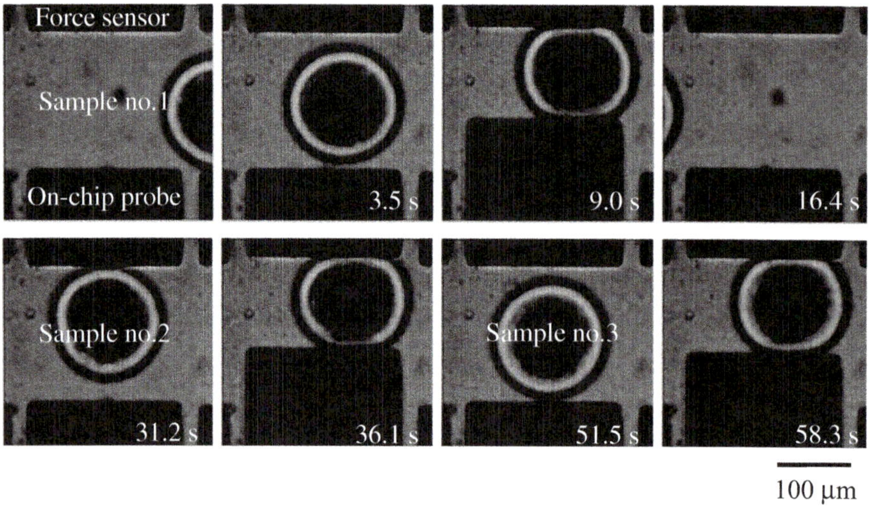

Fig. 1.14 Automated cell mechanical characterization

positioned at the manipulation point. The position of an oocyte is controlled by the syringe pump, and the oocyte at the manipulation point is deformed by the on-chip probe. The measurement time is approximately 10 s per oocyte, and the interval between oocytes is 5 to 10 s. Thus, the total measurement time is approximately 15 to 20 s per oocyte. We concluded that high-throughput cellular mechanical characterization is achieved.

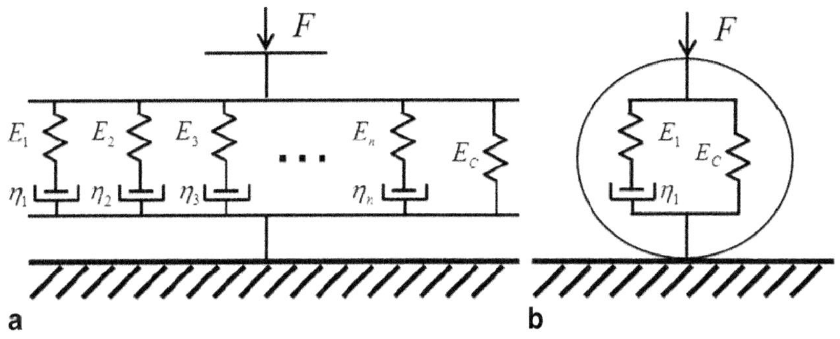

Fig. 1.15 Modeling of cellular deformation: **a** analytical model of deformation (generalized Maxwell model), and **b** deformation model of oocyte

1.3.4 Modeling of Cell Deformation

Our system is based on visual feedback control. Therefore, it is possible to measure the cellular force and the magnitude of any deformation as a function of time. This section describes the evaluation of the cell mechanical characteristics by using a deformation model that can be written as shown in Fig. 1.15. In this model, an oocyte is assumed to be a microparticle with viscoelastic properties, and the microparticle is pushed against a rigid plane. At first, an oocyte is regarded as a homogeneous isotropic elastic microparticle, and Hertz's theory is applied to the contact model between the microparticle and the plane. The applied force F can be described as a function of the displacement of a microparticle under the assumption that the deformation of the plane can be ignored, as shown in Eq. 1.6.

$$F = \frac{4}{3}\frac{R_0^{\frac{1}{2}}}{1-\nu^2}E'\delta^{\frac{3}{2}} \tag{1.6}$$

where R_0, ν, and E' are the radius, the Poisson's ratio, and Young's modulus of the microparticle, respectively. Then, considering the viscoelastic properties of the oocyte, we assume that E' can be expressed by the generalized Maxwell model as shown in Fig. 1.15a. In the case of Fig. 1.15a, the relaxation modulus $E(t)$ can be expressed as follows:

$$E(t) = E_c + \sum_i E_i e^{-\frac{t}{\tau_i}} \tag{1.7}$$

where E_c is the long term modulus once the material is totally relaxed, E_i is the constant value of the time dependence parameter, τ_i is the relaxation time defined as the time required until the initial value becomes *1/e*. By substituting Eq. 1.7 into Eq. 1.6 and ignoring the effect of the acceleration due to the mass of the microparticle, the unstable force can be expressed as follows,

$$F(t) = \frac{4}{3}\frac{R_0^{\frac{1}{2}}}{1-\nu^2}\left(E_c + \sum_i E_i e^{-\frac{t}{\tau_i}}\right)\delta^{\frac{3}{2}} \tag{1.8}$$

Now, to simplify the calculation, we regard the oocyte as having a single dependence term in our model ($i = 1$) as shown in Fig. 1.15b. To obtain the values of E_c and E_1, we measure the cellular force and the deformation of the oocyte as a function of time, using the images captured with the CCD camera. In this case, we can obtain the total number n of the equations for two unknown parameters, as shown in Eq. 1.9.

$$\begin{cases} F(t_1) = \frac{4}{3}\frac{R_0^{\frac{1}{2}}}{1-\nu^2}\left(E_c + E_1 e^{-\frac{t_1}{\tau_1}}\right)\delta_1^{\frac{3}{2}} \\ F(t_2) = \frac{4}{3}\frac{R_0^{\frac{1}{2}}}{1-\nu^2}\left(E_c + E_1 e^{-\frac{t_2}{\tau_1}}\right)\delta_2^{\frac{3}{2}} \\ F(t_3) = \frac{4}{3}\frac{R_0^{\frac{1}{2}}}{1-\nu^2}\left(E_c + E_1 e^{-\frac{t_3}{\tau_1}}\right)\delta_3^{\frac{3}{2}} \\ \vdots \\ F(t_n) = \frac{4}{3}\frac{R_0^{\frac{1}{2}}}{1-\nu^2}\left(E_c + E_1 e^{-\frac{t_n}{\tau_1}}\right)\delta_n^{\frac{3}{2}} \end{cases} \tag{1.9}$$

Here, defined matrices $\boldsymbol{A}$ and $\boldsymbol{B}$ are defined by the following equation,

$$A = \begin{bmatrix} 1 & e^{-t_1/\tau_1} \\ 1 & e^{-t_2/\tau_1} \\ 1 & e^{-t_3/\tau_1} \\ \vdots & \vdots \\ 1 & e^{-t_n/\tau_1} \end{bmatrix}, \qquad B = \begin{bmatrix} P(t_1)/\delta(t_1)^{3/2} \\ P(t_2)/\delta(t_2)^{3/2} \\ P(t_3)/\delta(t_3)^{3/2} \\ \vdots \\ P(t_n)/\delta(t_n)^{3/2} \end{bmatrix} \tag{1.10}$$

Eventually, the unknown parameters are obtained using Eq. 1.11.

$$\begin{bmatrix} E_C \\ E_T \end{bmatrix} = \frac{3}{4}\frac{1-\nu^2}{R_0^{1/2}}\left(A^T A\right)^{-1} A^T B \tag{1.11}$$

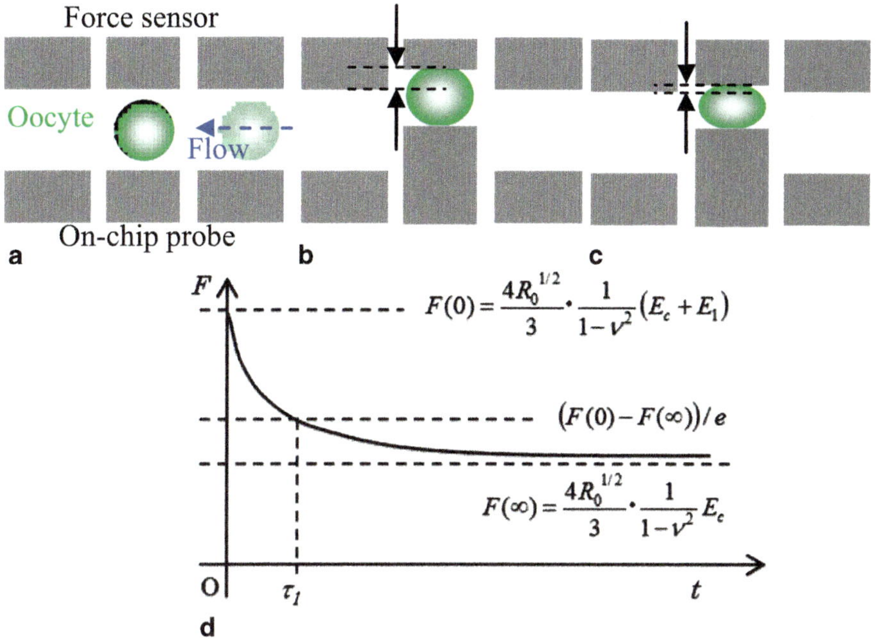

Fig. 1.16 Concept of on-chip cell mechanical characteristics: **a** cell positioning, **b**, **c** deformation relaxation, and **d** cellular force as a function of time

1.3.5 Cell Mechanical Characterization

Figure 1.16 shows the concept of the on-chip cell mechanical characterization. First, an oocyte is transported to the manipulation point of the robochip (Fig. 1.16a). Then, the oocyte is deformed by the on-chip probe (Figs. 1.16b and c. Finally, we evaluated the displacement and the cellular force as a function of time based on the images captured by the CCD camera (Fig. 1.16d). Figure 1.17a shows sequential photographs of a typical experimental result. The on-chip probe deforms the oocyte and the reaction force of the oocyte decreases with the passing of time. Figure 1.17b shows the measured cellular force as a function of time. The blue and red plots show the experimental values and the calculated values, respectively. The experimental values are measured by the images, and the analytical values are simulated using Eq. (1.3) with the measured cellular parameters. Then, E_c and E_1 are calculated using Eq. (1.6). In this case, E_c and E_1 are calculated as being 4.83 and 4.86 kPa. The experimental and simulated values are in good agreement and the results show that it is possible to adopt a deformation model to analyze the mechanical properties of an oocyte. Finally, the mechanical properties of the oocytes, which are cultured for 2 (group A), 14 (group B), and 18 (group C) days after harvesting, were measured to evaluate the quality of the oocyte, as shown in Fig. 1.18. Each group consisted of 10 samples. The average of E_c decreased with the number of culture days and the average of E_1 increased. These results indicate that the oocyte has viscoelastic properties and that it is important to analyze the mechanical properties of

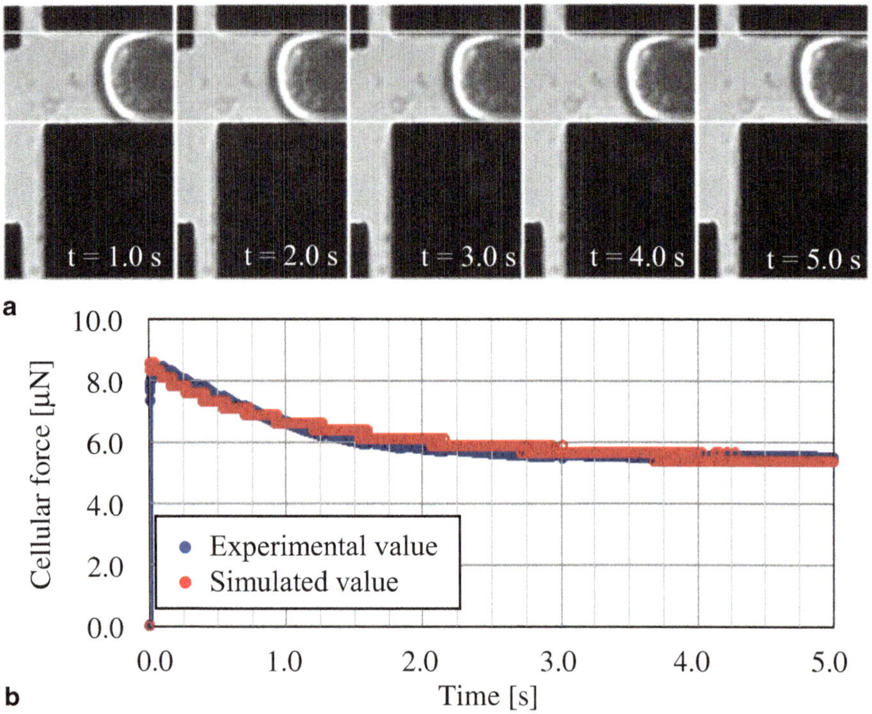

Fig. 1.17 On-chip cell mechanical characterization: **a** sequential photographs of demonstration and **b** results of cellular force of oocyte as a function of time

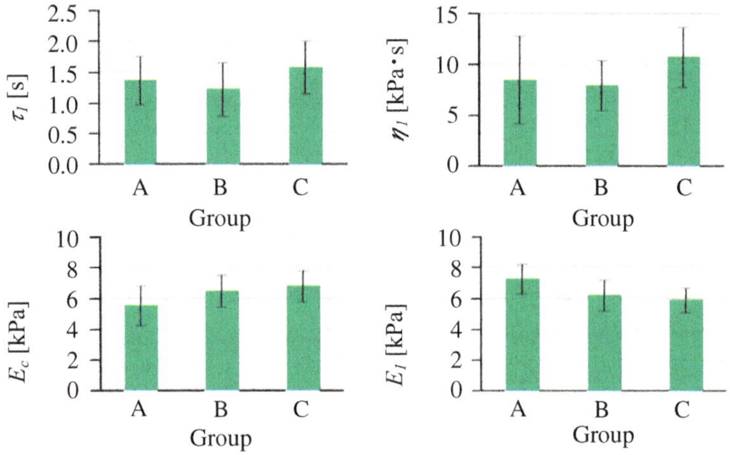

Fig. 1.18 Measured cell mechanical characters

an oocyte to evaluate the quality. Eventually, a quality evaluation of the cell will be possible by mechanical characterization. Therefore, the evaluation of the relationship between E_c and E_1 will be a promising technology for evaluating the quality of a cell.

1.4 Conclusion

We have described cell mechanical characterization based on on-chip robotics. To evaluate the mechanical properties of oocytes, we integrated an on-chip probe and an on-chip force sensor into a microfluidic chip as a robochip. This robochip has the advantages of both mechanical micromanipulators and lab-on-a-chip devices, which are, briefly, a high level of accuracy, a high output force, and high throughput. The characterization system is automated by visual feedback control of the flow in the microchannel and the on-chip probe; continuous mechanical characterization is achieved in this way. The throughput of our system is approximately 20 s per oocyte. Moreover, measurement of the viscoelastic properties is demonstrated as a means of evaluating the quality of an oocyte. We note that by applying robot technologies to microfluidic devices, we can evaluate cell characters quickly and quantitatively. Therefore, this method is a promising technology for contributing to the analysis of the physical and biological properties of cells.

References

1. Fauver ME, Dunaway DL, Lilienfeld DH, Craighead HG, Pollack GH (1998) Microfabricated cantilevers for measurement of subcellular and molecular forces. IEEE Trans Biomed Eng 45(7):891–898
2. Murayama Y, Constantinou CE, Omata S (2004) Micro-mechanical sensing platform for the characterization of the elastic properties of the ovum via uniaxial measurement. J Biomech 37:67–72
3. Sun Y, Wan K-T, Roberts KP, Bischof JC, Nelson BJ (2003) Mechanical property characterization of mouse Zona Pellucida. IEEE Trans Nanobioscience 2(4):279–286
4. Gauthier M, Piat E (2004) An electromagnetic micromanipulation system for single-cell manipulation. J Micromech 2(2):87–119
5. Wacogne B, Pieralli C, Roux C, Gharbi T (2008) Measuring the mechanical behaviour of human oocytes with a very simple SU-8 micro-tool. Biomed Microdevices 10(3):411–419
6. Dittrich PS, Tachikawa K, Manz A (2006) Micro total analysis systems. Latest advancements and trends. Anal Chem 78(12):3887–3907
7. Whitesides GM (2006) The origins and the future of microfluidics. Nature 442:368–373
8. Wu Z, Willing B, Bjerketorp J, Jansson JK, Hjort K (2009) Soft inertial microfluidics for high throughput separation of bacteria from human blood cells. Lab Chip 9:1193–1199
9. Lee NY, Yamada M, Seki M (2005) Development of a passive micromixer based on repeated fluid twisting and flattening, and its application to DNA purification. Anal Bioanal Chem 383(5):776–782
10. Moritani T, Yamada M, Seki M (2011) Generation of uniform-size droplets by multistep hydrodynamic droplet division in microfluidic circuits. Microfluid Nanofluid 11(5):601–610
11. Jang K, Xu Y, Tanaka Y, Sato K, Mawatari K, Konno T, Ishihara K, Kitamori T (2010) Single-cell attachment and culture method using a photochemical reaction in a closed microfluidic system. Biomicrofluidics 4(3):032208
12. Yamashita T, Tanaka Y, Idota N, Sato K, Mawarati K, Kitamori T (2011) Cultivation and recovery of vascular endothelial cells in microchannels of a separable micro-chemical chip. Biomaterials 32(10):2459–2465

13. Sakuma S, Arai F (2013) Cellular force measurement using a nanometric-probe-integrated microfluidic chip with a displacement reduction mechanism. J Robot Mechatron 25:277–284
14. Sakuma S, Arai F (2013) High throughput mechanical characterization of oocyte using robot integrated microfluidic chip. Proceeding of the 2013 IEEE/RSJ Int. Conf. on Intelligent Robots and Systems (IROS), pp 2047–2052
15. Papi M, Sylla L, Parasassi T, Brunelli R, Monaci M, Maulucci G, Missori M, Arcovito G, Ursini F, Spirito MD (2009) Evidence of elastic to plastic transition in the zona pellucida of oocytes using atomic force spectroscopy. Appl Phys Lett 94(15):153902
16. Arai F, Ogawa M, Fukuda T (1999) Selective manipulation of a microbe in a microchannel using a teleoperated laser scanning manipulator and dielectrophoresis. Adv Robotics 13(3):343–345
17. Onda K, Arai F (2012) Multi-beam bilateral teleoperation of holographic optical tweezers. Opt Express 20(4):3633–3641
18. Yamanishi Y, Sakuma S, Onda K, Arai F (2010) Powerful actuation of magnetized microtools by focused magnetic field for particle sorting in a chip. Biomed Microdevices 12(4):745–752
19. Yamanishi Y, Sakuma S, Kihara Y, Arai F (2010) Fabrication and application of 3-D magnetically driven microtools. J Microelectromech Syst 19(2):350–356
20. Hagiwara M, Kawahara T, Yamanishi Y, Masuda T, Feng L, Arai F (2011) On-chip magnetically actuated robot with ultrasonic vibration for single cell manipulations. Lab Chip 11:2049–2054

Chapter 2
Dimensionless Evaluation of Cell Deformability with High Resolution Positioning in a Microchannel

Chia-Hung Dylan Tsai, Shinya Sakuma, Fumihito Arai and Makoto Kaneko

Abstract This chapter covers dimensionless evaluation for the stiffness-based deformability of a cell using a high-resolution vision system and a microchannel. In conventional approaches, the transit time of a cell through a microchannel is often utilized for the evaluation of cell deformability. However, such time includes both the information of cell stiffness and viscosity. In this work, we eliminate the effect from cell viscosity, and focus on the cell stiffness only. We find that the velocity of a cell varies when enters a channel, and eventually reaches to equilibrium where the velocity becomes constant. The constant velocity is defined as the equilibrium velocity of the cell, and it is utilized to define the observability of stiffness-based deformability. The necessary and sufficient numbers of sensing points for evaluating stiffness-based deformability are discussed. Through the dimensional analysis on the microchannel system, three dimensionless parameters determining stiffness-based deformability are derived, and a new index is introduced based on these parameters. The experimental study is conducted on the red blood cells from a healthy subject and a diabetic patient. With the proposed index, we showed that the experimental data can be nicely arranged.

Keywords Dimensionless evaluation · Cell deformability · Microchannel system · High-speed vision

---

Part of the materials in this chapter is from C. Tsai, S. Sakuma, F. Arai and M. Kaneko, IEEE Transactions on Biomedical Engineering, vol.61, no.4, pp1187-1195, 2014. The permission of reuse is granted by IEEE.

C.-H. D. Tsai (✉) · M. Kaneko
Department of Mechanical Engineering, Osaka University, Osaka, Japan
e-mail: tsai@hh.mech.eng.osaka-u.ac.jp

S. Sakuma · F. Arai
Department of Micro-Nano Systems Engineering, Nagoya University, Nagoya, Japan
e-mail: sakuma@mech.nagoya-u.ac.jp

M. Kaneko
e-mail: mk@mech.eng.osaka-u.ac.jp

F. Arai
e-mail: arai@mech.nagoya-u.ac.jp

T. Arai et al. (eds.), *Hyper Bio Assembler for 3D Cellular Systems*,
DOI 10.1007/978-4-431-55297-0_2

2.1 Introduction

Cell stiffness is important for disease diagnosis based upon known relations with diseases, such as malaria, sepsis, diabetes and etc. [1–4]. For example, Brandao et al. showed that sickle cell anemia causes stiffening of red blood cells (RBCs) [5]. Among all the evaluation methods, microchannel-based methods have a great advantage of high-speed measurement, and it is useful and necessary for obtaining sufficient number of samples regarding the statistical significance of the results. The feature of high-speed measurement is especially important when it comes to practical applications.

In microchannel systems, a cell is deformed to pass through a microchannel whose diameter is smaller than the diameter of the cell, as shown in Fig. 2.1a. Different cell deformabilities result in different cell behaviors during the transits. For example, a cell with low deformability would move slower inside a channel because a greater resistance against the cell motion is generated on the contact between the cell and channel. On the contrary, a cell with high deformability would move faster upon the same consideration. According to this idea, the transit time for a cell through a microchannel is often used as a deformability index (DI) for the evaluation of cell deformability in conventional approaches [6–9].

However, we observed that cells with the same transit time sometimes exhibit very different characteristics in terms of motion profiles, as illustrated in Figs. 2.1b and c. Figure 2.1b shows a series of locations of two cells through microchannels. The motion profiles of them are plotted in Fig. 2.1c. The cell on the top of Fig. 2.1b moves at a constant velocity throughout the channel while the lower one starts with acceleration motion, and then moves at a constant velocity. These two cells apparently have different characteristics, but their deformability would be evaluated as the same by conventional approaches due to the same transit time. In order to distinguish these cells, we propose a new evaluation parameter, *equilibrium velocity*, by focusing on the cell motion when they are in equilibrium during the transit. Cell viscosity is a time-dependent property[1], so that the response damped out when a cell reaches equilibrium. Therefore, we can focus on only the time-independent properties, such as cell stiffness, by using equilibrium velocity. In the case of Fig. 2.1b and c, the equilibrium velocity of the cell on the top of Fig. 2.1b is slower than the lower one by comparing the corresponding slopes of the linear segment from their motion profiles. Assuming the same amount of deformation in the channel, we can say that the cell with greater equilibrium velocity has better deformability.

The observability of stiffness-based deformability is introduced based on the newly proposed parameter. Since the cell stiffness is directly related to the equilibrium velocity, we define the stiffness-based deformability of a cell as observable when its equilibrium velocity is measurable. The sufficient and necessary number of sensing points to evaluate cell stiffness are discussed.

Experimental study is conducted on the RBCs from a healthy subject and a diabetic patient. Two phases, the phases of deformation and equilibrium, were

[1] The time-dependent property of viscosity is sometimes referred as *rate-dependant*.

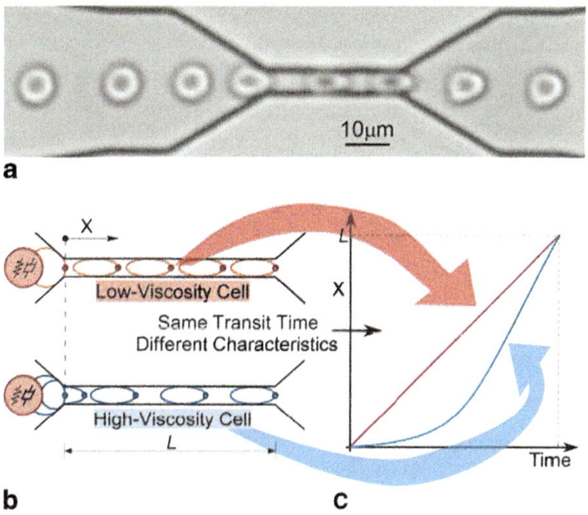

Fig. 2.1 **a** A series images of a RBC through a microchannel under a microscope. **b** Two cells pass through identical microchannels with the same transit time. The ellipsoids represent cell locations at the same time intervals. **c** The motion profiles of cells in Fig. 2.1b. The transit time for two RBCs through the channel are the same while the motion profiles are different

experimentally observed from the motion profiles. Dimensional analysis is performed on the basis of the stiffness-related parameters, and three dimensionless parameters for determining stiffness-based deformability are derived. Finally, a new dimensionless $\widehat{DI}$ is introduced by combining three dimensionless parameters, and the difference between the evaluation using transit time and $\widehat{DI}$ is compared and shown based on the experimental results.

The rest of the chapter is organized as follows: Firstly, a brief review on related works is presented in Sect. 22.2. The observation and physics behind the cell behavior through a microchannel is explained in Sect. 22.3. After that, the observability of cell stiffness is introduced in Sect. 22.4, and it is followed by the experimental study in Sect. 22.5. The experimental results are discussed by using the newly introduced dimensionless $\widehat{DI}$ in Sect. 22.6. Finally, the chapter is concluded in Sect. 22.7.

2.2 Related Works

The mechanical properties of cells have been used as biomarkers for cell biophysics [10]. Various methods have been developed for evaluating the deformability of biological cells [11–17]. These methods can be categorized into direct and indirect methods based on the nature of sensing techniques. The direct methods are the ones evaluating cell properties by the response of cells while directly applying an external force onto them with an instrument, such as an atomic force microscope (AFM) [17, 16] and optical tweezers [5, 12, 14]. For examples, Radmacher et al. measured the viscoelastic properties of human platelets with an AFM [18]. Dao et al. studied on the mechanics of human red blood cells by optical tweezers [19].

On the other hand, the indirect methods are the ones evaluating cell properties based on cell behavior without giving a direct excitation. One of the first works were done by Worthen et al. [20]. Isermann et al. measure nuclear mechanics in [21]. Tsukada et al. measured the deformability of RBCs in diabetes mellitus [4]. Adamo et al. assessed cell deformability by sensing the travel time of a cell through a narrow passage by the electrical impedance of cells [8]. Youn et al. studied the deformability of RBCs by using an array of orifices [22]. Zheng et al. developed a high-throughput measurement for RBCs [6]. Tomaiuolo et al. analyzed the membrane viscoelasticity of RBC membrane based on the relation between fluid pressure and cell velocity [13]. Hou et al. studied the deformability of breast cancer cells using a microchannel [7]. Byun et al. characterized the deformability and surface friction of cancer cells [23]. Gossett et al. evaluated cell mechanical properties based on hydrodynamic stretching and cell deformation [24]. In our former works, the transit time of a cell through a microchannel was used for the evaluation of cell stiffness [25]. An observation of two-phase motion of cells through a microchannel from experiments is shown in [26]. Later, a method for separating two phases was proposed in [27], and the comparison between cell stiffness evaluated by the equilibrium velocity and average velocity is discussed in [28]. The concept of observability is firstly mentioned in [29].

In this chapter, we conducted experiments on a healthy subject and a diabetic patient. Moreover, we present the dimensional analysis to obtain a dimensionless index of cell stiffness-based deformability for general microchannel system.

2.3 The Stiffness-Based Deformability

General motion profile of a cell through a microchannel is illustrated in Fig. 2.2. The profile can be separated into two parts based on the motion profile. The first part is when the cell just enters the channel, and cell velocity varies with time. The second part is when the cell eventually reaches a new equilibrium state, and the velocity becomes constant. The motion profiles in first and second part are illustrated by a curve and a line in Fig. 2.2. According to the difference of cell motion before and after reaching the equilibrium, we define the cell as in Phase I and Phase II during the transit, respectively.

It is well-known that most of biological cells have the properties of viscoelasticity [30]. Hence, a cell can be modeled by mechanical elements, such as springs and dampers, for representing cell stiffness and viscosity, respectively. From the viewpoint of mechanics, the force response of a deformed spring, F_k, is the function of its stiffness, k, and deformation, δ_k, which is

$$F_k = \mathrm{k}\delta_k \tag{2.1}$$

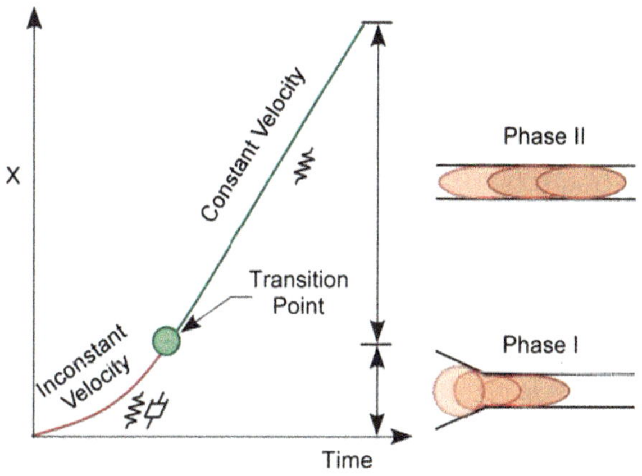

Fig. 2.2 General cell behavior inside a microchannel can be separated into two phases, Phase I and II. Two phases are represented by a exponentially increasing curve and a straight line, respectively

The force response of a deforming damper, F_c, is a function of the damping coefficient, c and deformation rate, $\dot{\delta}_c$ which is

$$F_c = c\dot{\delta}_c \text{ where } \dot{\delta}_c = \frac{d\delta_c}{dt} \tag{2.2}$$

The overall force response of a deformed viscoelastic material would be the resultant of both the responses in Eqs. (2.1) and (2.2). When a cell reaches equilibrium, every deformation, δ_k and δ_c, remains constant. Thus, the response of cell stiffness remains constant, and the response of cell viscosity becomes zero.

For example, Kelvin-Voigt model is one of classic viscoelastic models, and it represents viscoelastic response by a spring and a damper connected in parallel. In the case of Kelvin-Voigt model, the total force response would be the sum of Eqs. (2.1) and (2.2), which is

$$F = F_k + F_c \tag{2.3}$$

Also, the deformation of the spring and damper are synchronized because of parallel connection, and that gives $\delta_k = \delta_c = \delta$ While a cell is deformed to enter a narrow channel, $\delta \neq 0$ and $\dot{\delta} \neq 0$, thus $F_k \neq 0$ and $F_c \neq 0$. It means that the cell motion is affected by both cell stiffness and viscosity. When the deformation is concluded and the cell reaches to equilibrium, $\delta \neq 0$ and $\dot{\delta} = 0$, thus $F_k \neq 0$ and $F_c = 0$. As a result, the cell viscosity no longer affects the cell motion, and, its stiffness can be solely evaluated from the equilibrium state. Based on the idea, stiffness-based deformability is defined as follows:

Definition 1

Stiffness-based deformability is defined as the deformability of a cell when it is in equilibrium, where only the stiffness affects its motion but not the viscosity.

2.4 Observability of Cell Stiffness in Microchannel system

The observability of a microchannel system is defined as

Definition 2

Stiffness-based deformability of a cell is observable if equilibrium velocity of a cell inside a microchannel is measurable.

The purpose for introducing the concept of observability is to minimize the resources needed for the evaluation of stiffness-based deformability. To measure the equilibrium velocity of a cell requires continuous tracking of the cell through a microchannel as the method explained in [27]. The question then arises as to how many sensing points are necessary and sufficient in order to measure the equilibrium velocity of a passing cell? The *sensing point* here represents the locations where the position sensors are implemented as in an actual setup. The sensors are capable of detecting a passing cell, thus the time, T_i of a passing cell at the location, X_i, of the sensor can be recorded for position analysis, where i indicates the i-th sensing point.

It is not possible to measure the velocity of a cell in a microchannel with one sensor. Therefore, two sensing points are definitely necessary. Let's consider system variables shown in Fig. 2.3 and two cases shown in Fig. 2.4a and b. Since it is impossible to determine the transition point by just 2 points, as shown in Fig. 2.4a, the equilibrium velocity, u_{eq1}, cannot be obtained. It means that the cell stiffness-based deformability cannot be evaluated by two sensing points, and the cell stiffness-based deformability is not observable.

Nevertheless, two sensing points can still be enough to evaluate cell stiffness under a special case when

$$\frac{D_c}{\lambda_m} \to 0 \tag{2.4}$$

Equation (2.4) represents the channel length, λ_m, is much longer than cell diameter, D_c, as shown in Fig. 2.4b. The ratio between the duration of Phase I and Phase II is small enough so that Phase I can be neglected. It implies that

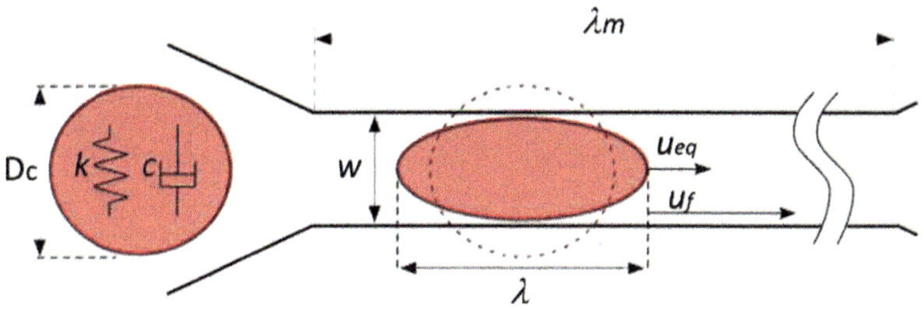

Fig. 2.3 The parameters of a microchannel system.

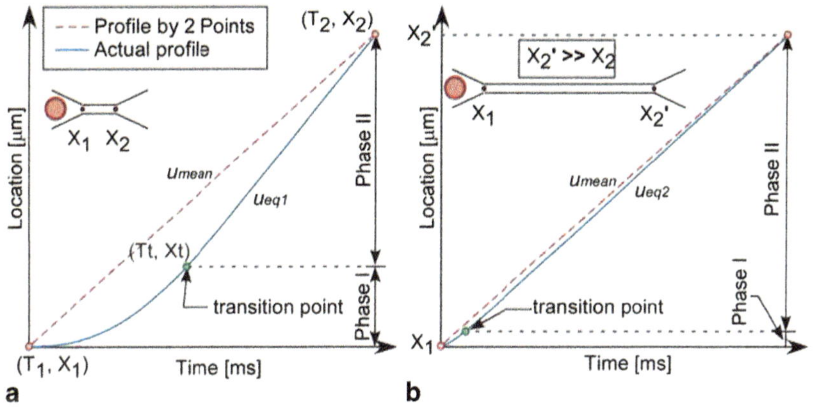

Fig. 2.4 **a** The mean velocity, u_{mean} and equilibrium velocity, u_{eq} are obviously different since the velocity varies significantly while the cell enters the channel. **b** $u_{mean} \rightarrow u_{eq}$ when the channel length approaches infinity, which makes the effect of transient responses insignificant [29]

$$\frac{X_t}{\lambda_m} \approx 0 \text{ and } (\mathrm{T} - T_t) \approx \mathrm{T} \tag{2.5}$$

Where T_t and X_t are the time and location of the transition point where a cell concludes to a new equilibrium state. The equilibrium velocity can be written as

$$\begin{aligned} u_{eq2} &= \lim_{\frac{X_t}{\lambda_m} \to 0} \frac{\lambda_m - X_t}{T - T_t} \\ &= \lim_{\frac{X_t}{\lambda_m} \to 0} \frac{1 - \frac{X_t}{\lambda_m}}{\frac{T}{\lambda_m} - \frac{T_t}{\lambda_m}} \\ &= u_{mean} \end{aligned} \tag{2.6}$$

From Eq. (2.6), we have that the equilibrium velocity equals the mean velocity, u_{mean} under the condition of Eq. (2.5). It shows that the two sensing points are sufficient in this special case. The results lead to the following Theorem

Theorem 1

Stiffness-based deformability of a cell is observable by two sensing points if $(D_c / \lambda_m) \to 0$

Next, in the case of three sensing points, an additional sensing point is included, and it can be used as a checking point for examining if three points,(T_1, X_1), (T_2, X_2) and (T_3, X_3), are on the same line. If the three points are on the same line, it say that the cell has reached equilibrium. Accordingly, another theorem about observability of cell stiffness with three sensing points is

Theorem 2

For cases, **except** $(D_c / \lambda_m) \to 0$, three sensing points are necessary and sufficient for judging whether a cell is in equilibrium or not.

As an extension to Theorem 2, three or more sensing points are sufficient for judging if the stiffness-based deformability of a tested cell is observable.

2.5 Experiments

2.5.1 Experimental Setup and Configuration

Figure 2.5 shows the experimental system. A microscope (Olympus Ltd.) is equipped with a high-speed vision system (Photron Inc.), where the shutter speed and the spatial resolution of the system are 500 *ns* and 0.24 μm, respectively. A microfluidic chip is fabricated with microchannels inside it. The dimension of the channel is 30 μm in length, 4 μm in height and 4 μm in width, as shown in Fig. 2.6. The velocity of the flow inside the channel is controlled by adjusting the pressure difference between the inlet and outlet of the microchannel as shown in Fig. 2.7. When the pressure difference increases, the flow rate increases, and vice versa.

RBCs are obtained from two subjects, a diabetic patient (male, 50 s) and a healthy subject (male, 30 s). The subjects have read and signed the consent for the experiment before the experiment. The blood was withdrawn from the subjects by a licensed doctor 30 min before the experiments.

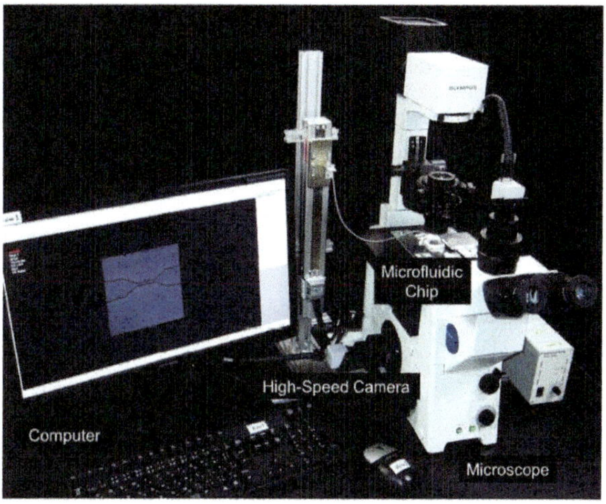

Fig. 2.5 The experimental setup includes a microfluidic chip, a microscope, a high-speed camera and a computer. A microchannel is fabricated inside the microfluidic chip

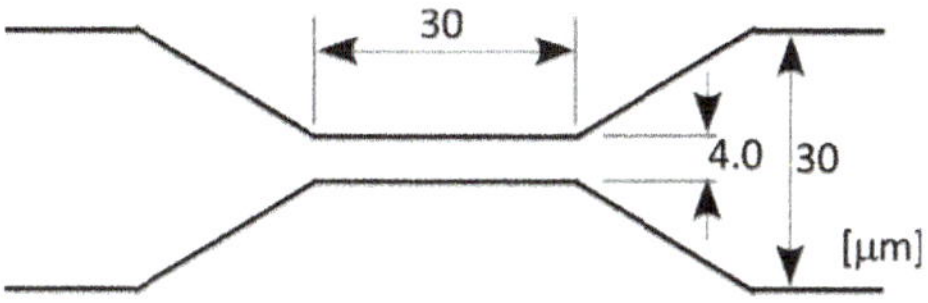

Fig. 2.6 The dimensions of the microchannel used in the chapter are shown here

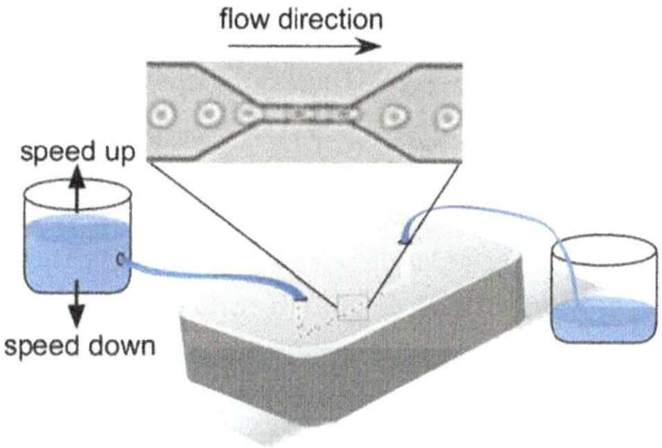

Fig. 2.7 The flow inside a microchannel is directly controlled by the pressure difference between the pressure at the inlet and outlet of the microfluidic chip [29]

2.5.2 *Experimental Procedure*

The step-by-step experimental procedure is listed as follows:

1. RBCs are acquired 30 min before the experiment. The blood is kept in vacuum tubes avoiding any interaction with air.
2. Standard saline solution washed through the microchannel before the experiment for making sure there is no air in the channel as well as for keeping the surface condition as similar as possible between different tests.

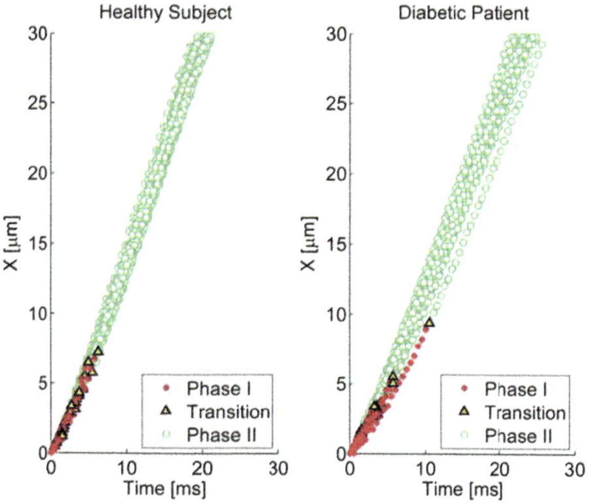

Fig. 2.8 The cell motion profiles of the healthy subject (*left*) and the patient (*right*). The cell motion inside the microchannel is separated into two phases by the method proposed in [27]. Data points in Phase I are in *red*, and the ones in Phase II are in *green*. *Black* markers indicate the transition point of each profile

3. The blood is diluted with the same saline solution at the ratio of 1/50. After that, the blood-saline mixture is injected into the microchannel from the inlet.
4. The flow inside the microchannels is directly controlled by the pressure different between the inlet and outlet of the microfluidic chip as shown in Fig. 2.7. The actual flow rate for each experiment is calculated based on cell motion using the method developed in [31].
5. The high-speed vision system is set to focus on the microchannel area where the channel width is 4.0 µm in the experiment[2], and cell motion through the channel is recorded at the prescribed frame rate, 2000 fps.
6. After finishing capturing images from the high-speed vision system, the images are processed by MATLAB with the toolbox of image processing for tracking the motion of RBCs. The information of the time and location of each RBC through the microchannel are measured and recorded.

2.5.3 Experimental Results

The motion profiles of RBCs through the microchannel for both the healthy subject and diabetic patient are plotted in Fig. 2.8. Each motion profile is divided into two phases by the method proposed in [27], and the red, green and black markers represent the data points in Phase I, II and the transition point, respectively. The cell flow-in velocity, as average cell velocity over the cross-sectional area before

[2] The width of 4.0 μm is chosen since RBCs are generally ranged from 6–8 μm in diameter. The microchannel should be narrow enough to deform the cell but not too narrow to cause damages. Moreover, based on our experience, RBCs are easily stuck in the microchannels with width 3 μm or less.

entering the channel [31], are 0.26 μm/ms and 0.20 μm/ms for the tests with the RBCs of the healthy subject and the patient, respectively. The difference between the two flow-in velocities is within the error of the system, and will be normalized for a fair evaluation in the later analysis.

2.6 Discussion

2.6.1 Dimensional Analysis

For evaluating stiffness-based deformability of a cell by a microchannel system, a functional relation among the physical quantities of the system and the cell is assumed as follows[3]:

$$\mathrm{F}(k, D_c, \lambda, u_{eq}, u_f, w, \mu) = 0 \tag{2.7}$$

where k, D_c, λ, u_{eq}, u_f, w and μ are cell stiffness, cell undeformed diameter, cell in-channel length, equilibrium velocity, fluid velocity[4], channel width and fluid viscosity, respectively. The friction coefficient of contact is not included in Eq. (2.7) because it is assumed no direct contact between a cell and channel wall, and the interaction between them is always through a thin layer of fluid, which is as known as *Thin-Film Lubrication Theory*. (see details in Appendix A and [32]) The quantities are indicated in Fig. 2.3. The units of these 7 dimensional quantities are:

$$\mathrm{k} \sim \left[FL^{-1}\right], D_c \sim \left[L\right], \lambda \sim \left[L\right], u_{eq} \sim \left[LT^{-1}\right], u_f \sim \left[LT^{-1}\right], w \sim \left[L\right], \mu \sim [FL^{-1}T]$$

where [F], [L] and [T] are fundamental dimensions, and are force, length and time, respectively. Buckingham π theorem [33] is adopted for solving the dimensionless parameters of the system. The dimensional matrix is

$$\mathbf{M} = \begin{bmatrix} 1 & 0 & 0 & 0 & 0 & 0 & 1 \\ -1 & 1 & 1 & 1 & 1 & 1 & -2 \\ 0 & 0 & 0 & -1 & -1 & 0 & 1 \end{bmatrix} \tag{2.8}$$

The rows correspond to the dimensions [F], [L] and [T], and the columns correspond to the dimensional quantities k, D_c, λ, u_{eq}, u_f, w and μ. Since the system has 7 dimensional quantities and the rank of the dimensional matrix is 3, we are looking for 4 dimensionless parameters[5] from Eq. (2.8). The dimensionless parameters can be obtained by letting

---

[3] see Appendix A for the details of how physical quantities are obtained.

[4] fluid velocity here represent the velocity of fluid inside the channel when there is no cell in the channel. The fluid velocity is estimated based on the velocity profile as the method described in [31].

[5] the value of 4 is obtained by subtracting the rank, 3, from the total number of dimensional quantities, 7.

$$\mathbf{Ma} = \mathbf{0} \tag{2.9}$$

where **a** is the kernel of the matrix and can be written as

$$\mathbf{a} = [a_1, a_2, a_3, a_4, a_5, a_6, a_7]^T \tag{2.10}$$

so that

$$k^{a_1}, D_c^{a_2}, \lambda^{a_3}, u_{eq}^{a_4}, u_f^{a_5}, w^{a_6}, \mu^{a_7} \sim \text{dimensionless} \tag{2.11}$$

After solving the above equations, we have

$$a_5 = -a_1 - a_4 \tag{2.12}$$

$$a_6 = -a_2 - a_3 \tag{2.13}$$

$$a_7 = -a_1 - a_4 \tag{2.14}$$

Substituting Eqs. (2.12)–(2.14) to (2.11) yields

$$\left(\frac{k}{u_f \mu}\right)^{a_1} \left(\frac{D_c}{w}\right)^{a_2} \left(\frac{\lambda}{w}\right)^{a_3} \left(\frac{u_{eq}}{u_f}\right)^{a_4} \sim \text{dimensionless} \tag{2.15}$$

Finally, we have

$$\left(\frac{k}{u_f \mu}\right) = \mathrm{g}\left(\frac{u_{eq}}{u_f}, \frac{D_c}{w}, \frac{\lambda}{w}\right) \tag{2.16}$$

Regarding the physical meaning of each dimensionless parameter, Eq. (2.16) can be rewritten as

$$\hat{\phi} = g^{-1}(\hat{u}, \hat{\delta}, \hat{\lambda}) \tag{2.17}$$

where

$$\hat{\phi} = \left(\frac{k}{u_f \mu}\right)^{-1} \tag{2.18}$$

$$\hat{u} = \frac{u_{eq}}{u_f} \tag{2.19}$$

$$\hat{\delta} = \frac{D_c}{w} \tag{2.20}$$

$$\hat{\lambda} = \frac{\lambda}{w} \tag{2.21}$$

Equations (2.17)–(2.21) basically tell that cell stiffness can be evaluated from its equilibrium velocity and amount of deformation when the flow condition is known.

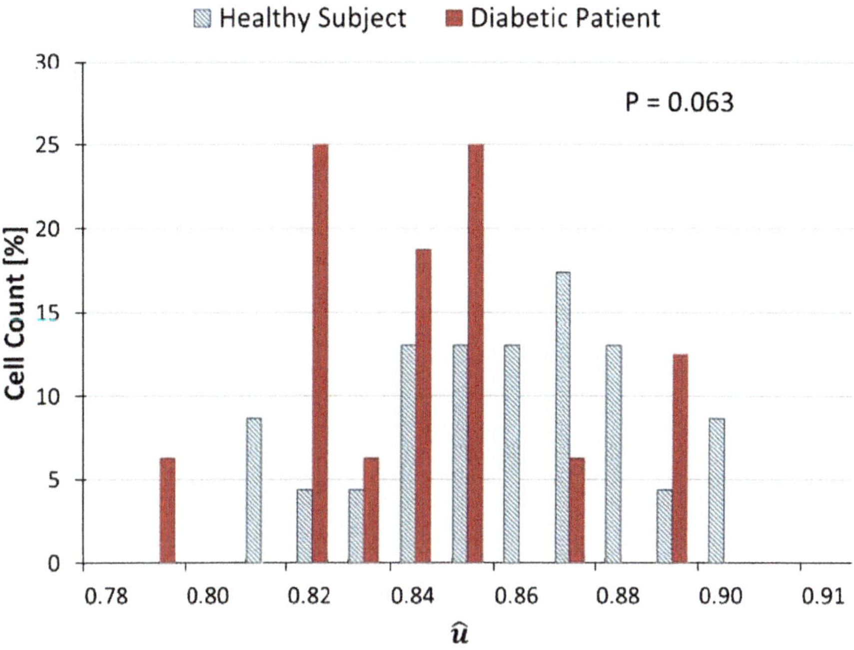

Fig. 2.9 The distribution of dimensionless parameter û for the healthy subject and the patient

The right of Eq. (2.18) is in inverse form for the consistency of all dimensionless parameters, so that the value of each parameter positively correlates cell deformability[6].

The experimental results are analyzed by the derived dimensionless parameters, and are shown in Figs. 2.9, 2.10 and 2.11. The P value of each result is calculated by a T-test. Figure 2.9 shows the distribution of û. We can find that the patient's RBCs have less deformability because the û are smaller, which means that the drop of velocity due to resistance force are greater. Figure 2.10 shows the distribution of $\hat{\delta}$. Since the patient's RBCs are smaller, in terms of normalized size $\hat{\delta}$, the amount of deformation for through the microchannel is less than the RBCs of the healthy subject. Figure 2.11 shows the distribution of $\hat{\lambda}$. We find that the patient has lower deformability of their RBCs in terms of the elongation ratio, $\hat{\lambda}$.

For evaluating cell stiffness-based deformability, all three dimensionless parameters, $\hat{\delta}$, $\hat{\lambda}$ and û, have to be considered as a whole. Therefore, we introduce a new dimensionless $\widehat{DI}$ by putting them together as

$$\widehat{DI} = \hat{\delta}\hat{\lambda}\hat{u} = \frac{D_c \lambda u_{eq}}{w^2 u_f} \tag{2.22}$$

[6] Here we consider cell deformability as the easiness of a cell to be deformed. Thus, high deformability reflects low stiffness and vice versa.

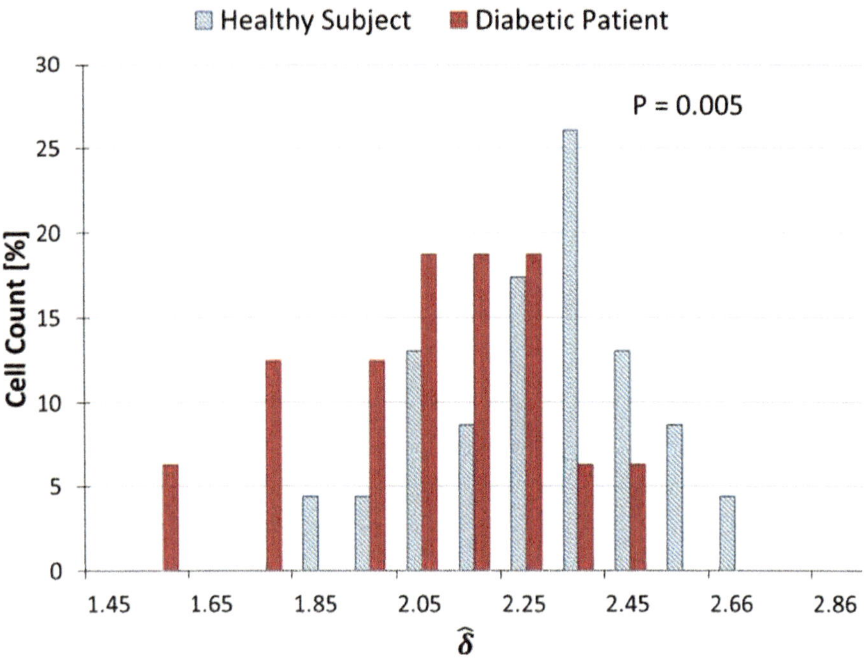

Fig. 2.10 The distribution of dimensionless parameter $\hat{\delta}$ for the healthy subject and the patient.

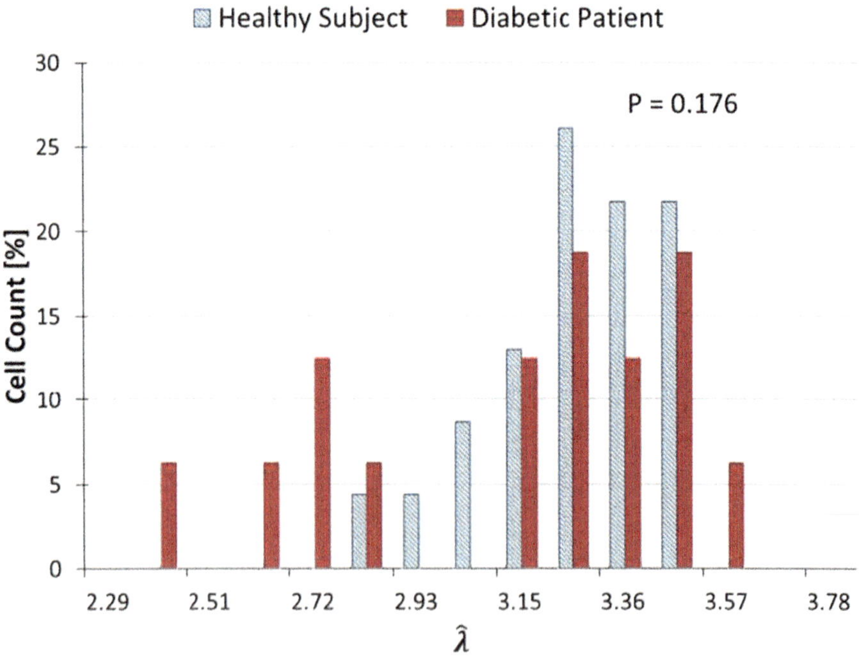

Fig. 2.11 The distribution of dimensionless parameter $\hat{\lambda}$ for the healthy subject and the patient

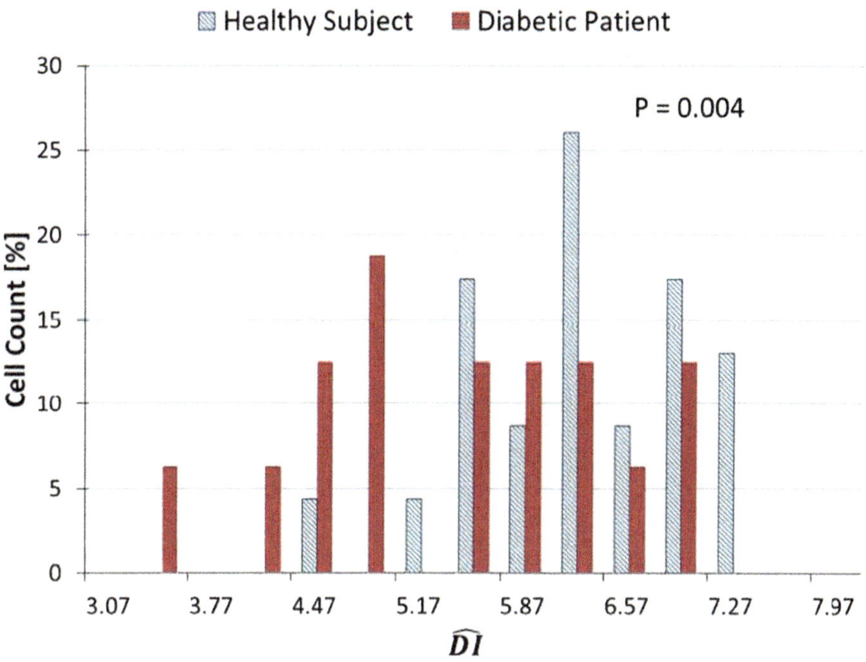

Fig. 2.12 The distribution of dimensionless stiffness index for the healthy subject and the patient

Figure 2.12 shows the distribution of $\widehat{DI}$ or both subjects. P value of 0.004 is obtained from T-test, and it tells that the difference between the patient and healthy subject is significant. Among all dimensionless parameters, $\widehat{DI}$ gives the smallest value of P which shows that $\widehat{DI}$ can distinguish a patient from a healthy person most effectively.

2.6.2 *Comparison with Transit-time Method*

The same set of experimental results are evaluated by transit-time method for comparing with the newly proposed $\widehat{DI}$ in Fig. 2.12. For compensating different flow-in velocities of cells, we normalize the transit time, *T*, by the fluid velocity, u_f,

$$\widehat{T} = \frac{T}{\lambda_m / u_f} \tag{2.23}$$

Where $\widehat{T}$ is a comparable index of cell deformability of transit-time method, and is dimensionless. Figure 2.13 shows the distribution of $\widehat{T}$ for the two subjects with $P=0.014$. It is shown that the healthy subject's RBCs have better deformability than the patient from the viewpoint of transit time. However, it is hard to tell if the

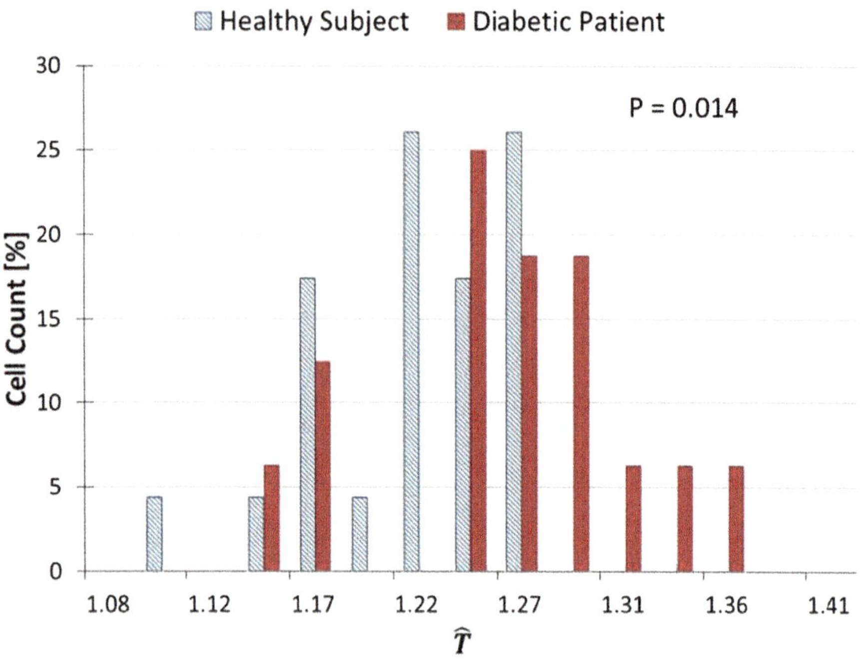

Fig. 2.13 The distributions of the dimensionless transit time, $\hat{T}$, in Eq. (2.24) for the healthy subject, in blue, and the diabetic patient, in red, are shown. $\hat{T}$ represents overall deformability, both cell stiffness and cell viscosity, of RBCs without considering individual sizes

difference of the distributions is caused by the cell stiffness, viscosity or both of the RBCs while these parameters are all considered in the newly proposed $\widehat{DI}$.

By comparing the results of the proposed dimensionless $\widehat{DI}$ in Fig. 2.12 and the one of the transit-time method in Fig. 2.13, there are two main advantages of the proposed $\widehat{DI}$ over the transit-time method used in conventional approaches:

1. $\widehat{DI}$ represents only the stiffness of a cell without the effect coming from its viscosity because only the cell motion in equilibrium is considered.
2. $\widehat{DI}$ takes into account not only cell motion but also the size variations among the tested cells while the conventional method includes no information of cell size, and is only valid for comparing the cells with similar size.

2.7 Conclusions

Four concluding remarks of this chapter are summarized as follows:

- A new parameter, equilibrium velocity, for evaluating cell stiffness using a microchannel is introduced. By using equilibrium velocity, we can eliminate the effect coming from time-dependent characteristics of a cell, such as cell viscosity, and evaluate cell stiffness alone.

- The observability of stiffness-based deformability is defined as a criteria for the development of future system. Cell stiffness-based deformability is said to be observable if the equilibrium velocity of a cell inside a microchannel is measurable.
- Dimensional analysis is performed for the microchannel system, and a new dimensionless $\widehat{DI}$ based on three derived dimensionless parameters is proposed. Experimental study on a diabetic patient and a healthy subject were conducted, and the results show that the deformability of patient's RBCs is lower than the ones of healthy subject.
- The proposed dimensionless $\widehat{DI}$ is demonstrated to be superior to the conventional one in the stiffness evaluation of the experimental results. The $\widehat{DI}$ eliminates viscosity effect, and also takes individual cell size into consideration.

Acknowledgement We would like to thank Dr. Sakata (M.D. Ph.D.), Dr. Ohtani (M.D. Ph.D.) and Dr. Taniguchi (M.D.) for their help in preparing blood samples from the subjects. This work is supported by The Ministry of Education, Culture, Sports, Science and Technology (MEXT) of Japan Grant-in-Aid for Scientific Research on Innovative Areas "Bio Assembler".

Appendix A Physical Quantities in Microchannel system

The model in *Thin-Film Lubrication Theory* [32] is adopted, and it is claimed that a very thin layer of fluid always exists between two objects. In other words, there is no direct contact between a cell and channel wall, and the interaction always through a thin layer of fluid between them. While a cell through a channel, the resistance, F_R, would be the shear force given by

$$F_R = \mu \frac{u_{eq}}{d_g} A_s \tag{2.24}$$

where μ, ,u_{eq}, d_g and A_s are fluid viscosity, equilibrium velocity, gap size, and the area where the force is exerted. By assuming the shape of deformed cell inside the channel is a cylinder, we have

$$A_s = \pi w \lambda \tag{2.25}$$

where λ is the cell length inside the channel (a.k.a. in-channel length). d_g can be regarded as a function of the compression force, F_c, acting on the cell

$$F_c = k\left(D_c - w\right) \tag{2.26}$$

where k, D_c and w are cell stiffness, undeformed diameter and channel width, respectively. Thus, we have

$$F_R = R\left(k, D_c, \lambda, u_{eq}, w, \mu\right) \tag{2.27}$$

On the other hand, the force pushing a cell forward (the pushing force, F_p) is

$$F_P = \Delta P A_c \tag{2.28}$$

where ΔP and A_c are the pressure difference between two sides of the channel and cross-sectional area, respectively. A_c is

$$A_C = \pi\left(\frac{w}{2}\right)^2 \tag{2.29}$$

Because it is difficult to directly measure the ΔP in a microchannel experimentally, fluid velocity is employed for the information of flow. We know that ΔP is a function of fluid velocity, u_f according to Hagen-Poiseuille equation [34], thus we have

$$\Delta P = H\left(u_f\right) \tag{2.30}$$

From Eqs. (2.28)–(2.30) we have

$$F_P = P\left(u_f, w\right) \tag{2.31}$$

While a cell reaches equilibrium, the force pushing cell forward, F_P and backward, F_R, are balanced, and can be represented by

$$P\left(u_f, w\right) = R\left(k, D_C, \lambda, u_{eq}, w, \mu\right) \tag{2.32}$$

In summary, we have

$$\mathcal{F}\left(k, D_C, \lambda, u_{eq}, u_f, w, \mu\right) = 0 \tag{2.33}$$

which shows that 7 physical quantities, k, , D_c, λ, u_{eq}, u_f, w and μ, determine how a cell moving inside a microchannel.

References

1. Bow H, Pivkin IV, Diez-Silva M, Goldfless SJ, Dao M, Suresh S, Niles JC, Han J (2011) A microfabricated deformability-based flow cytometer with application to malaria. Lab Chip 11:1065–1073
2. Glenister FK, Coppel RL, Cowman AF, Mohandas N, Cooke BM (2002) Contribution of parasite proteins to altered mechanical properties of malaria-infected red blood cells. Blood 99(3):1060–1063
3. Baskurt OK, Gelmont D, Meiselman HJ (2003) Red blood cell deformability in sepsis. Am J Respir Critical Care Med 157(2):421–427
4. Tsukada K, Sekizuka E, Oshio C, Minamitani H (2001) Direct measurement of erythrocyte deformability in diabetes mellitus with a transparent microchannel capillary model and high-speed video camera system. Microvasc Res 61:231–239

5. Brandao M, Fontes A, Barjas-Castro M, Barbosa L, Costa F, Cesar C, Saad S (2003) Optical tweezers for measuring red blood cell elasticity: application to the study of drug response in sickle cell disease. Eur J Hematol 70:207–211
6. Zheng Y, Shojaei-Baghini E, Azad A, Wang C, Sun Y (2012) High-throughput biophysical measurement of human red blood cells. Lab Chip 12:2560–2567
7. Hou HW, Li QS, Lee GYH, Kumar AP, Ong CN, Lim CT (2009) Deformability study of breast cancer cells using microfluidics. Biomed Microdevices 11:557–564
8. Adamo A, Sharei A, Adamo L, Lee B, Mao S, Jensen KF (2012) Microfluidics-based assessment of cell deformability. Anal Chem 84:6438–6443
9. Rosenbluth MJ, Lam WA, Fletcher DA (2008) Analyzing cell mechanics in hematologic disease with microfluidic biophysical flow cytometry. Lab Chip 8:102–1070
10. Carlo DD (2012) A mechanical biomarker of cell state in medicine. J Lab Autom 17(1):32–42
11. Roth KB, Eggleton CD, Neeves KB, Marr DWM (2013) Measuring cell mechanics by optical alignment compression cytometry. Lab Chip 13(8):1571–1577
12. Tan Y, Sun D, Wang J, Huang W (2010) Mechanical characterization of human red blood cells under different osmotic conditions by robotic manipulation with optical tweezers. IEEE Trans Biomed Eng 57(7):1816–1825
13. Tomaiuolo G, Barra M, Preziosi V, Cassinese A, Rotoli B, Guido S (2010) Microfluidics analysis of red blood cell membrane viscoelasticity. Lab Chip 11:449–454
14. Zhang H, Liu K (2008) Optical tweezers for single cells. J Royal Soci Interface 5:671–690
15. Lee GYH, Lim CT (2007) Biomechanics approaches to studying human diseases. Trends Biotechnol 25(3):111–118
16. Wojcikiewicz EP, Zhang X, Moy VT (2004) Force and compliance measurements on living cells using atomic force microscopy (afm). Biological Proc Online 6:1–9
17. Binnig G, Quate CF, Gerber C. Atomic force microscope. Phys Rev Lett 56(9):930–933
18. Radmacher M, Fritz M, Kacher CM, Cleveland JP, Hansma PK (1996) Measuring the viscoelastic properties of human platelets with the atomic force microscope. Biophys J 70:556–567
19. Dao M, Lim CT, Suresh S (2003) Mechanics of the human red blood cell deformed by optical tweezers. J Mech Phys Solids 51:2259–2280
20. Worthen GS, Schwab B 3rd, Elson EL, Downey GP (1989) Mechanics of stimulated neutrophils: cell stiffening induces retention in capillaries. Science 245(4914):183–186
21. Isermann P, Davidson PM, Sliz JD, Lammerding J (2012) Assays to measure nuclear mechanics in interphase cells. Curr Prot Cell Biol 56(22.16):1–21
22. Youn S, Lee DW, Cho Y (2008) Cell-deformability-monitoring chips based on strain-dependent cell-lysis rates. J Microelectromech Syst 17(2):302–308
23. Byun S, Son S, Amodei D, Cermak N, Shaw J, Kang JH, Hercht VC, Winslow MM, Jacks T, Mallick P, Manalis S (2013) Characterizing deformability and surface friction of cancer cells. Proc Natl Acad Sci PNAS 110(19):7580–7585
24. Gossett DR, Tse HTK, Lee SA, Ying Y, Lindgren AG, Yang OO, Rao J, Clark AT, Carlo DD (2012) Hydrodynamic stretching of single cells for large population mechanical phenotyping. PNAS 109:7630–7635
25. Hirose Y, Tadakuma K, Higashimori M, Arai T, Kaneko M, Iitsuka R, Yamanishi Y, Arai F (2010) A new stiffness evaluation toward high speed cell sorter. In Proceedings of the IEEE International Conference on Robotics and Automation, ICRA, pp 4113–4118, Anchorage, USA, May 2010
26. Tsai CD, Kaneko M, Sakuma S, Arai F (2012) Evaluation of cell impedance using a μ-channel. In Proceedings of the IEEE Engineering in Medicine & Biology Society, EMBC, pp 5518–5521, San Diego, USA, August 2012
27. Tsai CD, Kaneko M, Sakuma S, Arai F (2012) Phase decomposition of a cell passing through a μ-channel—a method for improving the evaluation of cell stiffness. In Proceedings IEEE International Conference on Mechatronics and Automation, ICMA, pp 138–143, Chengdu, China, August 5–8 2012

28. Tsai CD, Kaneko M, Sakuma S, Arai F (2012) Enhanced cell stiffness evaluation by two-phase decomposition. In Proceedings of the 16th International Conference on Miniaturized Systems for Chemistry and Life Sciences, μTAS12, pp 1009–1011, Okinawa, Japan, October 2012
29. Tsai CD, Kaneko M, Sakuma S, Arai F (2013) Observability of cell stiffness in micro-channel method. In Proceedings of the IEEE International Conference on Robotics and Automation, ICRA, pp 2792–2798, Karlsruhe, Germany, May 2013
30. Fung YC (1993) Biomechanics: mechanical properties of living tissues. Springer-Verlag, New York (http://www.springer.com/gp/book/9780387979472)
31. Tsai CD, Sakuma S, Kaneko M, Arai F (2013) Normalization of flow-in velocity for improving the evaluation on cell deformability. In *Proceedings* IEEE International Conference on Mechatronics and Automation, ICMA, pp 261–266, Takamatsu, Japan, August 2013
32. Elrod HG (1973) Thin-film lubrication theory for newtonian fluids with surfaces possessing striated roughness or grooving. J Lubrication Tech. 95(4):484–489
33. Gukhman AA (1965) Introduction to the theory of similarity. Academic, New York
34. Herrmann S, Klaus G (2003) Boundary-layer theory. Springer-Verlag Berlin, (http://www.springer.com/gp/book/9783540662709)

Chapter 3
Real-time Capillary-level Microchannel Flow Analysis Using a Full-pixel Frame-straddling Micro-PIV System

Idaku Ishii and Tadayoshi Aoyama

Abstract We develop a high throughput microscopic particle image velocimetry system that can compute flow vectors of 512 × 512 pixels in real time at 500 fps for fast microchannel flow. To compute many flow vectors at high speed, a gradient-based optical flow method is accelerated by calculating integral images of product sums of image brightness gradients and implementing them as parallel processes on a GPU-based frame-straddling high-frame-rate (HFR) vision system. Thus, the HFR vision system, having two cameras with a time delay function, can simultaneously compute hundreds of millions of flow vectors in a second, assuming a small image displacement between frames with a submillisecond delay. We conducted real-time flow measurement experiments to quantify capillary-level microchannel flow in a microfluidic chip with many 7-μm-width channels, and verified the high throughput performance of our system for long-term microchannel blood flow analysis.

Keywords Microscopic flow measurement · Particle image velocimetry (PIV) · Real-time image processing

3.1 Introduction

Microscopic particle image velocimetry (micro-PIV) [1, 2] is a well-known method of flow visualization that obtains the velocities and related properties of micro-channel flows from videos. Micro-PIV systems have been widely used in various applications such as pressure-driven microchannel flow [3], laminar flow and turbulence transition analysis [4], 3-D flow analysis based on stereo principles [5], and inkjet printhead analysis [6]. Such micro-PIV techniques enable microchannel flow measurement with high space resolution. Further, many studies on blood flow have been reported for understanding blood flow behavior in microcirculation:

I. Ishii (✉) · T. Aoyama
Department of System Cybernetics, Hiroshima University,
1-4-1, Kagamiyama, Higashi-Hiroshima, Japan
e-mail: iishii@robotics.hiroshima-u.ac.jp

T. Arai et al. (eds.), *Hyper Bio Assembler for 3D Cellular Systems*,
DOI 10.1007/978-4-431-55297-0_3

Sugii et al. [7] reported that blood flow in a straight microchannel had a blunt velocity profile, Chiu et al. [8] analyzed the effect of blood flow on monocyte adhesion to endothelial cells, Kim et al. [9] analyzed the blood behavior in a circular opaque microchannel using an X-ray PIV technique, Venneman et al. [10] measured the blood-plasma velocity in the beating heart of a chicken embryo, and Lima et al. [11] analyzed in vitro blood flow behavior with submicrometer optical thickness using a confocal micro-PIV system. Most these micro-PIV studies processed high-frame-rate (HFR) videos for the apparent fast microchannel flows, because the apparent velocity of microchannel flow in a microscopic image becomes larger as magnification is increased.

However, most micro-PIV systems processed the HFR videos offline for post microchannel flow analysis and human visualization because flow in these systems was estimated by cross-correlation based methods [12, 13], which require heavy computation for flow estimation. HFR video-based systems were limited to flow analysis for short time spans because offline HFR cameras had insufficient memory. Thus, current micro-PIV technology cannot always function as a real-time sensor to measure microchannel flow for long observation periods. To realize real-time micro-PIV, we developed a real-time frame-straddling HFR vision system [14] that can synchronize two camera inputs for the same camera view with a time delay on the order of microseconds. Assuming a small image displacement between frames with a tiny time delay, fast microchannel flow was simultaneously estimated using a gradient-based optical flow method, which is more suitable for real-time processing than cross-correlation-based PIV methods because flow distribution is estimated by calculating local brightness gradients, which do not need heavy computation. However, the number of measurement points per unit of time is limited by the performance of the personal computer (PC) in [14] because the optical flow estimation was executed in software on a PC. If we could accelerate the optical flow estimation to increase the number of measurement points in micro-PIV, full-pixel microchannel flow sensing could be conducted in real time at a high frame rate, enabling long-term microchannel flow analysis with high space and time resolution without the need for flow sensors to be physically attached.

This study develops a high-throughput micro-PIV system that can simultaneously compute hundreds of millions of flow vectors in a second for fast microchannel flow by accelerating a gradient-based optical flow algorithm on a GPU-based frame-straddling HFR vision system, which has two cameras with a submillisecond-delay function. The optical flow estimation was accelerated by computing integral images for product sums of brightness gradients of input images, which are required in the gradient-based optical flow method, and implementing them as parallel processes on the GPU-based frame-straddling HFR vision system.

Full-pixel flow vectors of 512×512 pixels can be computed for fast microchannel flow in real time at 500 fps. The performance of our system is verified using experimental results to quantify the spatio-temporal transitions of horse blood fast-flowing in a 7-μm-width microchannel array at the capillary level.

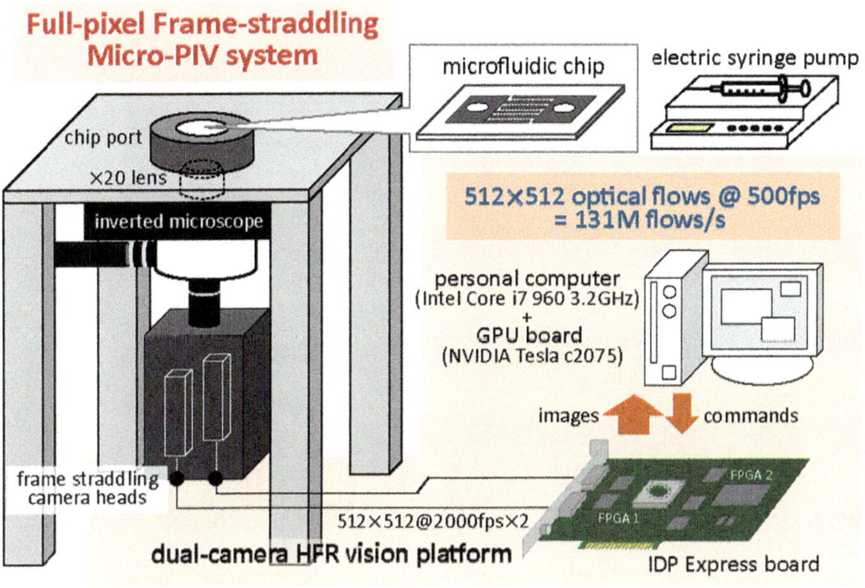

Fig. 3.1 System configuration

3.2 Frame-Straddling Micro-PIV System

Our micro-PIV system, an improvement on our previously developed frame-straddling micro-PIV system [14], comprises an HFR vision platform (IDP Express) [15] with a frame-straddling function for two camera inputs, a GPU board (Tesla c2075; NVIDIA Co., USA), a personal computer (PC), an inverted microscope with a microfluidic chip with 7-μm wide channels (MWA-MCFANbasic; Kikuchi Microtechnology Co., Japan), and an electric syringe pump (KDS200; KD Scientific Inc., USA). Figures 3.1 and 3.2 show its configuration and overview, respectively.

The IDP Express was designed to implement real-time video processing and recording of 512 × 512 images at 2000 fps. It consists of two camera heads and a dedicated FPGA image processing board (IDP Express board). The camera head has a CMOS image sensor of 512 × 512 pixels; its sensor and pixel sizes are 5.12 × 5.12 mm and 10 × 10 μm, respectively.

It can capture 8-bit gray-level 512 × 512 images at 2000 fps. The dimensions and weight of the camera head are 23 × 23 × 77 mm and 145 g. Figure 3.2b shows a camera housing mounted at the camera port of the microscope in which two camera heads of the IDP Express are arranged so that they can capture a common view via a prism. The IDP Express board has two camera inputs, so that two 512 × 512 images and their processed results can be mapped onto the standard memory of a PC at 2000 fps via a PCI-e bus. In this study, the IDP Express board was dedicated for dual-camera frame-straddling by improving the hardware logic for time delay

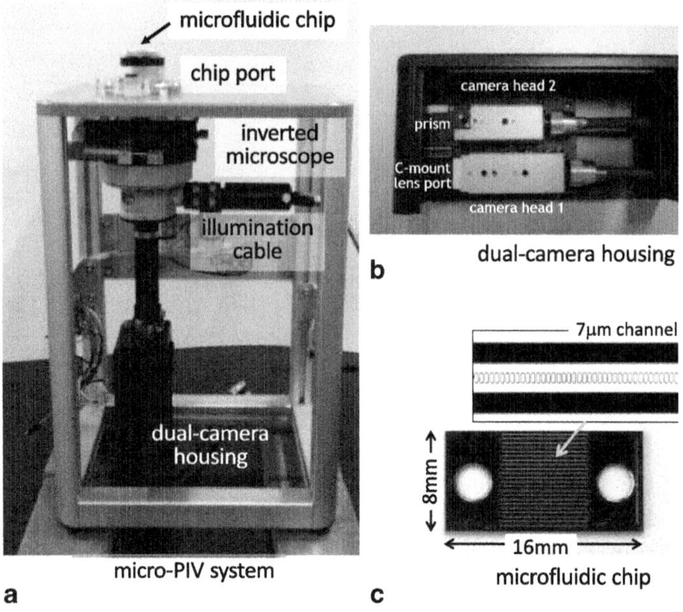

Fig. 3.2 System overview

control between the two camera heads; the image capture time delay is controlled via the PC from 0 to 0.5 ms in 9.9 ns steps.

The Tesla c2075 is a computing processor board accelerated by a NVIDIA Fermi GPU GF110. It has a processing performance of 1.03 TFlops using 448 ~ processor cores operating at 1.15 GHz, a bandwidth of 144 GB/s, an inner global memory of 6 GB, and fast shared memory of 64 kB. A PC with 16-lane PCI-e 2.0 buses and a processor chipset with DMA were adopted to transfer memory-mapped data between standard memory and the Tesla c2075 via the PCI-e bus. We used a PC with an ASUS P6T7 WS SuperComputer motherboard, Intel Core ™ i7 960 3.2 GHz CPU, and 3 GB RAM. We used a CUDA IDE provided by NVIDIA for coding the algorithms that had dedicated API functions for IDP Express in Windows 7 (32 bit), enabling us to access memory mapped data.

In this study, we observed microchannel flows in many 7-μm-width channels in an 8 × 16 mm-size microfluidic chip device of 0.5 mm thickness [16], as shown in Fig. 3.2c. It was fabricated of single crystal silicon for flow assessment at the capillary level. The microfluidic chip had 7854 channels, and the width, length, and depth of the channel were 7, 30, and 4.5 μm, respectively. Thirteen channels of them were observed under a 20X objective lens on the microscope, illuminated in the microscopic view by a metal-halide light source (PCS-MH375RC; Optical Garden Co., Japan). The measurement area was 256 × 256 μm, and one pixel corresponded to 0.5 μm in the 512 × 512 pixels microscopic view on the 10 μm-pixel-pitch image sensor.

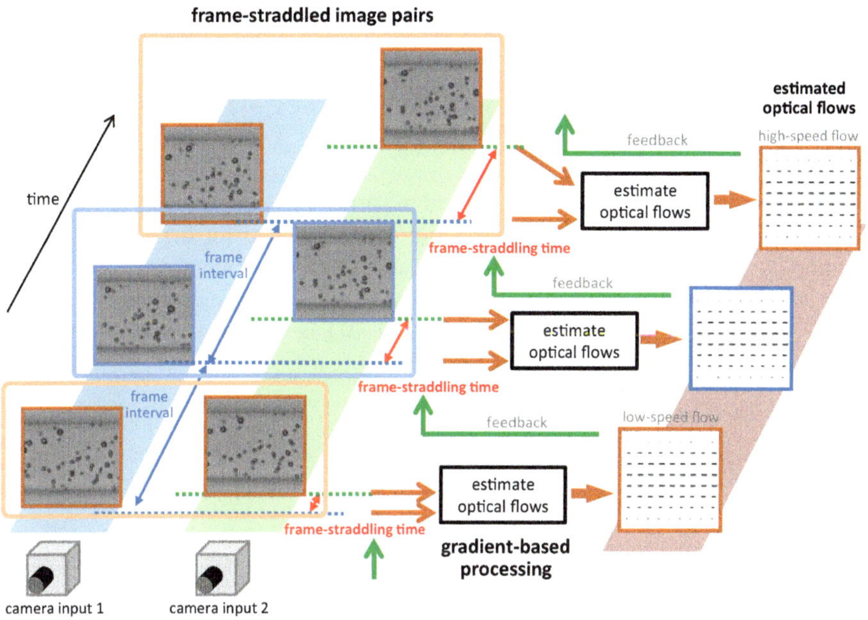

Fig. 3.3 Frame-straddling optical flow concept

3.3 Frame-Straddling Optical Flow Estimation Using Integral Gradient Images

3.3.1 Frame-Straddling Optical Flow

We proposed a VFS-OF algorithm [14] that improves on the Lucas-Kanade method [17] and is designed to estimate optical flow accurately for both high- and low-speed flows in the same system.

Figure 3.3 shows the VFS-OF method. Given a dual-camera high-speed vision system that synchronizes two camera inputs for the same view field with a time delay, the VFS-OF method adjusts its time delay for accurate, real-time gradient-based flow estimation so image displacement between frame-straddled images is controlled optimally to be small—around one pixel per frame on the subpixel order.

Compared to single-camera optical flow estimation at a fixed frame rate, VFS-OF improves the measurable velocity range by adjusting frame-straddling time to the amplitude of measured flow vectors. The VFS-OF method thus simultaneously estimates microchannel flow vectors at a variety of speeds while microchannel flow velocity is fluctuating rapidly in time and undergoing large changes in magnitude.

Here we assume that the two camera input images of a dual-camera high-speed vision system, ${}^1I(x,y,t)$ (camera input (1) and ${}^2I(x,y,t)$ (camera input 2), match perfectly:

$$ {}^1I(x,y,t) = {}^2I(x,y,t). \tag{3.1} $$

Product sums S_{xx}, S_{xy} and S_{yy} at t related to partial derivatives of the space direction use image of ${}^1I(x,y,t)$ (camera input 1) the same as for the Lucas-Kanade method. When the dual-camera frame interval is Δt, product sums related to partial derivatives of the time direction are calculated by dual-camera input ${}^1I(x,y,t)$, ${}^2I\left(x,\ y,\ t+\tau(t-\Delta t)\right)$ with time delay $\tau(t-\Delta t)$ determined by $t-\Delta t$.

$$ \begin{aligned} S_{xx} &= \sum_{\Gamma(x,y)} {}^1I_x {}^1I_x, \\ S_{xy} &= \sum_{\Gamma(x,y)} {}^1I_x {}^1I_y, \\ S_{yy} &= \sum_{\Gamma(x,y)} {}^1I_y {}^1I_y, \\ S'_{xt} &= \sum_{\Gamma(x,y)} {}^1I_x \frac{{}^1I(x,y,t) - {}^2I(x,y,t-\tau(t-\Delta t))}{\tau(t-\Delta t)}, \\ S'_{yt} &= \sum_{\Gamma(x,y)} {}^1I_y \frac{{}^1I(x,y,t) - {}^2I(x,y,t-\tau(t-\Delta t))}{\tau(t-\Delta t)}. \end{aligned} \tag{3.2} $$

Partial derivatives of space direction ${}^1I_x$, ${}^1I_y$ related to ${}^1I(x,y,t)$ are calculated as follows:

$$ \begin{aligned} {}^1I_x &= \frac{1}{2}\left({}^1I(x+1,y,t) - {}^1I(x-1,y,t)\right), \\ {}^1I_y &= \frac{1}{2}\left({}^1I(x,y+1,t) - {}^1I(x,y-1,t)\right). \end{aligned} \tag{3.3} $$

Frame-straddling time between the two camera inputs $\tau(t-\Delta t)$ is adaptively adjusted using flow vectors $(v_x(x,y,t-k\Delta t), v_y(x,y,t-k\Delta t))$ measured at time $t-k\Delta t (k=1,\cdots,K)$ calculated for K frames. This time-averaged flow vector distribution is calculated as follows:

$$ \begin{aligned} &(\tilde{v}_x(x,y,t-\Delta t), \tilde{v}_y(x,y,t-\Delta t)) \\ &\qquad = \left(\frac{1}{K}\sum_{k=1}^{K} v_x(x,y,t-k\Delta t), \frac{1}{K}\sum_{k=1}^{K} v_y(x,y,t-k\Delta t) \right). \end{aligned} \tag{3.4} $$

The time delay for a dual camera is determined by $(\tilde{v}_x(x,y,t-k\Delta t),\tilde{v}_y(x,y,t-k\Delta t))$ after calculating time-averaged flow vector distribution as follows:

$$\tau(t-\Delta t) = \frac{A}{\max_{x,y}\left(\sqrt{\tilde{v}_x^2(x,y,t-\Delta t)+\tilde{v}_y^2(x,y,t-\Delta t)}\right)}, \tag{3.5}$$

where A is a constant that indicates the control target to estimate flow velocity. A is perfectly set to a small value of around one pixel or less on the subpixel order in the case of optical flow estimation using the Lucas-Kanade method.

Based on the amplitudes of measured flow vectors, frame-straddling time $\tau(t-\Delta t)$ is adjusted in Eq. (3.5), a small frame-straddling time is set for high-speed flow, and a large frame-straddling time is set for low-speed flow. Flow vector $(v_x(x,y,t),v_y(x,y,t))$ at time t is thus estimated by controlling frame-straddling time based on flow vectors $(v_x(x,y,t-k\Delta t),v_y(x,y,t-k\Delta t))$ $(k=1,\cdots,K)$ measured by time $t-\Delta t$ using the following equation:

$$\begin{pmatrix} v_x \\ v_y \end{pmatrix} = \frac{1}{S_{xx}S_{yy}-S_{xy}^2}\begin{pmatrix} S_{yy}S'_{xt}-S_{xy}S'_{yt} \\ -S_{xy}S'_{xt}+S_{xx}S'_{yt} \end{pmatrix} \tag{3.6}$$

Using the frame-straddling optical flow method with a time delay on the order of microseconds, we measured fast microchannel flow accurately in real time [14], however, the number of measurement points and the output rate of flow vectors were limited by PC performance. Two limited modes were implemented when $\Gamma(x,y)$ was set to a neighbor region of 16×16 pixels (m = 16): (1) intersection mode (16 points, 1000 fps), and (2) whole-image mode (16×16 points, 50 fps). This is because the frame-straddling optical flow method was implemented in software on a PC, and its computational complexity depended on the size of $\Gamma(x,y)$. The frame-straddling optical flow method in [14] requires a computational complexity of $O\left(fm^2N^2\right)$ when flow velocities for all pixels of an $\mathrm{N}\times\mathrm{N}$ image are estimated using an $\mathrm{m}\times\mathrm{m}$ cell to calculate the product sums of brightness gradients at frame rate f.

Thus, to realize full-pixel flow estimation for real-time microchannel flow analysis with high space and time resolution, it is necessary to reduce the computational complexity of the gradient-based optical flow method so that it is independent of the cell size used in the brightness gradient calculation. It is also important to increase the computational power available for optical flow estimation by a parallel implementation on a GPU-based HFR vision system.

3.3.2 Integral Gradient Images

In this study, the computational complexity needed to calculate the product sums of brightness gradients in Eq. (3.2) is reduced by calculating the following five integral gradient images ii_{xx}, ii_{xy}, ii_{yy}, ii_{xt}, and ii_{yt} for the product sum of the brightness gradients, I_x^2, I_xI_y, I_y^2, I_xI_t and I_yI_t, respectively, as follows:

$$s_{\xi\eta}(x,y) = s_{\xi\eta}(x-1,y) + M(x,y)I_\xi(x,y,t)I_\eta(x,y,t), \tag{3.7}$$

$$ii_{\xi\eta}(x,y) = ii_{\xi\eta}(x,y-1) + s_{\xi\eta}(x,y) \quad (\xi,\eta = x,y,t), \tag{3.8}$$

where $s_{\xi\eta}(x,y)$ is the cumulative row sum, the boundary conditions are $s_{\xi\eta}(-1,y)=0$ and $ii_{\xi\eta}(x,-1)=0$, and $M(x,y)$ is a one-bit mask image consisting of preassigned observable regions. The integral image of a gray-level image has been used to accelerate the computation of Haar-like features for fast face detection in [18], and our integral gradient image is its expansion for brightness gradient images.

When $\Gamma(x,y)$ is a region of $m \times m$ pixels, whose starting point is (a,b), the product sums of brightness gradients, $S_{\xi\eta}$, can be computed using the values of integral gradient images on the four vertices of $\Gamma(x,y)$ as follows:

$$S_{\xi\eta} = ii_{\xi\eta}(a,b) + ii_{\xi\eta}(a+m-1,b+m-1) - ii_{\xi\eta}(a+m-1,b) - ii_{\xi\eta}(a,b+m-1). \tag{3.9}$$

Using the integral gradient images, we can accelerate the computation of the product sums of brightness gradients, even for computations that uses a large cell-size because the computational complexity is independent of cell size. We can estimate flow vectors at all pixels of $N \times N$ images in the frame-straddling optical flow method with a computational complexity of $O(N^2)$, dependent only on image size.

3.3.3 *Implemented Algorithm and Its Specifications*

For full-pixel level optical flow estimation at high speed, we implemented the improved frame-straddling optical flow method as parallel logic on a GPU board in our dual-camera HFR vision system.

The algorithm implemented on the dual-camera HFR vision system has the following steps:

1. Image acquisition
   Two 8-bit gray-level 512×512 input images, ${}^1B(x,y,t)$, and ${}^2B(x,y,t-\tau)$, captured from the two camera heads with time delay τ, are memory-mapped into PC memory.
2. Transfer of input images to GPU memory

The two PC-memory mapped images are transferred to the global memory on the GPU board.

3. Image correction

To align the locations and brightnesses of the two input images, the input image of camera 2 ${}^{2}B(x,y,t-\tau)$ is corrected based on that of camera 1 ${}^{1}I(x,y,t)={}^{1}B(x,y,t)$ as follows:

$$ {}^{2}I(x,y,t)=\mathrm{Blut}(\mathrm{Warp}({}^{2}B(x',y',t))), \tag{3.10} $$

where $\mathrm{Blut}(\cdot)$ is a lookup table that shows the relationship between the brightnesses of cameras 1 and 2. $\mathrm{Warp}(\cdot)$ is a function to transform the non-integer coordinates of a pixel to integers after bilinear interpolation. The non-integer coordinates $(x',y')=(a_1x+a_2y+a_5, a_3x+a_4y+a_6)$ of camera 2 are given by integer coordinates (x,y) of camera 1 along with the affine parameters of the geometric relationship between the images of cameras 1 and 2.

The correction process expressed in (3.10) is conducted in parallel on the GPU board for 16×16 blocks of 32×32 pixels.

The brightness lookup table $\mathrm{Blut}(\cdot)$ and the affine parameters $a_i (i=1,\cdots,6)$ are calculated a priori.

4. Calculation of integral gradient images

Using the two corrected images, ${}^{1}I(x,y,t)$ and ${}^{2}I(x,y,t)$, five products of brightness gradients, I_x^2, I_xI_y, I_y^2, I_xI_t and I_yI_t, are calculated in parallel on the GPU board for 16×16 blocks of 32×32 pixels.

Each of the five brightness gradient image $I_\xi I_\eta$ is scanned in the x direction to accumulated its cumulative row sum $s_{\xi\eta}(x,y)$ in parallel on the GPU board for 32×32 blocks of 16×16 pixels.

Next, each cumulative row sum $s_{\xi\eta}(x,y)$ is accumulated in parallel in the y direction to form its integral gradient image $ii_{\xi\eta}(x,y)$ on the GPU board for 32×32 blocks of 16×16 pixels.

5. Calculation of flow vector images

For each integral gradient image, the product sum of brightness gradient $S_{\xi\eta}$ is calculated by selecting the values of $ii_{\xi\eta}(x,y)$ on the four vertices of the $m\times m$ region $\Gamma(x,y)$, as expressed in Eq. (3.9). This process is conducted in parallel on the GPU board for 16×16 blocks of 32×32 pixels.

A full-pixel 512×512 flow vector image that gives the flow vectors (v_x,v_y) for all pixels of the input images is estimated using the five product sums of the brightness gradients, as expressed in Eq. (3.6). This process includes a time-averaging process using the flow vectors estimated at the ten previous frames for noise reduction; it is conducted in parallel on the GPU board for 16×16 blocks of 32×32 pixels.

Table 3.1 shows the execution times of the improved frame-straddling optical flow method when the flow vector images were calculated on our dual-camera HFR vision system; the execution times were independent of cell size when calculating the product sums of brightness gradients. The steps (2)-(5) were accelerated by executing them in parallel on the GPU board, and the total execution time for

Table 3.1 Execution times

Process	Time [ms]
Image acquisition	0.20
Image transfer CPU to GPU	0.18
Image correction	0.10
Integral gradient image calculation	0.72
Flow vector image calculation	0.40
Total	1.60

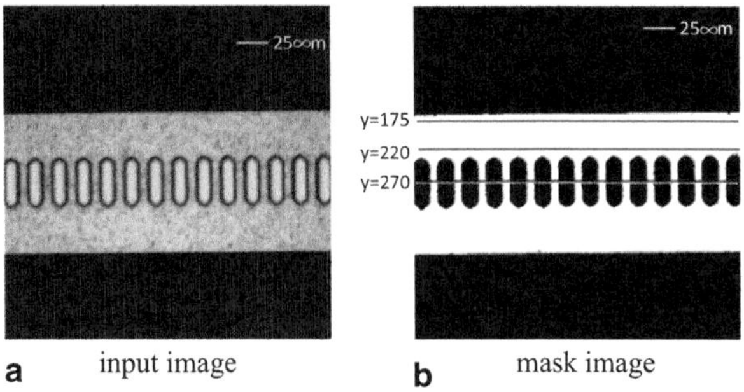

Fig. 3.4 Microchannels to be observed. **a** input image, **b** mask imgae

acquiring a flow vector image of 512 × 512 pixels was within 1.60 ms, including the data transfer time from the PC memory to the GPU board. We confirmed that the flow vectors at 512 × 512 points were estimated in real time at 500 fps; our improved micro-PIV system can compute 131 million flow vectors per second, eight thousand times more quickly than our previously reported real-time micro-PIV measurement [14].

3.4 Experiments

To verify the performance of our system, several experiments were conducted using defibrinated horse blood, which were flowing in a 7-μm-width microchannel array under a 20 × objective lens.

The liquids to be observed were supplied to the microfluidic chip using an electric syringe pump, and flowed through the channels from the lower to upper portions of the image.

Figure 3.4a shows the microscopic view. Microchannel flows through thirteen channels can be observed in the sufficiently illuminated region in the middle of the image. However, it was too dim to observe microchannel flows in the upper and lower regions in the image because of the channel structure of the microfluidic chip.

We preassigned a one-bit mask image $M(x, y)$ for the observable regions as shown in Fig. 3.4b, excluding the microchannel walls and dark areas.

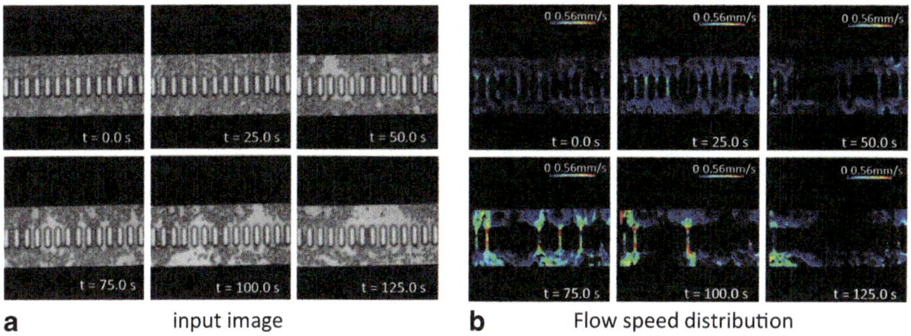

Fig. 3.5 Estimated microchannel blood flows. **a** input image, **b** Flow speed distribution

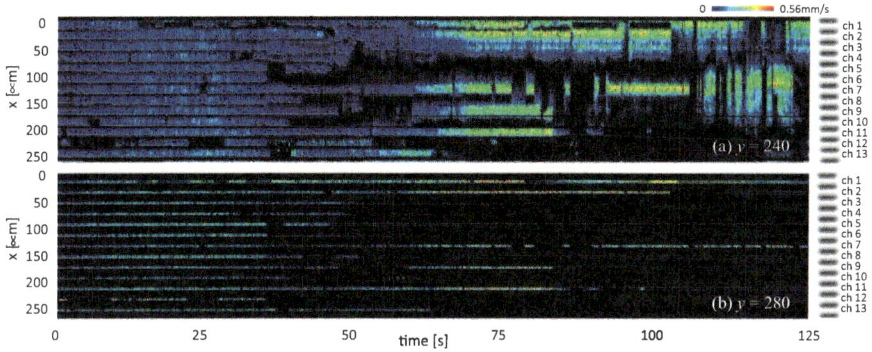

Fig. 3.6 Time-transient flow speeds for microchannel blood flow

The affine parameters and brightness lookup table between the two input images were calculated during prior dual-camera calibration. The exposure time of the camera heads was 50 μs.

The cell size for the optical flow estimation was set to 32 × 32 pixels ($m = 32$) in the experiments.

Time-transient microchannel flow distributions were measured when defibrinated horse blood was supplied to the microfluidic chip. The defibrinated horse blood used in the experiment was provided by Kojin Bio Co., Japan. The blood cells, having a diameter of 5–6 μm, were used as tracer particles for PIV measurements. The horse blood was supplied to the microfluidic chip at a flow rate of 25 μl/min. The frame-straddling time was =600 μs.

Figure 3.5 shows (a) a six image sequence of camera 1, and (b) the estimated flow speed distributions at intervals of 25 s. Here $t = 0$ was the start time of observation, and the duration of observation was 125 s. Figure 3.6 shows the time-transient profiles of flow speeds estimated on the intersection lines; $y = 240$ around the upper exits of the channels, and $y = 280$ at the centers of the channels. Figure 3.7 shows the temporal changes of flow speeds estimated at points at the centers of the channels.

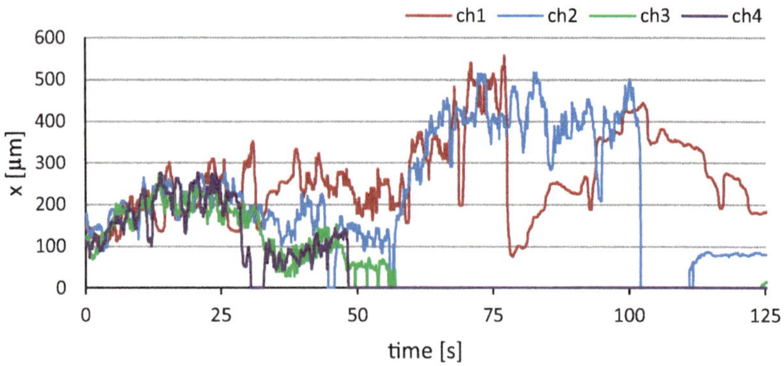

Fig. 3.7 Flow speeds at centers of microchannels $y=280$

In Fig. 3.7, the noise in the estimated flow speeds was reduced with median filter over sample data in a second.

It was observed that blood flowed in all the channels at the start of observation, however, the number of blocked channels increased as time passed; time-varying irregular flow paths were generated because of the blocked channels. The numbers of blocked or stagnant channels were 0, 1, 5, 8, 10 and 11 at $t=0$, 25, 50, 75, 100 and 125 s, respectively. This is because blood cells collected in the entrance of the channel and blocked it when they adhered to each other. In Fig. 3.7, the flow speeds estimated at channels 2, 3, and 4, ((62, 280), (100, 280), and (138, 280) in image (x, y) coordinates), in which blood flows were stopped in the middle of observation, are plotted, compared to those points estimated in channel 1, (27, 280), in which blood cells continuously flowed over 125 s. For channels 2, 3 and 4, the flow speeds estimated at $t=0$ s were 174, 115, and 112 µs, respectively, whereas blood flows were almost stopped at $t=102$, 57, and 48 s, respectively. For channel 1, the flow speed estimated at $t=0$ s was 169 µm/s, whereas the flow speed of channel 1 increased and irregularly changed as time passed; the flow speeds estimated at $t=75$ and 125 s were 503 and 182 µm/s, respectively. This is because the increase of blocked channels affected the flow speeds in the remaining non-blocked channels where blood smoothly flowed, and the number of blocked channels did not increase monotonically; blocked or non-blocked states were frequently perturbated in several channels.

3.5 Conclusion

In this study, we developed a real-time micro-PIV system that can simultaneously compute 131 million flow vectors per second; flow vectors of 512 × 512 pixels were estimated for fast microchannel flow in real time at 500 fps. Spatio-temporal transitions of blood flowing in a capillary-level microchannel array are quantified to verify the performance of our system for the long-term analysis of microchannel blood flow. We plan to develop an LOC-based long-term blood flow observation in a microchannel to monitor and quantify its spatio-temporal changing flow distribution, corresponding to the state of blood cells in capillary blood circulation.

References

1. Santiago JG, Wereley ST, Meinhart CD, Beebe DJ, Adrian RJ (1998) A particle image velocimetry system for microfluidics. Exp Fluids 25(4):316–319
2. Mielnik MM, Saetran LR (2005) Micro particle image velocimetry: an overview. Exp Fluids 38(3):1–8
3. Meinhart CD, Wereley ST, Santiago JG (1999) PIV measurements of a microchannel flow. Exp Fluids 27(5):414–419
4. Sharp KV, Adrian RJ (2004) Transition from laminar to turbulent flow in liquid filled microtubes. Exp Fluids 36(5):741–747
5. Klank H, Goranovic G, Kutter JP, Gjelstrup H, Michelsen J, Westergaard CH (2002) PIV measurements in a microfluidic 3D sheathing structure with three-dimensional flow behaviour. J Micromech Microeng 12(6):862–869
6. Meinhart CD, Zhang H (2000) The flow structure inside a microfabricated inkjet printhead. J Microelectromech Syst 9(1):67–75
7. Sugii Y, Nishio S, Okamoto K (2002) In vivo PIV measurement of red blood cell velocity field in microvessels considering mesentery motion. Physiol Meas 23(2):403–416
8. Chiu J, Chen C, Lee P, Yang C, Chuang H, Chien S, Usami S (2003) Analysis of the effect of distributed flow on monocytic adhesion to endothelial cells. J Biomech 36(12):1883–1895
9. Kim G, Lee S (2006) X-ray PIV measurements of blood flows without tracer particles. Exp Fluids 41(2):195–200
10. Vennemann P, Kiger KT, Lindken R, Groenendijk BCW, Stekelenburg-DeVos S, Ten Hagen TLM, Ursem NTC, Poelmann RE, Westerweel J, Hierck BP (2006) In vivo micro particle image velocimetry measurements of blood-plasma in the embryonic avian heart. J Biomech 39(7):1191–1200
11. Lima R, Wada S, Tanaka S, Takeda M, Ishikawa T, Tsubota K, Imai Y, Yamaguchi T (2008) In vitro blood flow in a rectangular PDMS microchannel: experimental observations using a confocal micro-PIV system. Biomed Microdevices 10(2):153–167
12. Hart DP (2000) Super-resolution PIV by recursive local-correlation. J Visual 3(2):13–22
13. Sugii Y, Nishio S, Okuno T, Okamoto K (2000) highly accurate iterative PIV technique using a gradient method. Meas Sci Technol 11(12):1666–1673
14. Kobatake M, Takaki T, Ishii I (2012) A real-time micro-PIV system using frame-straddling high-speed vision. Proceedings of the IEEE International Conference Robotics Automation pp 397–402, St. Paul, Minnesota, May 2012
15. Ishii I, Tatebe T, Gu Q, Moriue Y, Takaki T, Tajima K (2010) 2000 fps real-time vision system with high-frame-rate video recording. Proceedings of IEEE International Conference on Robotics and Automation pp 1536–1541, Anchorage, Alaska, May 2010
16. Kikuchi Y, Sato K, Mizuguchi Y (1994) Modified cell flow microchannels in a single-crystal silicon substrate and flow behavior of blood cells. Microvasc Res 47(1):126–139
17. Lucas B, Kanade T (1981) An iterative image registration technique with applications in stereo vision. Proceedings of the DARPA Image Understanding Workshop pp 121–130, Washington DC, April 1981
18. Viola P, Jones M (2001) Rapid object detection using a boosted cascade of simple features. Proceedings of the IEEE Conference on Computer Vision and Pattern Recognition, pp 511–518, Kauai, Hawaii, December 2001

Chapter 4
High-throughput Measurements of Single Cell Rheology by Atomic Force Microscopy

Kaori Kuribayashi-Shigetomi, Ryosuke Takahashi, Agus Subagyo, Kazuhisa Sueoka and Takaharu Okajima

Abstract The compliant mechanical properties of single cells have been extensively investigated and these properties are known to exhibit a strong dependence on the surrounding environments and also cell types, functions and conditions. An understanding of the cell behavior is important for applications of tissue engineering. Accurate rheological measurements are essential to elucidate the mechanisms of cell integrity and fluidity and are also key to mechanically identifying and separating single cells for cellular and tissue engineering. Of the various existing nano- and micro-rheology techniques, atomic force microscopy (AFM) shows great potential as a minimally invasive method. AFM allows mechanical measurements to be performed without the need for chemical modifications, *via* nano-scale contact between the AFM probe and the cell surface. In this chapter, we describe a recent advance in which micro-fabricated substrates are used for high-speed, automated AFM rheological measurements on size- and position-controlled cells.

Keywords Atomic force microscopy · Micro-fabricated substrates · Cell rheology · Single cell analysis · Complex shear modulus

---

T. Okajima (✉) · K. Kuribayashi-Shigetomi · R. Takahashi · A. Subagyo · K. Sueoka
Graduate School of Information Science & Technology, Hokkaido University, Kita-ku N14 W9, Sapporo 060-0814, Japan
e-mail: okajima@ist.hokudai.ac.jp

K. Kuribayashi-Shigetomi
e-mail: kaorik@ist.hokudai.ac.jp

R. Takahashi
e-mail: r-takahashi@ist.hokudai.ac.jp

A. Subagyo
e-mail: agus@ist.hokudai.ac.jp

K. Sueoka
e-mail: sueoka@ist.hokudai.ac.jp

T. Arai et al. (eds.), *Hyper Bio Assembler for 3D Cellular Systems*,
DOI 10.1007/978-4-431-55297-0_4

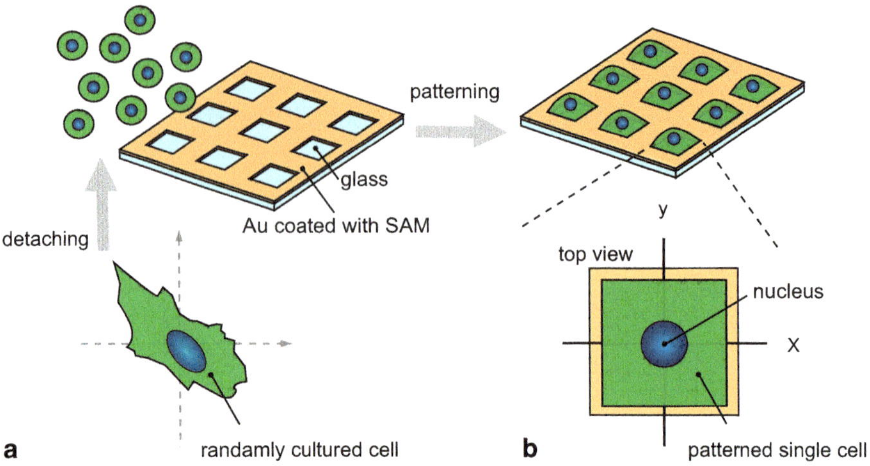

Fig. 4.1 Schematic of cells **a** before and **b** after adhering to glass regions of the patterned substrate. The Au surface is modified with a SAM to prevent cellular adhesion. The shape of cells grown on standard culture dishes is random, so it is difficult to define a coordinate in the cell for cell mechanical measurement (**a**). The shape of cells cultured on the micro-fabricated substrate is controlled, thus the position of the cells can be defined precisely

4.1 Introduction

Living cells can be classified into two types: adherent and non-adherent. Adherent cells attach to neighboring cells and/or extracellular matrices (ECMs) and thereby stabilize their structures that are associated with integrity, fluidity, and various cell functions. To understand their intrinsic biological properties, AFM has been extensively used to mechanically probe cells adhered to substrates. Detailed relationships between mechanical properties and cytoskeletal structures have been revealed by AFM at the single cell level [1].

However, less is known about the statistics of cellular mechanical properties, which is crucial to quantify and sort individual cells for collection. [2–5]. Although there are large spatial and temporal variations in the mechanical properties of adherent cells, there have been no techniques that provide accurate statistical information.

AFM can be used to acquire statistical information on adherent cell mechanics if the following two problems can be solved. The first problem is regarding the randomness of cell shapes adhering to the flat surfaces in standard culture dishes (Fig. 4.1a). In this case, cellular elastic moduli can vary at different positions within the same cell owing to structural heterogeneity. Thus it is important to define a uniform measurement position to quantify cell-to-cell mechanical variation. This problem can be solved by using a micro-fabricated substrate that arranges single cells in defined positions and controls their shape and cytoskeletal structures in a uniform geometric configuration (e.g., squares, Fig. 4.1b). Additionally, this cell patterning restricts lateral movement [6] and reduces temporal fluctuations of the cytoskeleton [7] so that highly reproducible measurements can be obtained. Two types of micro-fabricated substrates have been used: a commercial cell microarray

(LiveCell Array™, NUNC) composed of hexagonal microarray wells [8, 9], and a homemade micro-fabricated substrate discussed here.

The second problem is the limited scan range of commercial AFMs, which are optimized for nano-scale force measurements in local regions within a single cell [1]. To realize high-throughput AFM measurements of a large number of cells, the scan range must be expanded to the millimeter-scale [10]. By combining a large-scan-range AFM with cell micro patterning, as described below, we have been able to measure statistical distributions of cell moduli.

4.2 Micro-Fabricated Substrates Suitable for Single Cell Measurements by AFM

For high-throughput AFM rheology of single cells, the following features of micro-fabricated substrates are required: (i) Cells must be isolated and separately patterned, and (ii) regions where cells are placed on the substrates can be observed with an optical microscope to precisely place the AFM probe at the same measurement location of each cell. If these requirements are met, the mechanical properties of many cells over a large area can be acquired automatically.

Various methods of cellular pattering have been developed, including micro-contact printing [6, 7, 11, 12], photolithography [13–15] and stencil patterning [16, 17], in combination with a variety of cell-adhesive substrates, such as glass, metals, and plastics [18]. We use photolithography to fabricate micro-patterned gold (Au) on a glass substrate and then coat the Au with a self-assembled monolayer (SAM) of ethylene glycol (EG) to prevent cellular adhesion. The bare glass regions are sized for single cells and are arrayed uniformly with high resolution over a large area (Fig. 4.1a). These regions are easily identified with an optical microscope. Single cells are periodically arranged on the micro-fabricated substrate (Fig. 4.1b). Precise AFM positioning within each cell is achieved by observing the micro-fabricated patterns on the substrate instead of the patterned cells themselves.

The Au micro-patterned substrate was fabricated by standard photolithography techniques (Fig. 4.2). After the chromium (Cr)/Au thin film was deposited on the glass substrate (Fig. 4.2a), a photoresist (PR) was patterned with ultraviolet light and dissolved with a developer (Fig. 4.2b). Then, both Au and Cr layers were etched away (Fig. 4.2c). After removing PR, the substrate was finally immersed in an EG solution in ethanol (Fig. 4.2d). Before culturing cells on the patterned substrate, we coated the glass regions with fibronectin (FN) of ECMs to promote adhesion (Fig. 4.2e). Mouse fibroblast NIH/3T3 cells were plated onto the substrates at a density of 1×10^4 cells/cm^2. Non-adhered cells were washed out after a 4-h culture period (Fig. 4.2f).

An array of single NIH/3T3 cells cultured on the micro-patterned substrate is shown in Fig. 4.3a, b. We found that 900–1600 μm^2 square areas were suitable to pattern the single cells. The distance between the patterned edges is 30–40 μm,

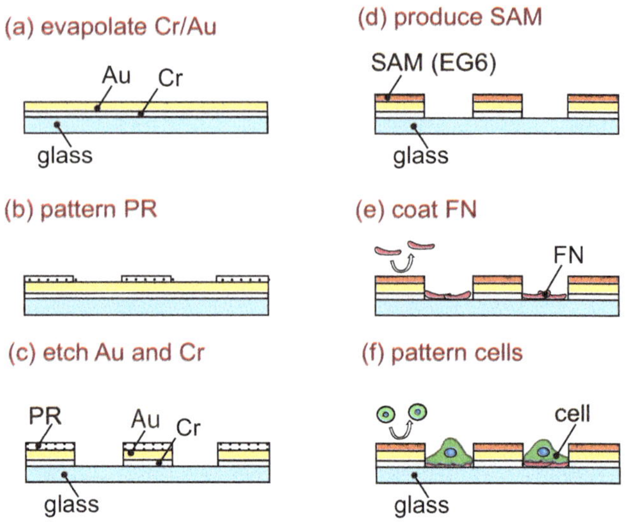

Fig. 4.2 Fabrication of micro-patterned Au substrates for single cell AFM analysis. The Cr/Au layers with 3 nm/25 nm in thickness were patterned by photolithography, and Au surface was coated by an EG6 SAM to resist cellular adsorption. The cells adhered in the ordered bare glass regions

which is a sufficient separation to prevent for cells bridge between the micropatterns [19, 20]. Phase contrast images merged with fluorescence images of labeled actin filaments (green) and nuclei (blue) (Fig. 4.3b) indicate that the cells stably adhered and fully spread into the squares. Moreover, the nuclei are located close to the center of each square cell, and the actin filaments align diagonally, indicating

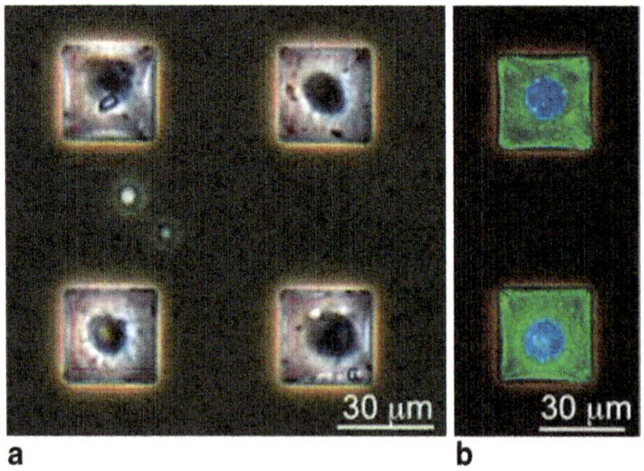

Fig. 4.3 Optical images of NIH/3T3 cells cultured after 12 h on the 30 μm × 30 μm etched regions of the Cr/Au patterned glass substrate. **a** Phase contrast image. **b** Phase contrast image merged with fluorescence image. Actin filaments and nuclei were labeled with Alexa Fluor Phalloidin 488 conjugate and Hoechst 33258, respectively

that the micro-fabricated substrate controls both the cell shape and the intracellular cytoskeletal structures [4]. Because the cells are highly stable, lateral movement is restricted, reducing temporal fluctuation of the cytoskeleton.

The micro-patterned substrates can be recycled by cleaning them ultrasonically in detergent and by UV ozone exposure. This is a distinct advantage when compared with other micro-fabricated substrates.

4.3 Coarse and Fine Scans for AFM Probing of Cells on Micro-patterned Substrates

Adherent cells can be patterned on the micro-fabricated substrates over centimeter-scale areas. To perform high-throughput AFM measurements over these distances, we need coarse positioning of the AFM probe over the cell array, followed by fine (nanoscale) scans for locations within the cell. This operation can be achieved by using two pulse motors for coarse x and y positioning that is synchronized to the AFM fine scanner system [9] where there is no limit on the coarse scanning range (Fig. 4.5b). Matching the coarse and fine positioning was performed by capturing sample images *via* a charge-coupled device (CCD) on the inverted optical microscope (Fig. 4.4a, b, c). Images of the substrates were binarized (Figs. 4.4a', b', c') and thus the absolute position was readily determined by image analysis of the Au-etched patterns. As a result, the position error caused from a slight tilt of the sample and from the mechanical drift of the AFM system could be precisely calibrated and corrected during large sample scans.

4.4 Statistics of Single-Cell Rheology Measured by AFM

The rheological properties of each cell were characterized by probing it locally with an AFM cantilever probe that was vertically oscillating with a small amplitude at frequency f. This force modulation method is based on measuring sample-induced changes in amplitude and phase with a lock-in amplifier (Fig. 4.5b) [21–24]. To reduce cell damage, a colloidal bead was often used [8, 9, 25]. From the AFM data, the storage G' and loss G'' moduli of the cell were determined as a function of f.

The complex shear modulus G^* is estimated from a Hertz model expressed as [1]:

$$F_0 = \frac{4E_0 R^{1/2}}{3(1-\nu^2)} \delta_0^{3/2}, \tag{4.1}$$

where F_0 is the loading force, E_0 is Young's modulus of the cell, R is the radius of the probe, δ_0 is the static indentation depth of the probe, and ν is Poisson's ratio of

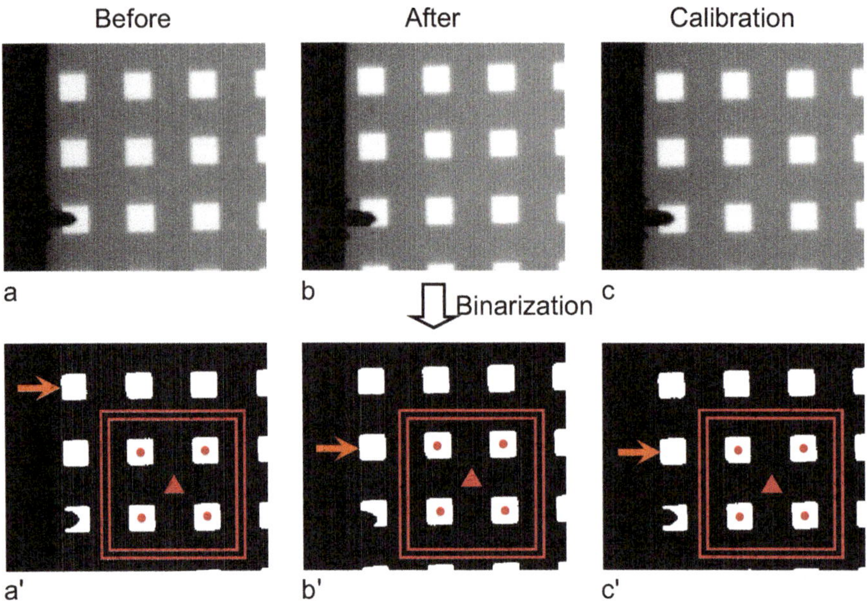

Fig. 4.4 **a–c** Bright field images of micro-fabricated substrates. **a** Before and **b** after moving the sample stage with pulse motors and **c** after calibrating the position error with the AFM scanner. **a'–c'** Respective binarized images of bright field images. The same glass region is indicated by *arrows*. The region of interest indicated by a *square* with a *double line* includes four glass regions. The dots represent the centroid of each pattern and the triangles represent the middle point of four centroids. The sample position was determined from the point of the triangle. The position error was estimated from the displacement of the triangles **a'** before and **b'** after moving the sample stage. The relative position of the cantilever tip to glass region after calibration was nearly identical to that before moving (**a** and **c**)

the cell (assumed here to be 0.5). In force modulation mode, the indentation depth δ is the sum of δ_0 and the small transient indentations δ^*_1 due to the oscillating cantilever (Fig. 4.5a):

$$\delta = \delta_0 + \delta_1^* e^{i\omega t}. \tag{4.2}$$

The asterisk indicates a complex number, $e^{i\omega t}$ is Euler's formula, and ω is the angular frequency of the cantilever oscillation. By using Eqs. 4.1 and 4.2, the loading force F^* is approximately expressed as [21–24]:

$$F^* = F_0 + F_1^* = \frac{4R^{1/2}}{3(1-\nu^2)}\left(E_0\delta_0^{3/2} + \frac{3}{2}E_1^*\delta_0^{1/2}\delta_1^*\right), \tag{4.3}$$

where F_1^* is the frequency-dependent term, and the frequency-dependent E_1^* is given by $2(1+\nu)G^*$ [23, 26].

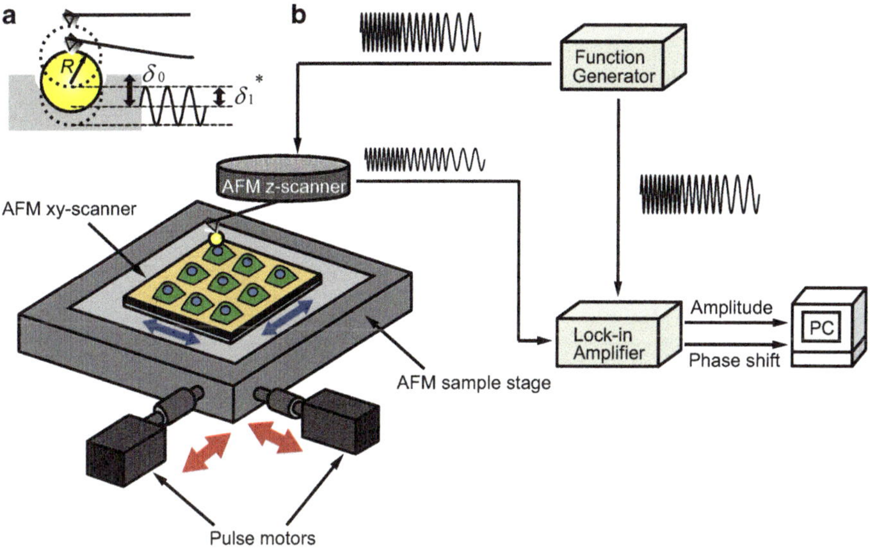

Fig. 4.5 **a** Schematic of indentation by pressing an AFM cantilever probe with a colloidal bead to the sample. The probe was relatively and vertically modulated to the sample. **b** Schematic of the AFM force modulation measurements on cells on micro-patterned substrates. Big *double-headed arrows* represent x and y coarse positioning using two pulse motors and the small arrows represent fine positioning using the AFM scanner

Assuming a viscous effect from the surrounding liquid, G^* is given by:

$$G^* = G' + iG'' = \frac{1-\nu}{4(R\delta_0)^{1/2}}\left[\frac{F_1^*}{\delta_1^*} - ib(0)f\right], \tag{4.4}$$

where $F_1^* = 2(R\delta_0)^{1/2}E_1^*\delta_1^*/(1-\nu^2)$. The drag factor $b(0)$ at contact was determined by extrapolating values of $b(h)$, where h is the measured distance between the probe and the cell [27].

Figure 4.6 shows cellular G' and G'' measured at the center of the square-shaped patterns. The results shown here are consistent with previous studies [8, 9, 28–30] where the cell modulus exhibited a log-normal distribution, the geometric mean increased, and the variation decreased with increasing frequency.

Figure 4.7a plots the geometric mean of G' and G'' of the NIH/3T3 cells as a function of f over 5-200 Hz. The G' exhibited a linear power-law behavior, while the G'' exhibited a nonlinear power-law in the high frequency range. The frequency dependence of G^* was observed previously [31, 32], where the cells are regarded as a soft glassy material.

The power-law behavior of cell rheology has been interpreted in terms of a structural damping model with the addition of Newtonian viscosity [31, 32], as given by

$$G^* = G_0 g(\alpha)\{1 + i\eta(\alpha)\}\left(\frac{f}{f_0}\right)^{\alpha} + i\mu f, \tag{4.5}$$

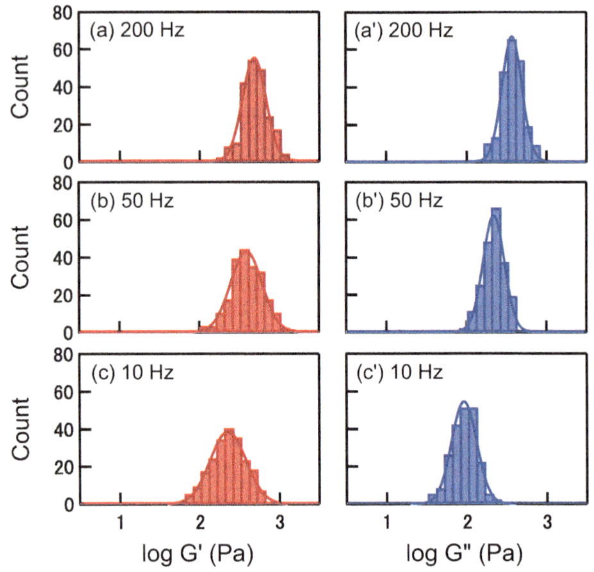

Fig. 4.6 Distributions of the storage G' (*left*) and loss G'' (*right*) moduli of NIH/3T3 cells ($N > 200$) on micro-patterned substrates at different frequencies: (**a** and **a**') 200 Hz, (**b** and **b**') 50 Hz and (**c** and **c**') 5 Hz. The solid lines represent fits to log-normal distribution functions

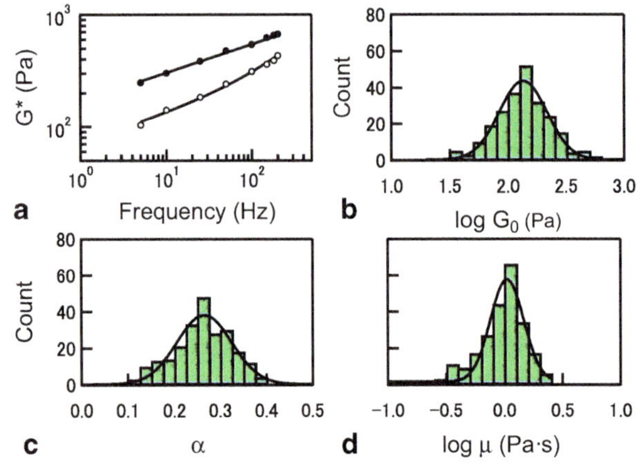

Fig. 4.7 **a** Frequency dependence of G' (*filled*) and G'' (*open*) of NIH/3T3 cells. Solid lines represent fits to Eq. 4.5, where the frequency scale factor, f_0, was arbitrarily set to 1 Hz. Distributions of **b** G_0 on a logarithmic scale. **c** α on a linear scale, and **d** μ on a logarithmic scale. The solid lines represent fits to log-normal distribution functions (**b** and **d**) and a normal distribution function (**c**)

where α is the power-law exponent, $g(\alpha) = \Gamma(1 - \alpha)\cos(\pi\alpha/2)$, and Γ is the gamma function. G_0 is a scale factor of the modulus at a frequency scale factor f_0. The hysteresivity $\eta(\alpha)$ is equivalent to $\tan(\pi\alpha/2)$, and μ is the Newtonian viscous damping coefficient.

Using this model, we can estimate rheological parameters of cells from data acquired with our AFM system (Fig. 4.7b, c, d). We can see that the distributions of G_0 and μ are fit with log-normal functions, whereas the distribution of α is a normal Gaussian. These ensemble distributions are in good agreement with those reported previously [8, 9, 28–30]. Our AFM technique can be applied to any adherent cells because our cell patterning technique and AFM operation are flexible. To enhance and speed up the measurement time, fast lock-in detection is being developed.

4.5 Conclusion

In this chapter, we described an AFM technique for high-throughput rheology of single adherent cells. Micro-fabricated substrates allow us to rapidly characterize frequency-dependent mechanical properties, and to quantify cell-to-cell variability, by confining cell culturing to well defined patterns and shapes. We emphasize that bio-AFM techniques are still rapidly evolving and will be applied more broadly to explore rheological properties of cells, to characterize tissues and organs for bioengineering, and for medical applications.

References

1. Morris VJ, Kirby AR, Gunning AP (2009) Atomic force microscopy for biologists, 2nd edn. Imperial College Press, London
2. Guck J, Schinkinger S, Lincoln B, Wottawah F, Ebert S, Romeyke M, Lenz D, Erickson HM, Ananthakrishnan R, Mitchell D, Käs J, Ulvick S, Bilby C (2005) Optical deformability as an inherent cell marker for testing malignant transformation and metastatic competence. Biophys J 88 (5):3689–3698
3. Cross SE, Jin YS, Rao J, Gimzewski JK (2007) Nanomechanical analysis of cells from cancer patients. Nat Nanotechnol 2(12):780–783
4. Plodinec M, Loparic M, Monnier CA, Obermann EC, Zanetti-Dallenbach R, Oertle P, Hyotyla JT, Aebi U, Bentires-Alj M, Lim RYH, Schoenenberger CA (2012) The nanomechanical signature of breast cancer. Nat Nanotechnol 7(11):757–765
5. Gossett DR, Tse HT, Lee SA, Ying Y, Lindgren AG, Yang OO, Rao J, Clark AT, Di Carlo D (2012) Hydrodynamic stretching of single cells for large population mechanical phenotyping. Proc Natl Acad Sci U S A 109(20):7630–7635
6. Singhvi R, Kumar A, Lopez G, Stephanopoulos G, Wang D, Whitesides G, Ingber D (1994) Engineering cell shape and function. Science 264:696–698.
7. Kandere-Grzybowska K, Campbell C, Komarova Y, Grzybowski BA, Borisy GG (2005) Molecular dynamics imaging in micropatterned living cells. Nat Methods 2(10):739–741

8. Hiratsuka S, Mizutani Y, Tsuchiya M, Kawahara K, Tokumoto H, Okajima T (2009) The number distribution of complex shear modulus of single cells measured by atomic force microscopy. Ultramicroscopy 109:937–941
9. Cai P, Mizutani Y, Tsuchiya M, Maloney JM, Fabry B, Van Vliet KJ, Okajima T (2013) Quantifying cell-to-cell variation in power-law rheology. Biophys J 105(5):1093–1102
10. Takahashi R, Ichikawa S, Subagyo A, Sueoka K, Okajima T (2014) Atomic force microscopy measurements of mechanical properties of single cells patterned by microcontact printing. Adv Robot 28:449–455
11. Chen CS, Mrksich M, Huang S, Whitesides GM, Ingber DE (1997) Geometric control of cell life and death. Science 276(5317):1425–1428
12. Théry M, Pépin A, Dressaire E, Chen Y, Bornens M (2006) Cell distribution of stress fibres in response to the geometry of the adhesive environment. Cell Motil Cytoskeleton 63(6):341–355
13. Bhatia S, Yarmush M, Toner M (1997) Controlling cell interactions by micropatterning in co-cultures: hepatocytes and 3T3 fibroblasts. J Biomed Mater Res 34:189–199
14. Kuribayashi K, Tsuda Y, Nakamura H, Takeuchi S (2010) Micro-patterning of phosphorylcholine-based polymers in a microfluidic channel. Sens Actuators B 149(1):177–183
15. Tseng Q, Wang I, Duchemin-Pelletier E, Azioune A, Carpi N, Gao J, Filhol O, Piel M, Théry M, Balland M (2011) A new micropatterning method of soft substrates reveals that different tumorigenic signals can promote or reduce cell contraction levels. Lab Chip 11(13):2231–2240
16. Carter S (1967) Haptotactic islands: a method of confining single cells to study individual cell reactions and clone formation. Exp Cell Res 48:189–193
17. Selvarasah S, Chao S, Chen C, Sridhar S, Busnaina A, Khademhosseini A, Dokmecia M (2008) A reusable high aspect ratio parylene-C shadow mask technology for diverse micropatterning applications. Sens Actuators A 145–146(1):306–315
18. Folch A, Toner M (2000) Microengineering of cellular interactions. Annu Rev Biomed Eng 02:227–256
19. Kuribayashi-Shigetomi K, Onoe H, Takeuchi S (2012) Cell origami: self-folding of three-dimensional cell-laden microstructures driven by cell traction force. PLoS One 7(12):e51085
20. Teshima T, Onoe H, Kuribayashi-Shigetomi K, Aonuma H, Kamiya K, Ishihara H, Kanuka H, Takeuchi S (2014) Parylene mobile microplates integrated with an enzymatic release and handling of single adherent cells. Small 10(5):912–921
21. Radmacher M, Tillmann RW, Fritz M, Gaub HE (1992) From molecules to cells – imaging soft samples with the atomic force microscope. Science 257(5078):1900–1905
22. Radmacher M, Tilmann RW, Gaub HE (1993) Imaging viscoelasticity by force modulation with the atomic force microscope. Biophys J 64(3):735–742
23. Alcaraz J, Buscemi L, Grabulosa M, Trepat X, Fabry B, Farre R, Navajas D (2003) Microrheology of Human lung epithelial cells measured by atomic force microscopy. Biophys J 84:2071–2079
24. Mahaffy RE, Park S, Gerde E, Kas J, Shih CK (2004) Quantitative analysis of the viscoelastic properties of thin regions of fibroblasts using atomic force microscopy. Biophys J 86:1777–1793
25. Ducker WA, Senden TJ, Pashley RM (1991) Direct measurement of colloidal forces using an atomic force microscope. Nature 353:239–241
26. Landau LD, Lifshiz EM (1986) Theory of elasticity, 3rd edn. Pergamon Press, Oxford
27. Alcaraz J, Buscemi L, Puig-de-Morales M, Colchero J, Baro A, Navajas D (2002) Correction of microrheological measurements of soft samples with atomic force microscopy for the hydrodynamic drag on the cantilever. Langmuir 18:716–721
28. Balland M, Desprat N, Icard D, Fereol S, Asnacios A, Browaeys J, Henon S, Gallet F (2006) Power laws in microrheology experiments on living cells: comparative analysis and modeling. Phys Rev E 74(2 Pt 1):021911
29. Massiera G, Van Citters KM, Biancaniello PL, Crocker JC (2007) Mechanics of single cells: rheology, time depndence, and fluctuations. Biophys J 93(10):3703–3713

30. Desprat N, Richert A, Simeon J, Asnacios A (2005) Creep function of a single living cell. Biophys J 88(3):2224–2233
31. Fabry B, Maksym GN, Butler JP, Glogauer M, Navajas D, Fredberg JJ (2001) Scaling the microrheology of living cells. Phys Rev Lett 87:148102
32. Fabry B, Maksym GN, Butler JP, Glogauer M, Navajas D, Taback NA, Millet EJ, Fredberg JJ (2003) Time scale and other invariants of integrative mechanical behavior in living cells. Phys Rev E 68:041914.

Chapter 5
Discrimination of Cells with Specific Antigens Expressed on a Membrane Based on the Dielectrophoresis

Tomoyuki Yasukawa and Fumio Mizutani

Abstract Formation of line pattern with cells based on dielectrophoresis (DEP) was applied to simple and rapid distinction of cells with specific surface antigens from a cell population. Dispersed cells were accumulated to gap areas of the interdigitated band array (IDA) electrode modified with antibody within 5 s by negative DEP (n-DEP) and captured by immunoreactions. Unbounded cells without the specific antigen on the membrane were removed to form another pattern by switching the applied voltage of the band electrode. The time required for the assay was substantially short, 60 s for forcing and 60 s for the separation of unbounded cells. Furthermore, the present method does not require pretreatment such as target labeling or washing of unbound cells.

Keywords Cell patterning · Surface antigen · Label free · Phenotyping

5.1 Introduction

Particles placed in a spatially inhomogeneous electric field experience dielectrophoresis (DEP) force by the interaction of a polarization effect induced in the particles. DEP is attractive for the manipulation of micro- and nano-objects including biological living cells and bacteria [1] and has been used in the wide range of applications, such as a separation and sorting [2–4], trapping [5–7] and patterning. Cell arrays can be fabricated using positive DEP (p-DEP), which is a force directed particles towards the regions of electric field maxima [8–10]. Generally, p-DEP patterning is utilized to attract in array formats with the pair of electrodes at every elements modified with cell adhesive layers to produce cell patterns. Negative DEP (n-DEP) force which is a force to repel the particles from the high electric field regions has also been used to pattern cells [11–13]. The strategy for n-DEP cell patterning is the construction of the localized position enclosed with strong electric fields and allows

T. Yasukawa (✉) · F. Mizutani
Graduate School of Material Science, University of Hyogo, 3-2-1, Kouto, Kamigori, Ako 678-1297, Japan
e-mail: yasu@sci.u-hyogo.ac.jp

T. Arai et al. (eds.), *Hyper Bio Assembler for 3D Cellular Systems*,
DOI 10.1007/978-4-431-55297-0_5

to cell trapping by the repulsion force balanced from every direction. We have also previously contributed to the fabrication of periodic and alternate cell lines incorporating two cell types of cells using n-DEP [14]. Furthermore, by using both p-DEP and n-DEP, individual cell types with different dielectrophoretic properties were separately patterned at the different positions [15, 16]. We can easily and reversibly fabricate two different patterns with particles by controlling the direction of the DEP force with applied frequency. However, one feature was only used in most case of cell patterning. We recently developed rapid and simple sandwich-typed immunosensing systems using accumulation and re-dispersion of microparticles based on n-DEP [17]. The use of n-DEP manipulation of microparticles brought about the separation-free sensing of unreacted target molecules and microparticles within 3 min [18–20].

In this chapter, we introduce the useful application of the DEP manipulation for the rapid and simple discrimination of cells which were expressed a specific surface antigen [21, 22]. Surface antigen patterns expressed on living cells depend on a linage, differentiation and maturation. Phenotyping, which can be identified the presence and proportion of pathogenic cell populations, are useful for early medical diagnosis and prognosis prediction. A common approach is to use fluorescence labeling via antigen-binding to relieve specific antigens at cell surfaces; however, the methods are often qualitative and low throughput, and involve several complex steps for a modification and washing. The cell binding assay to the immobilized antibody was accelerated by n-DEP cell accumulation and the discrimination of the unbound cells was performed with spatial separation by switching the formation of electric field. We demonstrate here that Human promyelocytic leukemia (HL60) cells with surface antigen, CD33, can be trapped on the gap region modified with anti-CD33 antibody.

5.2 Particle Patterning at Different Positions in a Single Device

We demonstrated the formation of aligned structures with microparticles at the different positions in the single device [23]. Interdigitated band array (IDA, 2 mm long, 12 μm wide, and gap 50 μm) electrode was fabricated by indium tin oxide (ITO) layer which is a transparent conductive material. Another ITO electrode was mounted on the ITO-IDA electrode via a 30 μm thick polyester film to fabricate the DEP device with fluidic channel. Top and cross section of DEP device were shown in Fig. 5.1a, b, respectively. The application of the AC voltage with the frequency region for n-DEP formed a non-uniform electric field in the device.

A suspension of the polystyrene particles (6-μm diameter) was injected into the device. The patterns with particles were formed by applying AC voltage in the frequency region for n-DEP (1.0 MHz) to the band A and band B. Figure 5.2a, b show images of particle patterns obtained by use of the glass and ITO as upper substrates, respectively. The application of AC voltage (20 V_{pp}) with an identical frequency and

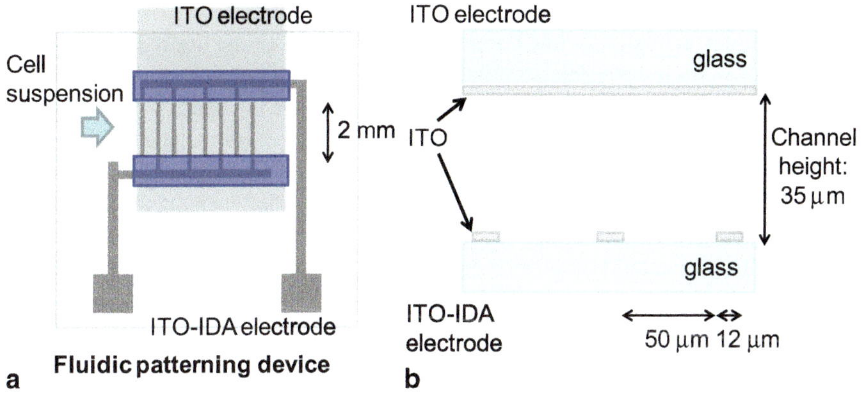

Fig. 5.1 **a** Top and **b** cross sectional views of fabricated DEP device

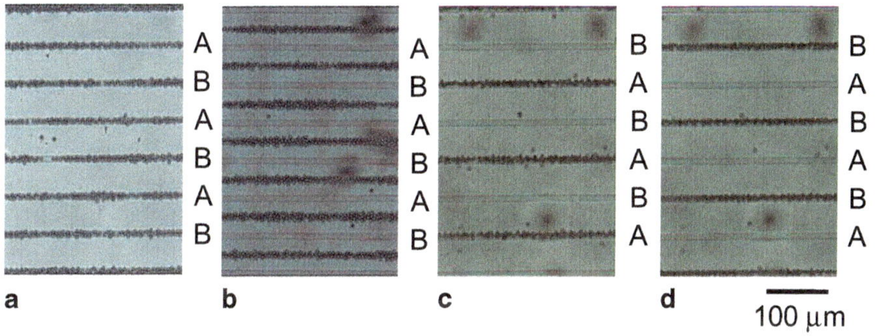

Fig. 5.2 Images of particle patterns obtained by use of **a** glass and **b** ITO as upper substrates. The AC voltage (20 V_{pp}) with an identical frequency and opposite phase to the band A and B was applied and upper ITO was connected to the ground. Images of patterns obtained by applying AC voltage with identical phase to **c** band A and upper ITO, and **d** band B and upper ITO. (Modified from [23])

opposite phase to the band A and B forced particles to form the line pattern along both bands within 1 s. The particles were guided to the glass surface to form periodical lines of positive projection above the band electrodes of the IDA template (Fig. 5.2a). On contrast, when we used the conductive ITO electrode connected to the ground as an upper substrate, the particles were aligned on the ITO surface above the gap of IDA electrodes (Fig. 5.2b).

The use of the conductive electrode as an upper substrate allowed to the formation of various patterns using the same design of device. Figure 5.2c shows an image for the particle formation. In this case, the AC voltage was applied to the upper ITO and band A with the same intensity (20 V_{pp}), frequency (1.0 MHz) and phase, while the AC voltage with opposite phase was applied to band B. Dispersed particles were guided to the areas above band A to form the alternating line pattern with the half of the positive projection. The electric field formed between the upper ITO and band A is relatively low compared to that between the upper ITO and band B. Thus, no particles were accumulated between the upper ITO and band B. On the other hand, the AC voltage was applied to the upper ITO and band B with the same intensity, frequency and phase to form another line pattern. The particles also rapidly accumulated at the areas above band B due to the repulsive force of n-DEP by the distribution of the strong electric fields generated between the upper ITO and band A and between each band (Fig. 5.2d). Thus, the line patterns with same design can be easily fabricated at the different position in the single DEP device. We applied the techniques to accumulate and capture cells with specific surface antigen via immunoreactions.

5.3 Formation of Two Different Patterns with Cells

DEP line patterning with HL60 cells was studied using a DEP patterning device consisted of the upper ITO and lower ITO-IDA electrode. HL60 cells (4.0×10^7 cells/mL) suspended in the DEP medium, which consisted of 250 mM sucrose and 250 mM HEPES buffer (pH 7.4), adjusted the conductivity to 400 mS/m with PBS were introduced into the device to manipulate the cells with the n-DEP. The AC voltage (15 V_{pp}, 100 kHz) was applied between upper ITO and the lower ITO-IDA electrode to accumulate the cells in the gap region between bands by the repulsive force of n-DEP, because the strong and weak electric fields were formed on the band electrodes and at the gap region. Figure 5.3a, b show the optical microscopic images before and 5 s after the voltage was turned on, respectively. The cells dispersed randomly in the channel (Fig. 5.3a) were rapidly accumulated in the gap region within 5 s (Fig. 5.3b). When the voltage applied to the band B was switched to zero, the cells moved on the band B with the repulsive force generated between the upper ITO and band A, and band A and B, resulting in the formation of another line pattern within 5 s (Fig. 5.3c). Moreover, the formation of the first line pattern in Fig. 5.3b reproduced as the voltage for the band B was again switched on. The results clearly indicated that the different patterns with cells could be easily created by applying the voltages with the different intensity to bands. Figure 5.3d, e show the cross-sectional plots of the electric field formed in the patterning device calculated from the digital simulation when a voltage is applied to both IDA bands and band A, respectively. The distribution of the electric field strength for the DEP patterning device was calculated by the finite element method (FEM) solver (COMSOL Multiphysics, Stockholm, Sweden). The regions with high electric field were found in the areas between the upper ITO and lower both bands (Fig. 5.3d). Thus, the suspended cells moved in the gap region by n-DEP. In contrast, when the voltage is applied to the band B, the regions with high electric field were found between the upper ITO and band A (Fig. 5.3e). Therefore, low electric field regions were formed on the band B. These results indicated that the different applications of the voltage for

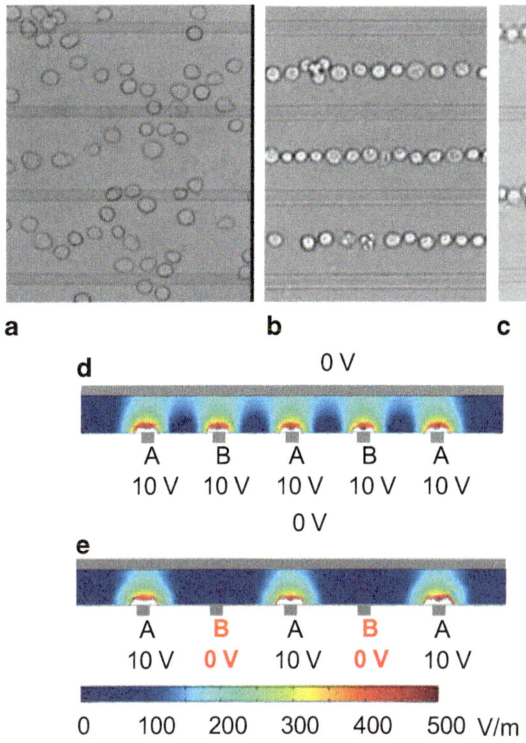

Fig. 5.3 Optical images of cells manipulated by n-DEP **a** before and **b** and **c** 5 s after the application of the ac voltage (intensity, 15 Vpp; frequency, 100 kHz). **b** The voltage was applied between the upper ITO and lower band A and B. **c** The voltage was applied to band A, while the upper ITO and band B was connected to the ground. **d** and **e** Cross-sectional views of the numerically calculated electric field formed in the patterning device. Upper and lower gray bars show the ITO and band electrodes, respectively. Applied voltages for bands A and B and the upper ITO were (D) 10, 10, and 0 V and (E) at 10, 0, and 0 V, respectively. (Modified from [22])

band B forced the cells to move to areas in the gap region or on the band B, which were in good agreement with the experimental results (Fig. 5.3a–c).

5.4 Capture of HL60 Cells with CD33 Antigen by n-DEP

We have investigated the number of cells accumulated in the gap region using the cell binding efficiency. Figure 5.4 depicts the cross sectional view of the DEP device and the present procedure for detecting the surface antigen. Cells were forced to move toward the gap regions between the bands and captured with the antibodies via immunoreactions. Uncaptured cells were removed and separated from the gap region to the band B by switching the applied voltage for band B to zero. As a result, the ratio of captured cell density can be easily calculated from the images

Fig. 5.4 Cross-sectional view of the DEP device and the method for discriminating cells with specific surface antigens using n-DEP [22]

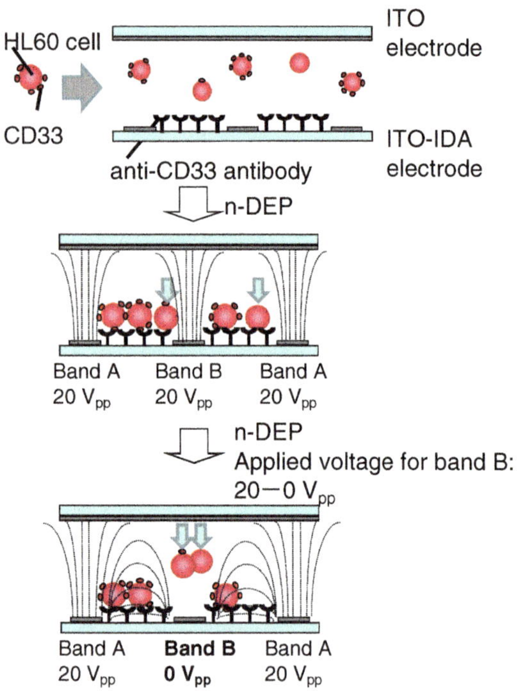

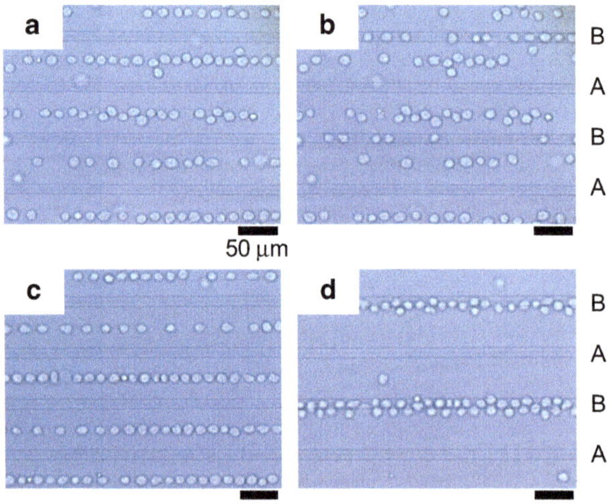

Fig. 5.5 Optical images of cells patterned by n-DEP. **a** Cells accumulated in the gap region modified with the anti-CD 33 antibody by applying AC voltage of same intensity (20 V_{pp}) and frequency (100 kHz) as bands A and B. **b** Cell pattern captured in the gap region modified with the anti-CD 33 antibody after separating unbound cells by switching off the band B voltage. **c** Cells accumulated in the gap region modified with the anti-mouse IgG antibody. **d** Cells removed from the gap region modified with the anti-mouse IgG antibody after the band B voltage was turned off. Duration of voltage application for accumulation and removal with n-DEP: 60 and 60 s. (Modified from [22])

of the line structure of cells in the gap region. Figure 5.5 shows the optical images of cells accumulated in the gap region (Fig. 5.5a) and captured cells after separating the unbound cells (Fig. 5.5b). Cells were accumulated in the gap region with n-DEP for 60 s and then the voltage for the band B was switched off to remove the cells from the gap region. Again, the uniformly dispersed cells initially start to move toward the gap region due to the strong repulsive force for n-DEP to form clear line patterns (Fig. 5.5a). After the voltage of band B was switched to zero, some cells accumulated in the gap region modified with anti-CD33 antibody were remained on the gap region even after the voltage was switched off, while the other moved on the band B (Fig. 5.5b). In contrast, almost all cells moved on band B after switching off the voltage to the band B when cells were accumulated in that with anti-mouse IgG antibody (Fig. 5.5d). The results suggested that the cells with CD33 surface antigens were reacted with the antibodies immobilized in the gap region and irreversibly captured at that position. The number of cells in the gap region quickly decreased in 30 s after switching off the voltage, and then reached at steady-state value. The cell binding efficiency is estimated from the steady-state value and found to be $68.3 \pm 3.2\,\%$. The slight undesired binding originated from the non-specific adsorption was observed on the bands modified with anti-mouse IgG antibody ($4.2 \pm 1.4\,\%$). The time as short as 30 s was required for removing unbound cells. Therefore, the cells with CD33 cell surface antigen can be rapidly identified from the cell suspension by the spatial separation based on the immunoreactions and DEP manipulation.

5.5 Specific Cell Binding in a Mixture Containing Nonspecific Cells

Non-specific cells used as a CD33 negative cell was prepared by treating HL-60 cells with anti-CD33 antibody. Cells were also labeled with fluorescent molecules (CFDA SE) to distinguish the antibody-treated non-specific cells. The suspensions of each specific and non-specific cell were mixed at different ratios to study the relationship between the content of CD33 positive cells in suspensions. The mixtures were introduced into the device and accumulated in the gap region by n-DEP for 60 s. The binding efficiency of the cells was calculated after switching off the voltage for band B for 60 s. The efficiency was defined as a ratio of average density of cells in the gap region immediately before and 60 s after AC voltage for band B was switched off to eliminate the unbound cells.

Mixtures of HL-60 cells specific and nonspecific to anti-CD33 were used to determine the number of cells captured in the gap area modified with anti-CD33. Figure 5.6a shows the image obtained by combining optical and fluorescent images, which were obtained 60 s after the applied voltage to band B was switched to zero. The initial concentration ratio of specific cells was set to 50 %. Almost all nonspecific cells with a fluorescent signal moved to band B, whereas the specific cells were captured in the gap between the band electrodes, even in the presence of

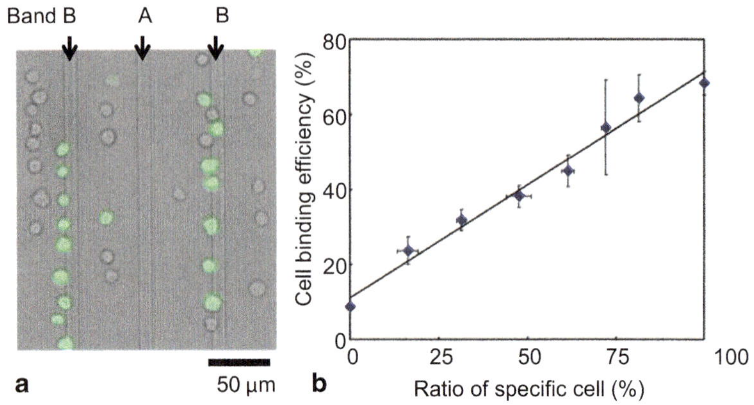

Fig. 5.6 Cells captured in the gap region from mixed suspensions of specific and nonspecific HL-60 cells. **a** Photographs were obtained by combining the optical and fluorescent images. Initial ratios in the original suspensions were set to 50 % of specific cells. **b** Ratio of cells captured in the gap. [22]

treated nonspecific cells (Fig. 5.6a). Figure 5.6b shows the ratio of cells captured in the gap between the band electrodes. The ratio of captured cells increased linearly with the increasing ratio of specific cells in the prepared mixture suspension. These results indicate that the presence of cells without the target antigen did not obstruct specific cell binding for detecting cells with surface antigens. The binding efficiency obtained in the present study improved compared to that obtained in our previous study using a combination of p- and n-DEP.

5.6 Conclusion

We propose a novel procedure for rapid and simple cell binding assay based on the manipulation of cells by DEP. On applying an AC voltage, cells were forced to accumulate in the gap region yielding the line formations. On the other hand, cells were repelled from the gap region and formed another line pattern on the band B by switching off the voltage for band B. Therefore, we can easily prepare the different pattern with cells by controlling the intensity of the applied voltage. The formation of second line pattern with cells was markedly inhibited by immunoreactions between the surface antigen and specific antibody immobilized in the gap region. HL60 cells with surface antigen were irreversibly captured in the gap region modified with specific anti-CD33 antibody, as a result, captured cells did not remove even after the voltage for the band B was switched off. Importantly, the capture of cells via immunoreactions was assisted by rapid accumulation with n-DEP. The total time as short as 90 s was required for detecting the antigen expressed on the cell surface. Moreover, no any pre-treatment, such as cell staining with fluorescent molecules was requited for the proposed method.

References

1. Morgan H, Green NG (2003) AC Electrokinetics: colloids and nanoparticles. Research Studies Press, Baldock
2. Voldman J (2006) Electrical forces for microscale cell manipulation. Annu Rev Biomed Eng 8:425–454
3. Vahey MD, Voldman J (2008) An equilibrium method for continuous-flowcell sorting using dielectrophoresis. Anal Chem 80:3135–3143
4. Yasukawa T, Suzuki M, Sekiya T, Shiku H, Matsue M (2007) Flow sandwich-type immunoassay in microfluidic devices based on negative dielectrophoresis. Biosens Bioelectron 22:2730–2736
5. Hughes MP, Morgan H, Rixon FJ, Burt JPH, Pethig R (1998) Manipulation of herpes simplex virus type 1 by dielectrophoresis. Biochim Biophys Acta Gen Subj 1425:119–126
6. Grom F, Kentsch J, Müller T, Schnelle T, Stelzle M (2006) Accumulation and trapping of hepatitis A virus particles by electrohydrodynamic flow and dielectrophoresis. Electrophoresis 27:1386–1393
7. Sebastian A, Buckle AM, Markx GHJ (2006) Formation of multilayer aggregates of mammalian cells by dielectrophoresis. J Micromech Microeng 16:1769–1777
8. Taff BM, Voldman J (2005) A scalable addressable positive-dielectrophoretic cell-sorting array. Anal Chem 77:7976–7983
9. Ho C-T, Lin R-Z, Chang W-Y, Chang H-Y, Liu, C-H (2006) Rapid heterogeneous liver-cell on-chip patterning via the enhanced field-induced dielectrophoresis trap. Lab Chip 6:724–734
10. Chiou PY, Ohta AT, Wu MC (2005) Massively parallel manipulation of single cells and microparticles using optical images. Nature 436:370–372
11. Mittal N, Rosenthal A, Voldman J (2007) nDEP microwells for single-cell patterning in physiological media. Lab Chip 7:1146–1153
12. Frenea M, Faure SP, Le Pioufle B, Coquet P, Fujita H (2003) Positioning living cells on a high-density electrode array by negative dielectrophoresis. Mater Sci Eng C 23:597–603
13. Puttaswamy SV, Sivashankar S, Chen R-J, Chin C-K, Chang H-Y, Liu CH (2010) Enhanced cell viability and cell adhesion using low conductivity medium for negative dielectrophoretic cell patterning. Biotechnol J 5:1005–1015
14. Suzuki M, Yasukawa T, Shiku H, Matsue T (2008) Negative dielectrophoretic patterning with different cell types. Biosens Bioelectron 24:1043–1047
15. Huang Y, Joo S, Duhon M, Heller M, Wallace B, Xu X (2002) Dielectrophoretic cell separation and gene expression profiling on microelectronic chip arrays. Anal Chem 74:3362–3371
16. Pethig R, Huang Y, Wang XB, Burt JPH (1992) Positive and negative dielectrophoretic confinement of bioparticles using interdigitated castellated electrodes. J Phys D 25:881–888
17. Lee HJ, Yasukawa T, Shiku H, Matsue T (2008) Rapid and separation-free sandwich immunosensing based on accumulation of microbeads by negative dielectrophoresis. Biosens Bioelectron 24:1000–1005
18. Lee HJ, Lee SH, Yasukawa T, Ramón-Azcón J, Mizutani F, Ino K, Shiku H, Matsue T (2010) Rapid and simple immunosensing system for simultaneous detection of tumor markers based on negative-dielectrophoretic manipulation of microparticles. Talanta 81:657–663
19. Ramón-Azcón J, Yasukawa T, Lee HJ, Matsue T, Sánchez-Baeza F, Marco M-P, Mizutani F (2010) Competitive multi-immunosensing of pesticides based on the particle manipulation with negative dielectrophoresis. Biosens Bioelectron 25:1928–1933
20. Ramón-Azcón J, Yasukawa T, Mizutani F (2011) Sensitive and spatially multiplexed detection system based on dielectrophoretic manipulation of DNA-encoded particles used as immunoreactions platform. Anal Chem 83:1053–1060
21. Hatanaka H, Yasukawa T, Mizutani F (2011) Detection of surface antigens on living cells through incorporation of immunorecognition into the distinct positioning of cells with positive and negative dielectrophoresis. Anal Chem 83:7207–7212

22. Yasukawa T, Hatanaka H, Mizutani F (2012) Simple detection of surface antigens on living cells by applying distinct cell positioning with negative dielectrophoresis. Anal Chem 84:8830–8836
23. Yasukawa T, Yoshida Y, Hatanaka H, Mizutani F (2013) Line patterning with microparticles at different positions in a single device based on negative dielectrophoresis. J Robot Mechatron 25:650–656

Chapter 6
Analysis of Physical Characteristic of Hematopoietic Cells

Shoichi Iriguchi, Tomoyuki Yamaguchi and Hiromitsu Nakauchi

Abstract Circulating tumor cells (CTCs) or fetal nucleated erythrocytes in maternal peripheral blood are extremely rare cell population ($\sim 1/10^5$) in circulating blood, though these cells are useful for diagnose diseases. Flow cytometry enabled us to detect such a rare cell populations. However, in order to sort those rare cell populations by fluorescent activated cell sorter (FACS), it is necessary to stain the cells with fluorescent conjugated antibody *ex vivo* and to obtain theoretically few-hundred milliliters of blood. Therefore, high-speed closed system for analysis and isolation of rare cell populations within the blood without cell staining is desirable. In this chapter, we described analysis of physical characteristic of hematopoietic cells by microfluidics-based devices for efficient detection of rare cell population in circulating blood.

Keywords High-speed closed system · Microfluidics-based devices · Rare cell population in circulating blood

6.1 Introduction

An organ (or tissue) that constitutes as a whole an individual is consisted itself of cells of distinct types. For example, the adult bone marrow is occupied with mature hematopoietic cells, including granulocytes, monocytes, macrophages, dendritic cells, B cells, natural killer cells, T cells, erythrocytes, and megakaryocytes, immature hematopoietic progenitor cells, and non-hematopoietic cells, such as mesenchymal cells and perivascular cells. Numerous efforts had been made to enrich and isolate distinct cell type. Density gradient centrifugation separates cells in organs based on distinct size and widely used in biological researches. Magnetic beads separation exploits antigen-antibody reactions and has been an effective method to purify cell populations of interest [1]. The MACS technology

T. Yamaguchi (✉) · S. Iriguchi · H. Nakauchi
Division of Stem Cell Therapy, Center for Stem Cell Biology and Regenerative Medicine, Institute of Medical Science, University of Tokyo, Tokyo, Japan
e-mail: tomoyama@ims.u-tokyo.ac.jp

T. Arai et al. (eds.), *Hyper Bio Assembler for 3D Cellular Systems*,
DOI 10.1007/978-4-431-55297-0_6

(a magnetic separation method) employed the principle that each cell types within a heterogeneous population express distinct surface markers (or surface antigens) on the cell membrane. The development of the flow cytometer, which is originally developed by Leonard Herzenberg at Stanford University in California and Marvin Van Dilla at the Los Alamos National Laboratories in New Mexico, revolutionized our approach to cell biology research, enabling isolation of cells with a particular antigen expression pattern from a wide variety of biological source. It has significantly contributed to provide new insights into our understanding of molecular cell biology, particularly in the fields of immunology, hematology, and stem cell biology. These researches also have highlighted the importance of identifications and investigations of rare populations within a heterogeneous population.

Flow cytometry is an analytical technique that measures the fluorescence intensities of a single cell stained with antibodies against different antigens tagged with different fluorophores. Another outstanding feature with the flow cytometer include isolation of viable cells expressing the proteins of interest while simultaneously analyze 10 parameters (proteins) or more from a heterogeneous population. A flow cytometer is consisted of two systems: a fluidic system and a measuring system. The fluidic system is designed to vertically align and transport cells in a sample to the flow cells by generating a laminar flow of sheath flow surrounding the sample flow from the sample loading port. A flow cytometer with the cell-sorting feature is also equipped with an electrostatic deflection system. The measuring system acquires Forward scatter (FSC) and Side scatter (SSC) as the aligned cell pass through lasers one by one. At the same time, fluorescence emitted from the fluorophores conjugated with antibodies bind to the cell as a result of excitation by the lasers is directed to a series of dichroic mirrors and optical filters to filter out noise and arrives at photomultiplier tube (PMT). The PMTs effectively amplify photons of the fluoresence by converting it to electrons, allowing measurement of the quantity of photons as a form of voltage.

With this technology, we have previously identified a population with a defined antigen expression pattern that highly purified murine hematopoietic stem cells, which represent only one in 300,000 cells within an adult bone marrow in the steady state. Hematopoietic stem cells give rise to all mature blood cells including erythrocytes, platelets, and leukocytes through step-wise differentiation processes with distinct intermediate progenitor cells while maintaining its population via self-renewal and sustain the blood cell populations throughout an adult's life. Hematopoietic stem cells reside in bone marrow cells, spleen, and blood in adult, and fetal liver in early embryo. Although the existence of the stem cells and its importance have been widely recognized by the bone marrow transplantation, direct experimental proof was difficult to perform due to the lack of the technology to isolate a subpopulation within a heterogeneous population at that time. Furthermore, competitive dilution assay postulated hematopoietic stem cell concentration to be 0.7–1.1 and 1.0–1.6 per 10^5 bone marrow cells, suggesting its isolation extremely difficult [2, 3]. The development of flow cytometers made it possible for researchers to prospectively isolate any cells of interest in the bone marrow of mice and examine its long-term ability to reconstitute myeloid and lymphoid cells in recipients' body, followed by long-term serial transplantation [4]. The first report appeared in 1988 by Spangrude et al. using specific cell surface markers (Thy-1^{lo}Lineage markers^{-}Sca-1^{+}) [5]. Inspired by these seminal studies, we have demonstrated the identification of a set of surface markers that highly enrich murine hematopoietic stem cells (Lineage markers^{-}c-Kit^{+}Sca-1^{+}CD$34^{low/-}$) capable of long-term reconstitution of lympho-hematopoietic system in recipients when transplanted with a single donor cell [6]. Further characterization of the population by single cell transplantation assays have shown that the most primitive hematopoietic stem cells compartment present with distinct functional capabilities based on their CD150 expression levels [7]. These findings have highlighted the impact of clonal analysis on stem cell biology using flow cytometers, which is directly translatable to other fields of basic biology as well as human studies. Figure 6.1 illustrates a schematic of a flow cytometer and an example of flow diagrams in purifying murine hematopoietic stem cells.

6.2 Circulating Tumor Cells (CTCs)

Activating invasion and metastasis are one of the hallmarks of cancer and account for nearly 90% of cancer-associated mortality [8]. However, it remains the most poorly understood aspects of cancer biology. Recent research have raised the notion that circulating tumor cells (CTCs) bridges a primary site to a distant organ and therefore understanding it would provide new therapeutic targets and diagnostic biomarkers [9]. In accordance with the hypothesis, a number of studies have shown correlations between high CTC numbers and severity of the disease, increased metastasis, and decreased time to relapse [10]. Moreover, CTCs have been detected even in cancer patients at the early stage, suggesting its diagnostic values [11]. In spite of these potential, researchers in this field have been encountering many obstacles [9]. One such obstacle includes technical issues inherited to the detection system currently employed in the clinic. Although the development of an automated immunomagnetic enrichment and staining system (CellSearch™) have accelerated our understanding of CTCs, the use of epithelial-specific markers and enrichment processes may result in loss of subpopulations of cancer cells, particularly cancer stem cells that had undergone epithelial-to-mesenchymal transition [12, 13]. These findings pose the need for the development of a novel detection and isolation systems to further evaluate the utility and biological significance of CTCs.

6.3 Prenatal Diagnosis for Detection of Chromosomal Abnormality of Embryo

Prenatal diagnosis is the test for detection of congenital abnormality of embryo. There are two types of diagnosis methods including definitive diagnosis such as amniocentesis, chorionic villus sampling and non-definitive diagnosis such as

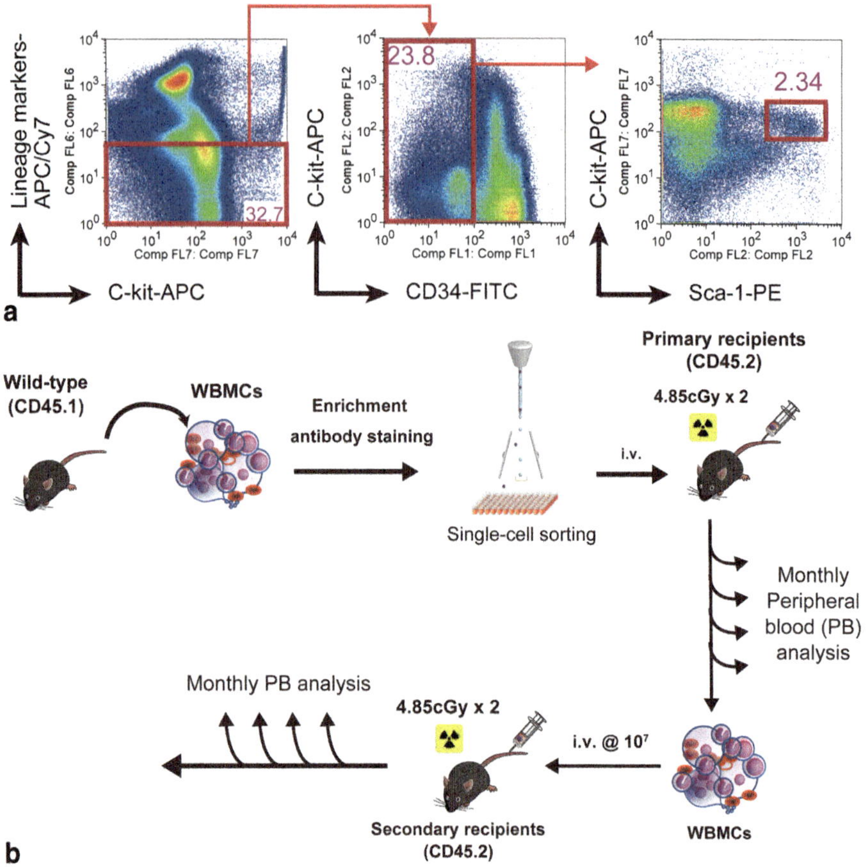

Fig. 6.1 The workflow of a single cell transplantation of murine hematopoietic stem cells in the Nakauchi Lab. **a** Representative flowdiagrams showing a gating strategy to isolate a murine hematopoietic stem cell population (Lineage marker$^-$/CD34^{low}/Sca-1^{Sc}/C-kit^{C-}). The antibodies used and its fluorescent conjugate is indicated. **b** A schematic illustrating the workflow of single cell transplantation assays. A bulk of whole bone marrow cells (WBMCs) is isolated from legs and pelvis of donor mice and subsequently magnetically selected against c-kit positivity. After the selection, the cells are stained with a cocktail of antibodies unique to a hematopoietic stem cell population and a single cell of the population is sorted into a well of a 96-well plate via a flow cytometer. The sorted cell is transplanted into a lethally irradiated recipient mouse together with a defined number of competitor cells. The fates of the donor cells are analyzed monthly using peripheral blood (PB) of each recipient. The self-renewal ability of the donor cells is assessed by serial transplantation, followed by monthly PB analysis at least up to 4 month after transplantation. Parts of image are adopted from [22]

ultrasonography, maternal serum screens. In order to perform definitive diagnosis, it is necessary to obtain *amniotic fluid*, villus or fetal blood by invasive strategy with the risk of abortion. On the other hand, ultrasonography and maternal serum screens are non-invasive technique; however, those are the inspection for the morbidity rate not for the definitive diagnosis. Recently Lo et al. reported that there are large quantities of cell free embryonic DNA in maternal circulating blood

Table 6.1 Prenatal diagnosis for detection of chromosomal abnormality of fetus

	Ultrasonography	Maternal serum screening	Non-invasive prenatal testing	Amniotic diagnosis	Chorionic villus sampling
Non-definitive/definitive	Non-definitive			Definitive	
Non-invasive/invasive	Non-invasive			Invasive	
Detection rate (%)	80–85		99.1	99.9	

and the concentration of cell free DNA fragment derived from each chromosome are proportional to the mass of each chromosome [14, 15]. The ratio of embryonic cell free DNA fragments in maternal circulating cell free DNA fragments is about 10~15% and this percentage increase as the week number in pregnancy increase [16]. From these observations, non-invasive prenatal testing (NIPT) which can detect abnormality in chromosomal number was developed [17]. In order to detect the chromosomal abnormality of fetus, over 10^7 embryonic cell free DNA fragments are sequenced by next generation sequencer and analyzed by comparing with human genome resources to determine which chromosome is the origin of fragment. Then the number of fragments in each chromosome is counted and compare with the number of fragments derived from embryo with normal karyotype to estimate the quantity of the number of each chromosome in specimen. Although the detection rate of chromosomal abnormality is high (ch13 trisomy; 91.7%, ch18 trisomy; 100%, ch21 trisomy; 99.1%) by NIPT, NIPT is not definitive diagnosis because only three chromosomes (ch13, 18, 21) can be analyzed currently and chromosome abnormality other than chromosome number (chromosomal translocation or mosaicism) cannot be detected [18] (Table 6.1). Therefore non-invasive definitive diagnosis technology for detection of chromosomal abnormality is still required. It has been identified the existence of fetal nucleated erythrocytes in maternal peripheral blood. Although these cells are rare cell population (one in tens of thousands of circulating blood cells), utilization of these cells for non-invasive prenatal diagnosis has been actively studied [19–21].

6.4 Hurdles and Solutions in Search for Rare Populations in Blood

Despite its significant contribution, ex vivo cell staining with antibodies and requirement for highly skilled operators have been major hurdles with flow cytometric analysis. Furthermore, in order to obtain a sufficient number of such a rare cell population for identification as well as down-stream analysis with high robustness, precision, and accuracy, one needs to obtain theoretically few-hundred milliliters of blood from a specimen. To overcome these issues, we aim to develop a novel high-speed closed system for analysis and isolation of rare cell populations within

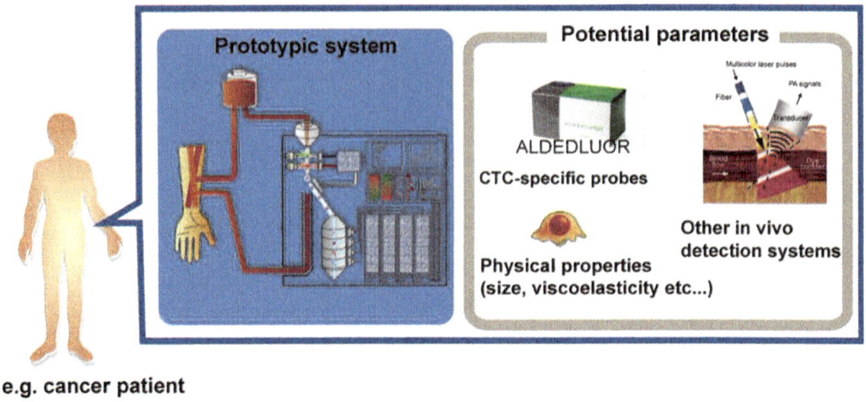

Fig. 6.2 A conceptual design for isolating a rare cell population within blood. A schematic of the system is illustrated on the *left*. Detection systems that measure potential parameters are integrated in an extracorporeal circulation from a patient (e.g. a cancer patient). The cell isolation system is located after the detection system to sort out cells of interest based on the measured parameters. Potential parameters include commercially available CTC-specific probes (ALDEFLUOR kit; StemCell Technologies, Vancouver, Canada), physical properties such as cell size, viscoelasticity and weights

the blood. We propose to analyze each cell in the blood at an ultra-high speed while they are in an extracorporeal circulation. The prototype of the system is illustrated in Fig. 6.2.

6.5 Analysis of Physical Properties of Hematopoietic Cells with a Microfluidic Device on an Ultra-High Speed Image Acquisition System

Since ex vivo antibody staining is not applicable in the proposed system, development of alternative parameters that characterize and distinguish cell types at ease is a prerequisite for this project. Although it has been long known that each cell type possesses distinct physical properties, such as weight, elasticity, dimension, and viscosity, a few studies have attempted such measurements in part due to technological hurdles. We have developed a novel microfluidic channel and an ultra high speed online image capturing system, respectively and attempted to measure physical properties of various cells, including murine bone marrow cells (mBMCs), human peripheral blood mononuclear cells (hPBMCs), a human cancer cell line, a mouse cancer cell line, and erythroblast cell lines at different stages of maturation established pluripotent stem cells, passing through the microfluidic channel while acquiring its real time images. This system allows us to measure the passing time of a cell while it travels through the region with a narrowed diameter deforming. With the passing times and diameters, we can obtain viscoelasticity of a cell. The schematic of the prototypic system is depicted in Fig. 6.3.

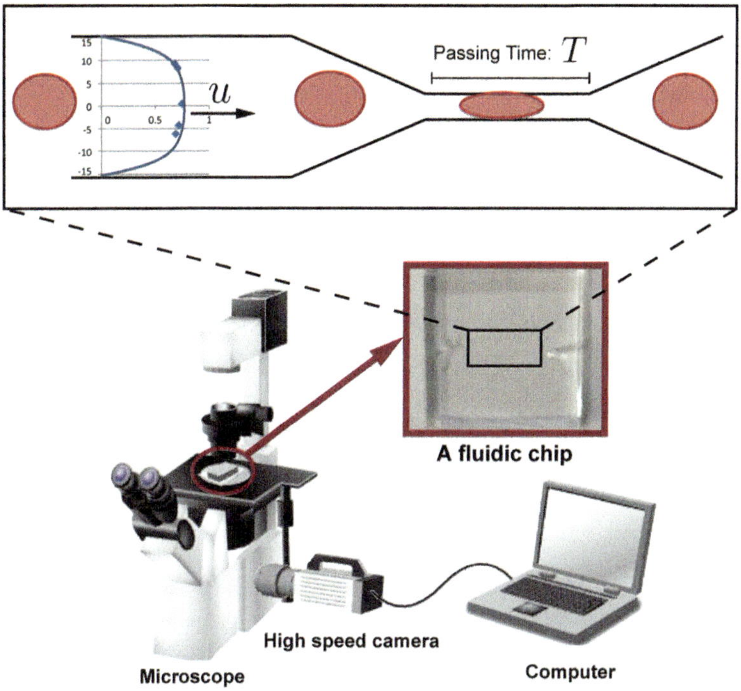

Fig. 6.3 Prototypic design of image acquisition system. Sequential images of a cell traveling through the channel are acquired by a high-speed camera connected to a conventional microscopy and to a computer. The high-speed camera is operated by software via the computer. An illustration of the microfluidic chip is shown on the top. The software processes the acquired images and generates a 2-D plot as shown in Fig. 6.4. Images are provided by the Kaneko lab

We performed the following experiments by using the ultra high-speed image acquisition system that we have developed. At the one day before or at the day of experiments, we prepared freshly isolated bone marrow cells in a femur of a C57BL/6 mouse, human PBMCs from a healthy volunteer, a human leukemia cell line (EOL-1: chronic eosinophilic leukemia), a mouse cancer cell line (4T1: mammary gland tumor), and the erythroblast cell line. The prepared cells were diluted to a density of 1×10^7 cell/ml, if possible, using either saline or phosphate-buffered saline with 3 % Fetal bovine serum (vol/vol) immediately before running the experiment. We used human erythrocytes prepared in the lab as a reference control. Subsequently, the cell suspension was drawn through the polyvinyl tubing to which a 2.5 ml syringe is attached at the other end and the tip was inserted to the inlet of the microfluidic channel. The cells were introduced into the channel by applying a pressure with the syringe piston. When the flow rate around the region with a narrowing diameter within the channel reached to a constant value, we started acquiring images for one minute. We repeated the experiments several times switching microfluidic channels with different diameters (6–10 μm). Cell diameters and passing times were automatically extrapolated by analyzing the images via a tailor-made program. The

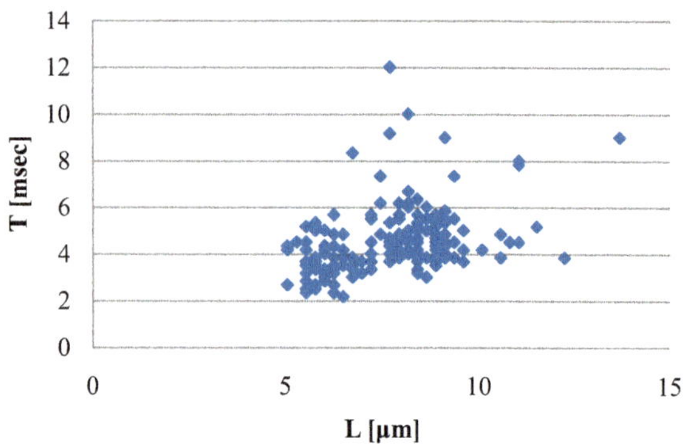

Fig. 6.4 A representative plot that can be obtained from an experiment. The *horizontal axis* represents the diameter of a cell at the entrance of the narrowed region in micrometer (*µm*), whereas the *vertical axis* shows the passing time of a cell in millisecond (*msec*). Each dot represents a cell passed through the channel

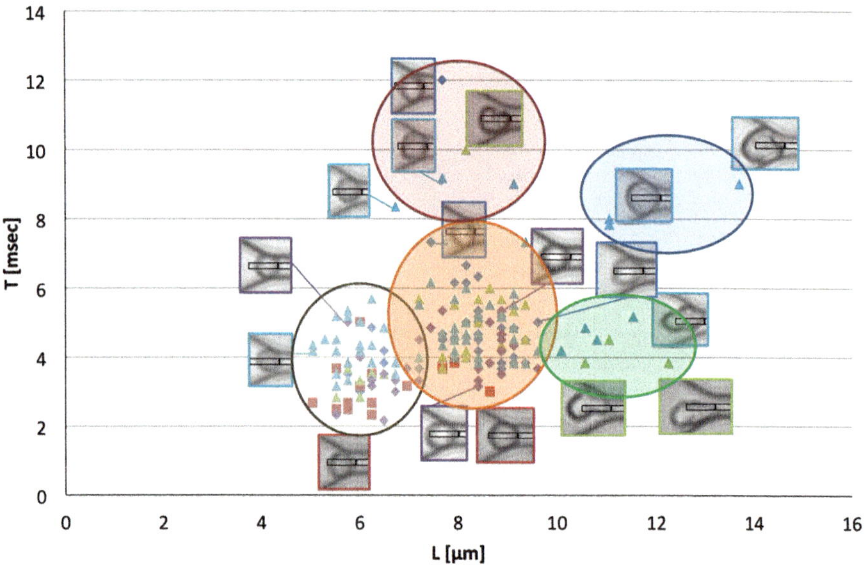

Fig. 6.5 A representative 2-D plot obtained with mouse bone marrow cells. Combined data from five independent measurements is shown. There is a positive correlation between passing time (*T*) and the length (*L*). Note two large populations on the plot (the *gray circle*; the *lower left* and the *orange circle*; *lower middle*). Some cells showed unexpected behaviors (the *red circle*; *top middle* and the *green circle*; *lower right*). *L* length in micrometer (or diameter of a cell), *T* the passing time in millisecond

results were exported as a excel spreadsheet containing a 2-D plot (Passing time v.s. cell diameter). Figure 6.4 shows representative data from the experiments.

The plots obtained with mBMCs experiments clearly indicated a linear correlation between passing time and cell diameter, whereas no such trend was observed from the plots using human erythrocytes (the reference control). Moreover, we observed clusters of cells on the plot that do not fit onto the regressed line, indicating that they behave differently in the channel. Some of those cells may be dying as postulated from its morphology (Fig. 6.5). Experiments with tumor cell lines

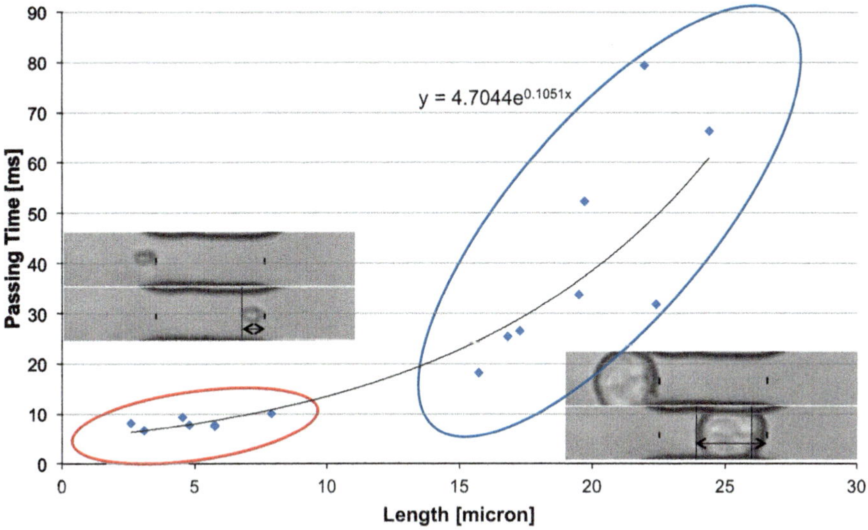

Fig. 6.6 A representative 2-D plot obtained with a mixture of human peripheral blood and the 4T1 cancer cell line. The plot from one measurement is shown here. The cluster on the left represents human peripheral blood cells (hPBCs) and the right cluster designates the 4T1 cancer cells. Note the distribution of the cancer cells spreading widely between 15–25 μm through 20–80 msec., although they all derived from a clonal origin. It appears possible to roughly discriminate hPBC from the cancer cells. *L* length in micrometer (or diameter of a cell), *T* the passing time in millisecond

unexpectedly demonstrated heterogeneous behaviors although they are clonal cell lines. Importantly, when human PBMCs and a tumor cell line were mixed together and simultaneously investigated for their passing times and diameters, it was possible to distinguish the two from the plots, which was generated from an unbiased method (Fig. 6.6). These results suggest the utility of the proposed system and physical properties as an alternative parameter for isolating a rare population in the blood.

Furthermore, another experiments conducted with PSC-derived erythroblasts under maturation and matured erythrocytes revealed a positive correlation between nucleus sizes and the passing times (Fig. 6.7). Since high elastic deformability of erythrocytes is attributed to its nucleation during the maturation stages and is a crucial property in vivo as it travel through the blood vessel, the result further strengthen the notion that it is of importance to use enucleated and fully maturated erythrocytes for an alternative blood transfusion source. Overall, the current system may also be useful for the quality assessment of the erythrocytes generated from the PSC technology upon the clinical application.

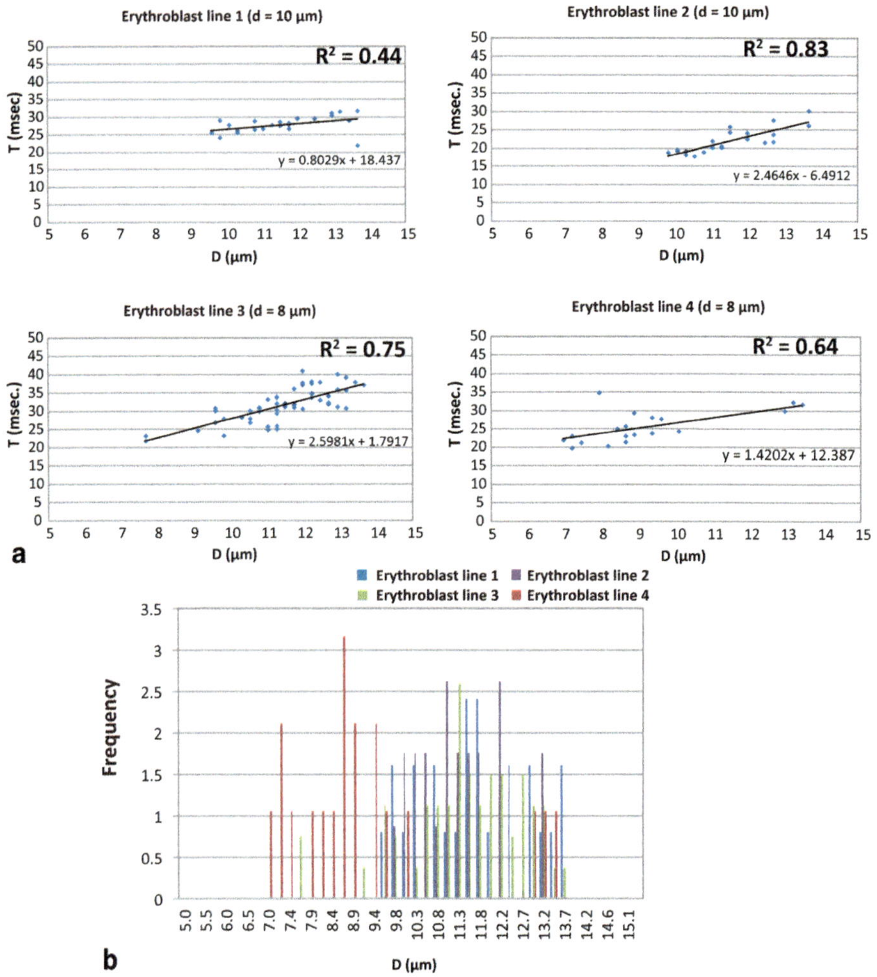

Fig. 6.7 2-D plots obtained with erythroblast cell lines at different stages of maturation. **a** Representative plots obtained with erythroblast cell lines at their maturation from proerythroblasts. The maturity of the cell lines is descending from 1, being erythroblast line 1 around the proethryroblast stage. The most matured line of the four (erythroblast line 4) comprised of the cells with smaller cell diameter. *D* length in micrometer (or diameter of a cell), *T* the passing time in millisecond; d the width of the narrowed region of the channel, where images are taken. **b** A histogram of the combined data of all four lines. Frequencies are plotted against diameters of the cells corresponding to (**a**). Most of the cells from line 4 appears on the *left* side of the histogram

6.6 Conclusion

The current study has demonstrated the utility of the proposed system to distinguish distinct cell types within a heterogeneous population with ease and high throughput. However, we observed some potential problems with the current system for further developmental processes. We sometimes encountered the situation where the entrance speeds of incoming cells, which are monitored at the entrance of the narrowing region, vary overtime during a data acquisition, making the data inconsistent. One possible cause of the problem may arise from the fluidic system exploited to the current system, which is generated by the potential energy. If so, we could overcome this by adopting an external sheath flow system utilized to flow cytometer. Another issue worthwhile to mention is related to surface finishing of the microfluidic chips. As we process high number of cells at a high speed, some debris from cells or media accumulate at the bottom of the channel, making a plug in between the entrance and the exit. Considering that blood contains a wide range of sticky material, this issue should be resolved. We may avoid this by coating the chip with serum, silane-coupling agents for glass bottom or by locating a filter before the chip. Although further improvement is necessary, we suppose microfluidics-based devices described here have a potential for efficient detection of rare cell population in circulating blood.

References

1. Miltenyi S, Müller W, Weichel W, Radbruch A (1990) High gradient magnetic cell separation with MACS. Cytometry 11:231–238
2. Harrison DE, Jordan CT, Zhong RK, Astle CM (1993) Primitive hemopoietic stem cells: direct assay of most productive populations by competitive repopulation with simple binomial, correlation and covariance calculations. Exp Hematol 21:206–219
3. Zhong RK, Astle CM, Harrison DE (1996) Distinct developmental patterns of short-term and long-term functioning lymphoid and myeloid precursors defined by competitive limiting dilution analysis in vivo. J Immunol 157:138–145
4. Morrison SJ, Weissman IL (1994) The long-term repopulating subset of hematopoietic stem cells is deterministic and isolatable by phenotype. Immunity 1:661–673
5. Spangrude GJ, Heimfeld S, Weissman IL (1988) Purification and characterization of mouse hematopoietic stem cells. Science 241:58–62
6. Osawa M, Hanada K, Hamada H, Nakauchi H (1996) Long-term lymphohematopoietic reconstitution by a single CD34-low/negative hematopoietic stem cell. Science 273:242–245
7. Morita Y, Ema H, Nakauchi H (2010) Heterogeneity and hierarchy within the most primitive hematopoietic stem cell compartment. J Exp Med 207:1173–1182
8. Hanahan D, Weinberg RA (2011) Hallmarks of cancer: the next generation. Cell 144:646–674
9. Plaks V, Koopman CD, Werb Z (2013) Circulating tumor cells. Science 341:1186–1188
10. Chaffer CL, Weinberg RA (2011) A perspective on cancer cell metastasis. Science 331:1559–1564
11. Rhim AD, Mirek ET, Aiello NM, Maitra A, Bailey JM, McAllister F, Reichert M, Beatty GL, Rustgi AK, Vonderheide RH, Leach SD, Stanger BZ (2012) EMT and dissemination precede pancreatic tumor formation. Cell 148:349–361
12. Mani SA, Guo W, Liao M-J, Eaton EN, Ayyanan A, Zhou AY, Brooks M, Reinhard F, Zhang CC, Shipitsin M, Campbell LL, Polyak K, Brisken C, Yang J, Weinberg RA (2008) The epithelial-mesenchymal transition generates cells with properties of stem cells. Cell 133:704–715
13. Thiery JP, Acloque H, Huang RYJ, Nieto MA (2009) Epithelial-mesenchymal transitions in development and disease. Cell 139:871–890
14. Lo YM, Corbetta N, Chamberlain PF, Rai V, Sargent IL, Redman CW, Wainscoat JS (1997) Presence of fetal DNA in maternal plasma and serum. Lancet 350:485–487.

15. Lo YM, Zhang J, Leung TN, Lau TK, Chang AM, Hjelm NM (1999) Rapid clearance of fetal DNA from maternal plasma. Am J Hum Genet 64:218–224
16. Lo YM, Tein MS, Lau TK, Haines CJ, Leung TN, Poon PM, Wainscoat JS, Johnson PJ, Chang AM, Hjelm NM (1998) Quantitative analysis of fetal DNA in maternal plasma and serum: implications for noninvasive prenatal diagnosis. Am J Hum Genet 62:768–775
17. Chiu RW, Chan KA, Gao Y, Lau VY, Zheng W, Leung TY, Foo CH, Xie B, Tsui NB, Lun FM, Zee BC, Lau TK, Cantor CR, Lo YM (2008) Noninvasive prenatal diagnosis of fetal chromosomal aneuploidy by massively parallel genomic sequencing of DNA in maternal plasma. Proc Natl Acad Sci U S A 105:20458–20463
18. Palomaki GE, Deciu C, Kloza EM, Lambert-Messerlian GM, Haddow JE, Neveux LM, Ehrich M, van den Boom D, Bombard AT, Grody WW, Nelson SF, Canick JA (2012) DNA sequencing of maternal plasma reliably identifies trisomy 18 and trisomy 13 as well as Down syndrome: an international collaborative study. Genet Med 14:296–305
19. Huang Z, Fong C-Y, Gauthaman K, Sukumar P, Choolani M, Bongso A (2011) Novel approaches to manipulating foetal cells in the maternal circulation for non-invasive prenatal diagnosis of the unborn child. J Cell Biochem 112:1475–1485
20. Bianchi DW, Williams JM, Sullivan LM, Hanson FW, Klinger KW, Shuber AP (1997) PCR quantitation of fetal cells in maternal blood in normal and aneuploid pregnancies. Am J Hum Genet 61:822–829
21. Hamada H, Arinami T, Kubo T, Hamaguchi H, Iwasaki H (1993) Fetal nucleated cells in maternal peripheral blood: frequency and relationship to gestational age. Hum Genet 91:427–432
22. Yamamoto R, Morita Y, Ooehara J, Hamanaka S, Onodera M, Rudolph KL, Ema H, Nakauchi H (2013) Clonal analysis unveils self-renewing lineage-restricted progenitors generated directly from hematopoietic stem cells. Cell 154:1112–1126

Part II
3D Cellular System Design

Chapter 7
Cell Manipulation and Cellular Parts Assembly for Constructing 3D Cellular Systems

Masaru Kojima, Yasushi Mae, Kenichi Ohara, Mitsuhiro Horade, Kazuto Kamiyama and Tatsuo Arai

Abstract Bio Assembler integrates the state-of-the-art micro robotics and tissue engineering to create 3D tissues *in vitro*. In Bio Assembler we have different key technologies in different complex structure for creating 3D cellular systems. In this chapter, we have reviewed our recent progress to construct 3D cellular systems. Especially, introduce three key technologies of micro robotics for 3D cellular construction, (1) Cell manipulation by microhand, (2) Construction of 3D lattice by hydrogel fiber and (3) Changeable Cell Culture mold for Advanced Cell Sheet. These proposed new technologies contribute to realize "Bio Assembler".

Keywords Multi-scalability · Parallel link mechanism · Micromanipulation · Alginate gel fiber · Cell culture device · Microactuator array

7.1 Introduction

Donor tissue transplantation for treating damaged tissue has had a serious problem: the smaller number of donors than of those needing transplantation. It has thus been advocated that tissue engineering (TE) studies should be promoted systematically [1, 2]. Tissue fabrication by TE is categorized into 2 main approaches: cellular and acellular. In principle, the cellular approach employs small units of living cells. During the past decade, many successful techniques using this approach for the fabrication of 3D engineered tissue have been reported, such as cell sheet stacking [3–5] and 3D cell accumulation [6, 7]. Recent studies have focused on technologies for generating active 3D tissues *in vitro*. Some previous studies have attempted to generate 3D tissues by laminating cultivated cell sheets [8, 9]. Others have developed bioprinting technologies for generating

T. Arai (✉) · M. Kojima · Y. Mae · M. Horade · K. Kamiyama
Graduate School of Engineering Science, Osaka University, 1-3 Machikaneyama-cho, 560-8531 Toyonaka, Osaka, Japan
e-mail: arai@sys.es.osaka-u.ac.jp

K. Ohara
Faculty of Science and Technology, Meijo University, 1-501 Shiogamaguchi, Tempaku-ku, 468-8502 Nagoya, Aichi, Japan

T. Arai et al. (eds.), *Hyper Bio Assembler for 3D Cellular Systems*,
DOI 10.1007/978-4-431-55297-0_7

tissue in arbitrary forms from liquid droplets containing cells, taking advantage of ink jet printer technology [10–12]. Yet another study dealt with vascular 3D network constructions aimed at generating biocompatible artificial 3D tissue by taking advantage of 3D printer technology [13]. Still other studies have focused on generating tissue in a variety of forms by producing hydrogel fiber in microfluidic channels [14–16]. The fabrication of complex tissue structures is limited, however, without the use of support scaffolds. In contrast, the acellular approach [17–20] generally employs biomaterials as support scaffolds to enhance the fabrication of complex tissue structures and to maintain the tissue structure during cell fusion. Use of a fixed scaffold would nevertheless inherently limit the variety of tissue structures fabricated. It is therefore difficult to rely solely on this approach for the fabrication of engineered organs that have a very complex structure.

Like when manufacturing a car, for example, an engineered kidney may be fabricated by the assembly of composites parts such as the ureter, capsule, cortex, and medulla. The *in vitro* assembly of tissue parts, cultivated under appropriate conditions, may later result in a functional organ. The only drawback to this assembly approach is that it requires many different parts that have different structures. To fabricate these parts, many individual fixed scaffolds, i.e., the current acellular approach, must be prepared, which requires a lot of effort and time.

In the above context, applying micro robotics to tissue fabrication is a promising way for automatic and high-speed fabrication of tissues. We have proposed a new methodology: Bio Assembler which integrates the state-of-the-art micro robotics and tissue engineering to create 3D tissues *in vitro*. In Bio Assembler we have different key technologies in different complex structure for creating 3D cellular systems. For example, we can say a cell manipulation technique is for a zero-dimensional structure, hydrogel fiber technique is for a one-dimensional structure, and cell sheet technique is for a two-dimensional structure. Micromanipulation technique can be also applied to assemble cellular parts with different structures. By combining these techniques, we will construct 3D cellular parts, and assemble them to realize complex 3D cellular system with functions. In this chapter, we introduce three key technologies of micro robotics for 3D cellular construction.

7.1.1 Cell Manipulation by Microhand

Various micromanipulation systems have recently been designed and used for applications such as manipulating biological cells and assembling cellar parts. The realization of a multi-scalable micromanipulator that can handle multisized micro objects and transport them with precise positioning in a large workspace is crucial. This system should have the ability to carry biological cells of different sizes for long distances (a few hundred μm) and position them precisely (sub-micron) to have them come in contact.

To achieve a multi-scalable microhand system that is compatible with various microapplications, i.e., a generalpurpose microhand, the following two features are required: first, the microhand end-effector should have a large workspace with a high resolution for grasping different-sized microobjects. Second, a large workspace with a precise positioning function is necessary for the transportation of microobjects. In order to realize a general-purpose micromanipulator that is compatible with various microapplications, however, it is critical to achieve manipulation of various-sized microobjects from a few microns to over a hundred microns. In this study, the workspace of the parallel mechanism has been optimized with the help of inverse kinematics by modifying the orientation of the end-effector, which is fixed to the end of the parallel mechanism, to meet the condition of handling target objects of various sizes.

Transportation of microobjects up to a millimeter-scale with sub-micron precision, creating a large workspace with high resolution is also important. One of the strategies available to do so is the combination of rough motion and fine motion functions in the same microsystem [21, 22]. In this study, the conventional rough-to-fine motion strategy has been improved to achieve fast precise positioning throughout a large workspace and to remove the time lag between rough and fine motions. These micromanipulation techniques will be available for not only cell manipulation but also cellular parts assembly in Bio Assembler [23].

7.1.2 Construction of 3D Lattice by Hydrogel Fiber

Research groups have proposed 3D tissue construction method using hydrogel fiber created by laminar flows in a combination of plural micropipettes [24, 25]. These approaches to 3D tissue generation *in vitro* still have the critical problem of continuously supplying oxygen and nutrients to tissue.

We describe the automated 3D tissue construction system in arbitrary forms from hydrogel fiber containing cells. Hydrogel fiber both supplies oxygen and nutrients to the cells it contains and is maneuverable by a robot system. We also describe the effectiveness of developed construction system through the experiments of 3D lattice structure construction from hydrogel fiber [26].

7.1.3 Changeable Cell Culture Mold for Advanced Cell Sheet

Instead of employing fixed scaffolds, we might need to consider a flexible scaffold, which makes many scaffold structures possible. A device provided with a changeable scaffold could be used to fabricate many different tissue structures. The simultaneous operation of such devices would certainly accelerate the production of engineered organs in helping meet the high demand of transplant tissue.

The changeable cell culture (C^3) mold is proposed for engineered tissue fabrication. This mold forms a temporary microscaffold. The displacement of each actuator varies according to the pressure of the working fluid inside its corresponding microchannel. The actuator is made of polydimethylsiloxane (PDMS), a material chosen for its biocompatibility, optical transparency, high elasticity, leakage prevention, and easy modification of actuator features. In our design, the C^3 mold consists of a 3 × 3 microactuator array with an actuator diameter of 500 μm and spacing of 650 μm. The physical characteristics of this polymeric actuator are discussed, and the possibility of using this device for real tissue fabrication is analyzed. Fabricated micropatterns on the agarose gel sheet are very similar to the tissue structures that would be obtained when real cells are used in future experiments [27].

7.2 Development of Multi-Scalable Microhand System with Precise Motion Ability

Numerous types of microhands have recently been designed to perform micromanipulation tasks that are crucial for micromachine assembly, microsurgery operations and biological cell analysis. Because most current microsystems are task-specific, the realization of a general-purpose microhand that is compatible with a wide range of applications is necessary. There are two problems to be solved in order to realize a general-purpose micromanipulation system however. First, creating a large workspace with high resolution in which to grasp multisized microobjects is still a challenging feature for available microhands. Second, precise motion throughout a large workspace for the transportation of microobjects in the limited space of a microscope is another arduous task.

In this section, we propose a microhand system that achieves multiscalability, i.e., a large workspace with precise positioning for the grasping and transportation of multisized microobjects. This system has been designed with an optimized parallel mechanism in which the manipulability of different-sized microobjects is improved from 1–45 to 1–132 μm. The proposed rough-to-fine motion strategy that allows us to achieve a large range with high resolution positioning ability for performing the transportation task moreover minimized error from 17 to 0.18 μm.

7.2.1 System Design: System Concept

At the beginning of the design process, the system concept has been defined as a micromanipulator that achieves large workspace and precise motion at the same time. This system is composed of two main parts: lower and upper modules. The lower module is for global motion, that is, the movement of both end-effectors that achieves the transportation of target objects and the positioning of the microhand. Two motorized stages—rough (Sigma-Koki, TSD-805S) and fine (Sigma-Koki, SFS-H60XYZ) stages—form the lower module and help realize the large workspace and precise motion necessary for the transportation task. Specifications of fine and rough stages are listed in Table 7.1. We use these two stages to move both end-effectors for long and rough motions, i.e., for traveling long distances (micro and millimeter-scales), and for short and fine motion (nano and micro-scales). This

Table 7.1 Motorized stages specifications

	Fine stage	Rough stage
Travel(mm)	0.1 (*X*, *Y*, *Z*)	25 (*X*, *Y*), 10 (*Z*)
Resolution(μm)	0.01	1
Accuracy(μm)	0.1	17
Repeatability(μm)	0.15	2

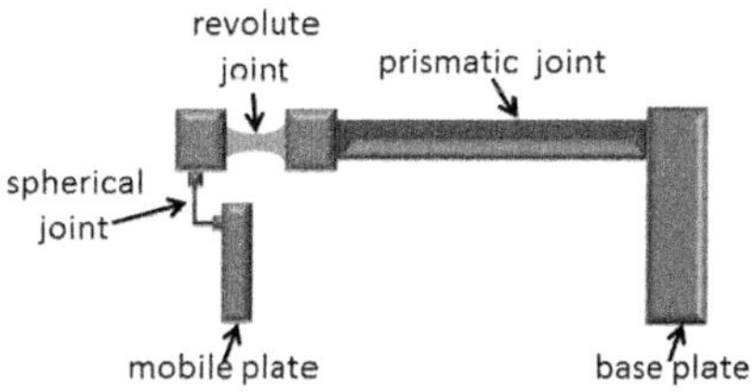

Fig. 7.1 Single kinematics chain of parallel mechanism including prismatic, revolute, and spherical joints

enables us to achieve precise positioning when grasping and releasing tasks are executed.

The upper module is for local motion, i.e., the movement of the upper end-effector. Local motion achieves the grasping and release of different-sized target objects. One manual stage for adjusting the upper end-effector with respect to the lower one and one parallel link mechanism for manipulating different-sized objects form the upper module and realize a large workspace for the grasping task. The manual stage moves 6 mm in the X-, Y-, and Z-directions with a resolution of 3 μm. The parallel link mechanism includes 3 piezoactuators as prismatic joints that can be extended up to 40 μm. Workspace optimization is realized by changing end-effector's orientation so that it has the largest feasible workspace for a fixed-size end-effector. In addition to its simple compact configuration, the parallel mechanism has the advantages of high speed, high accuracy, and high rigidity. It has simple joint mechanisms using a flexure hinge as a revolute joint and a wire as a spherical joint in each link, as shown in Fig. 7.1. The mechanism has a 3-DOF end-effector (two rotational and one translational motion), and has no singular point in its actual workspace. Three linear piezoelectric actuators (NEC TOKIN, AE0203D16) are arranged on a base plate so that their displacement is vertical to this plate. This arrangement has two advantages: first, the movable parts of the finger can be rather light, and, second, workspace along the Z-axis with reference to the base coordinate frame is its largest.

After the design concept is decided, a CAD model of the micromanipulator is prepared, as shown in Fig. 7.2.

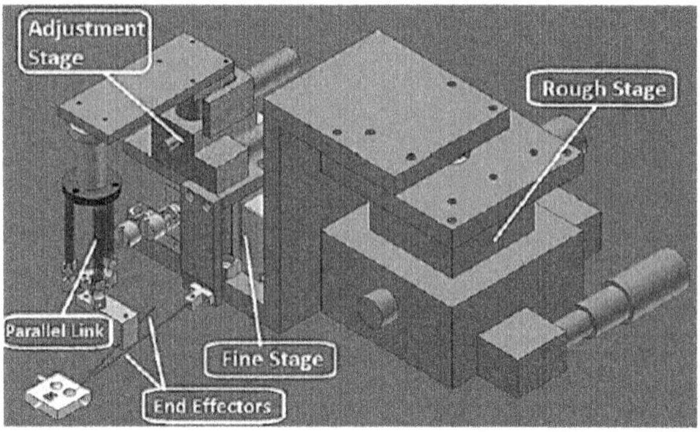

Fig. 7.2 CAD model concept

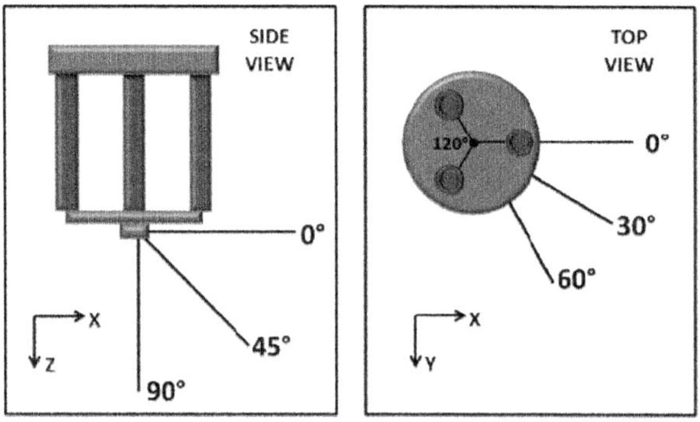

Fig. 7.3 Different orientations on X–Z (*left*) and X–Y (*right*) planes

7.2.2 System Design: Workspace Optimization

For the manipulation of different-sized microobjects, parallel mechanism workspace should be at least 100 μm for the X-direction, which is the direction of the motion of the end-effector when grasping and releasing tasks are carried out. The manipulation of large microobjects such as 97 μm diameter bovine cells will be thus achievable. In addition, accuracy should be at a sub-micron level for the manipulation of objects having a size of only a few microns, e.g., fibroblasts and bacteria. The upper end-effector is fixed to 3-DOF (prismaticrevolute-spherical (PRS) joints) parallel mechanism that was analyzed to understand upper end-effector workspace. Details of workspace analysis and the workspace optimization process are found in our previous work [28, 29]. We therefore discuss only workspace optimization results and ways of applying these results to the actual system. The purpose of workspace optimization is to decide the orientation of the end-effector for the largest horizontal workspace. Because inclined workspace cannot be used efficiently, we sought a flat workspace; whose X-axis can be used as a reference. To find the largest flat workspace, results were obtained about workspace of different orientations by checking two different planes, as shown in Fig. 7.3.

Table 7.2 Workspace results for different orientations of the end-effector on the X-Z plane and X-Y plane

X–Z plane	X (μm)	Y (μm)	*X–Y* plane	X (μm)	Y (μm)
0°	0.6	0.8	0°	530	480
45°	530	480	30°	510	480
90°	720	640	60°	540	480

Fig. 7.4 System constraints (*left*) and intermediate part solution (*right*)

On the X-Z plane, which is a side view of the parallel system, different orientations were checked, as shown in Fig. 7.3 (left). Workspace results for different orientations are listed in Table 7.2 (left). On the X-Y plane, which is the top view of the parallel system, different orientations were checked, as shown in Fig. 7.3 (right). Workspace results for different orientations are shown in Table 7.2 (right).

Workspace analysis results for different orientations are summarized as follows: if we change the orientation downward on the X-Z plane, workspace increases. The largest workspace is realized at 90°. If we change the orientation on the X-Y plane, the workspace size hardly changes. In addition, during the workspace analysis process, we found that the inclination of the workspace X-axis is inversely proportional to the orientation angle of the end-effector on the X-Z plane. For 0° orientation, the slope of the X-axis is 89.8° and for 90° orientation, it is 0.7°, which is close to a flat plane. Hence, to have a large flat workspace, the end-effector should form a 90° angle on the X-Z plane. If this is not feasible because of micromanipulation constraints, the orientation of the end-effector on the X-Z plane should be as close as possible to 90°.

Although a 90° orientation, normal to the ground, is the best angle for obtaining the largest flat workspace, there are two main constraints that should be considered. First, to observe the micromanipulation process with an inverted microscope, the end-effector should be placed between the light source and the lens, for which 90° orientation is not possible, as shown in Fig. 7.4 (left). Second, in order to avoid breaking the glass finger during 3D motion of the system, the end-effector should not be normal to the ground. When considering these two constraints with respect to

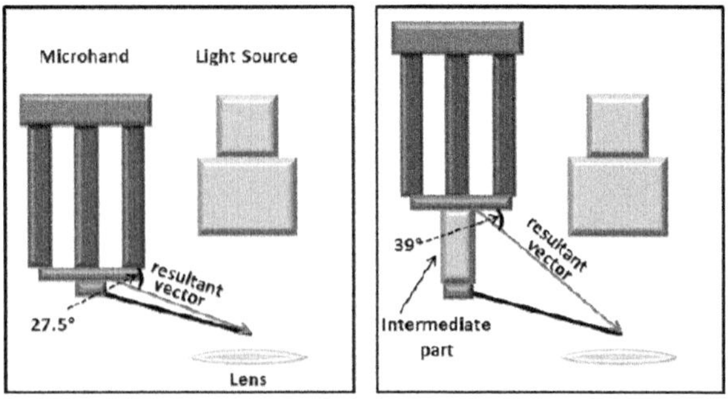

Fig. 7.5 Intermediate part solution to obtain a more inclined resulting vector

workspace optimization results, we propose using an intermediate part as a solution, as illustrated in Fig. 7.4 (right) and in Fig. 7.5, placed between the parallel link and the end-effector to obtain a more inclined resultant vector. A longer intermediate part means closer orientation to 90° for the resulting vector. Hardware limitations, system setup, and a vibration problem have to be considered, however, when deciding the maximum feasible length of the intermediate part. After these limitations were considered, the length of the intermediate part was set to 3.5 cm. With the optimized intermediate part, we were able to realize a 39° angle on the X-Z plane as the feasible degree closest to 90° orientation, as shown in Fig. 7.5.

After we consider workspace inclination for 39° orientation, if we keep the depth, which is the Z-direction, fixed, then we can move our end-effector 132 μm in the X-direction and 40 μm in the Y-direction with an observable resolution of 0.1 μm. Hence, with this workspace size, the ability of the microhand to manipulate different-sized microobjects was improved from 1–45 to 1–132 μm.

7.2.3 System Construction: System Configuration

Figure 7.6 shows the configuration of the micromanipulation system, in which, the main part is a two-fingered micromanipulator. Details of the actual system are shown in Fig. 7.7. The rough and fine stages are controlled by a Linux PC (Dell, XPS600, Pentium 4, 3.80 GHz) through commercially available stage controllers (Sigma-Koki: Omec-4BG, Fine-503). The parallel link mechanism is controlled by the same PC through a D/A board (Contec DA16-16(LPCI)L) and a drive amplifier (MATSUSADA, HJPZ-0.15Px3). Displacements are measured with a strain gauge attached to piezoelectric devices and sent to a PC through a strain amplifier (Kyowa MCD-16A) and an A/D board (Contec AD16-16(PCI)EV) for PI control, in order to compensate for the hysteresis effect of the piezoactuator. Parallel mechanism control in a closed loop is shown in Fig. 7.8, and the duration of one control cycle is about 1 ms.

The microhand and the target object are placed on the optical microscope stage. Images of end-effector ends are captured by the CCD camera (Point Gray Research,

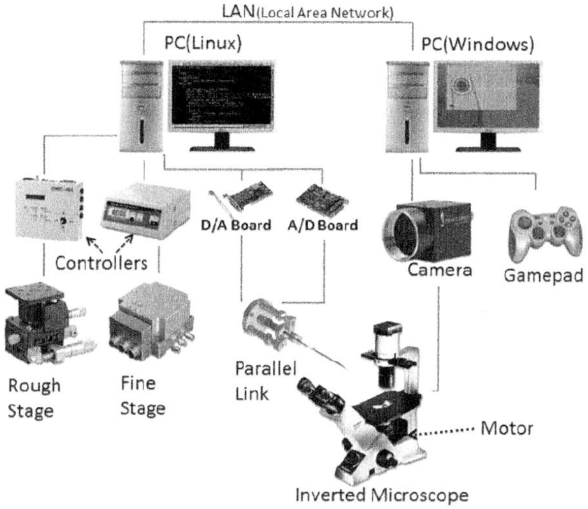

Fig. 7.6 System configuration

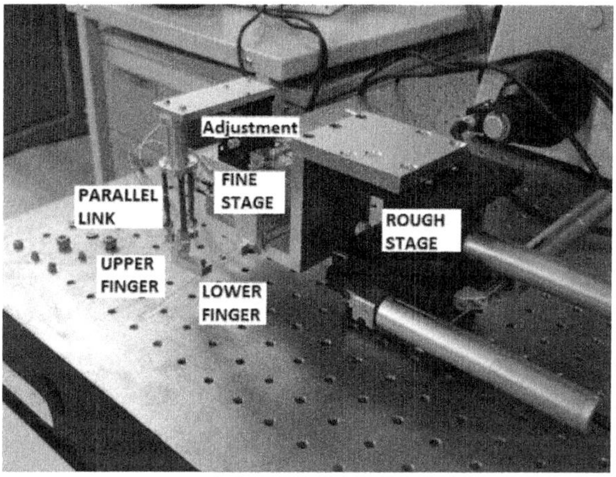

Fig. 7.7 Micromanipulator prototype

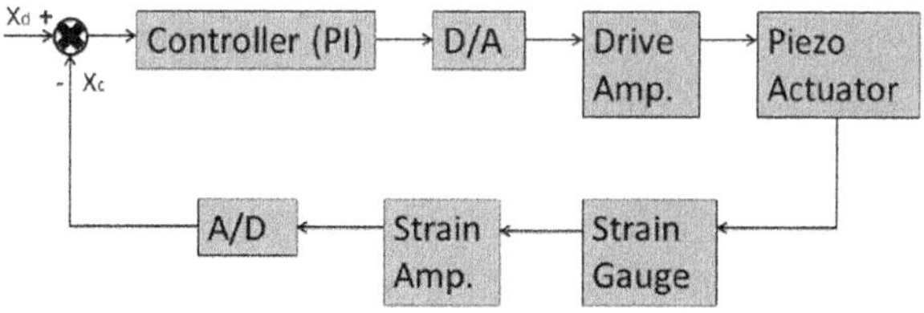

Fig. 7.8 Parallel mechanism control

Flea) and displayed on a Windows PC. The motor on the microscope controls the motion of the objective lens, used in keeping the target object in focus manually. The end-effectors—the upper and lower fingers—of this micromanipulator consist of two glass needles that have a 1 mm diameter and sharpened ends with less than 1 μm in curvature. These needles are attached to the manipulator so that they can be replaced easily with new ones if they break, become worn, or are dirty.

A joystick is more efficient than a keyboard for micromanipulation tasks. Teleoperation between the user and the micromanipulator was therefore realized by using a joystick. We controlled the lower and upper modules, which consist of two motorized stages and the parallel link part, simultaneously with a cordless gamepad (Logitech Rumblepad2).

The parallel link consists of three piezoactuators parallel to each other. An end-effector is placed at the end of the parallel link. To control upper module movement accurately, the relationship between the parallel link extension through the strain gauge and the end-effector position change through the camera was investigated. The parallel link mechanism was calibrated by obtaining the components of the calibration matrix that characterize the linear relationship between the displacement of the three piezoactuators and the change in the finger position in three directions. After calibration, absolute positioning accuracy is 2.1 μm for the 132 μm range in the X- and Y-directions. Average error is 0.9 μm and standard deviation is 0.7 μm.

In this microhand system, the grasping and release of microobjects can be realized by teleoperation through the parallel link mechanism, where the transportation of microobjects is achieved both by teleoperation and by automating through rough and fine stages.

7.2.4 System Construction: Realization of Multi-Scalability

The purpose of the lower module is to achieve longdistance transportation of microobjects with precise positioning. As the rough motion stage has large workspace with low accuracy and the fine motion stage has small workspace with high resolution, a combination of these two functions in the same system, which is the conventional rough-to-fine motion strategy, offers the key to achieve the purpose of the lower module [21, 22]. Because the conventional method has a time delay between rough and fine motions, however, we propose a rough-to-fine motion strategy without time lag.

7.2.5 System Construction: Realization of Multi-Scalability: Conventional Rough-to-Fine Motion Strategy

The conventional rough-to-fine motion strategy is employed to improve positioning accuracy. The strategy procedure is as follows: target position (Xt) in the visible area of the microscope camera of the end-effector is decided by the user. The system then sends the target position to the rough motion stage and the end-effector, which holds the microobject, moves to reach the target on the X-Y plane. Image resolution (Xr) depends on the magnification of the objective lens. By using a 60X objective lens, image resolution is 0.08 μm. The system should wait until the rough stage finishes its motion by checking the current and previous positions (Xc and Xp) of end-effectors through template matching as Eq. (1). Because the orientation of the end-effector does not change during manipulation, we apply the template matching method.

$$\begin{array}{ll} \text{If } |Xc - Xp| \geq Xr & : \text{rough motion continues.} \\ \text{Otherwise} & : \text{rough stage stops.} \end{array} \quad (7.1)$$

When the difference between current and previous positions is smaller than Xr, which means the rough motion stage has stopped, the system compares the target and current positions (Xt and Xc) of the end-effector through template matching. If the difference between them is less than Xr, the target position has been reached. If the difference is more than or equal to Xr, the difference is sent to the fine stage automatically to be compensated for by Eq. (7.2).

$$\begin{array}{ll} \text{If } |Xt - Xc| < Xr & : \text{motion is completed.} \\ \text{Otherwise} & : \text{compensate the error.} \end{array} \quad (7.2)$$

Last, the fine motion stage finely aligns the microobject with the target position on the X-Y plane with 0.1 μm error. The precision of template matching is 0.08 μm and maximum error of the fine motion stage is 0.1 μm, meaning that 17 μm error of the rough motion stage is minimized to 0.18 μm with the conventional rough-to-fine motion strategy. The algorithm is summarized in the part including time lag in Fig. 7.9 at left.

7.2.6 System Construction: Realization of Multi-Scalability: Proposed Rough-to-Fine Motion Strategy

Although large workspace with precise motion has been achieved using the above strategy, to decrease the duration of the transportation task, the strategy has been improved by considering time lag.

As seen in the part that does not include time lag in Fig. 7.9, the system does not wait for the rough motion stage to stop before compensating for error. When distance (Xdis) between the end-effector and the target position is close enough, i.e., in the range of workspace for fine motion stage (Xw), which should be 50 μm, the system sends distance (Xdis) to the fine stage in Eqs. (7.3)–(7.4).

$$Xdis = Xt - Xc \quad (7.3)$$

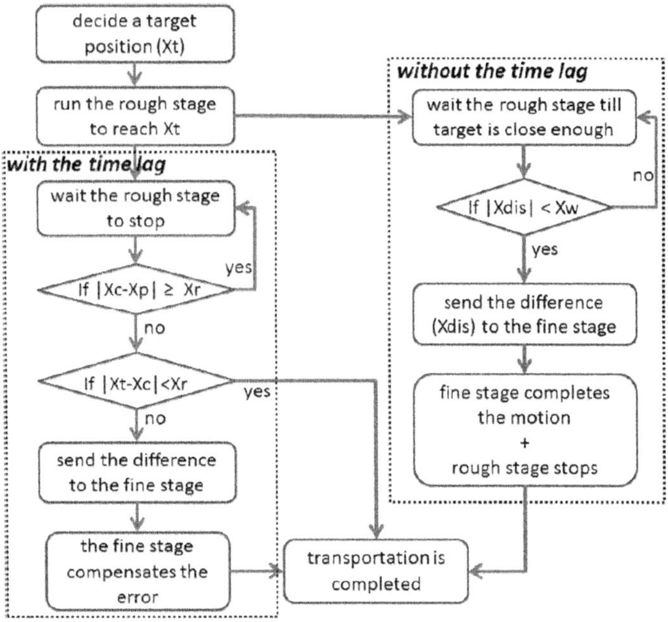

Fig. 7.9 Accurate positioning flowchart

$$\begin{array}{ll} \text{If} \mid \text{Xdis} \mid \geq \text{Xw} & : \text{rough motion continues.} \\ \text{Otherwise} & : \text{fine motion starts.} \end{array} \tag{7.4}$$

The fine stage moves the end-effector to the target position. When fine motion starts, the rough stage stops. Because the rough stage cannot stop suddenly, motion between the stop command and completely stopping the rough stage is considered when distance information is sent to the fine stage. When the fine stage finishes its motion, the transportation task is then completed with maximum error of 0.18 μm.

Conventional and proposed strategies are compared in Fig. 7.10. For the 150 μm travel distance in the X-direction, the system needs 2.76 s for rough motion, 1.31 s for time lag, and 0.005 s for fine motion, i.e., 4.08 s in total, to complete transportation with high precision (Fig. 7.10 (left)). Using the proposed strategy, in contrast, the system needs only 1.74 s to complete the transportation task as shown in Fig. 7.10 (right). Because the time lag depends on the system itself, e.g., the duration of communication between devices, even if the time lag is further shortened, our proposed strategy is still faster. With the first strategy, furthermore, the fine stage moves for a small distance, while with the second strategy, the fine stage moves for half of its workspace. Even if maximum error for both cases is 0.18 μm, error is therefore less for the first strategy most of the time.

As shown in Fig. 7.9, there are two routines for the system to continue: with and without the time lag. The user decides which routine is appropriate for the system with respect to priority such as time duration, speed, or accuracy.

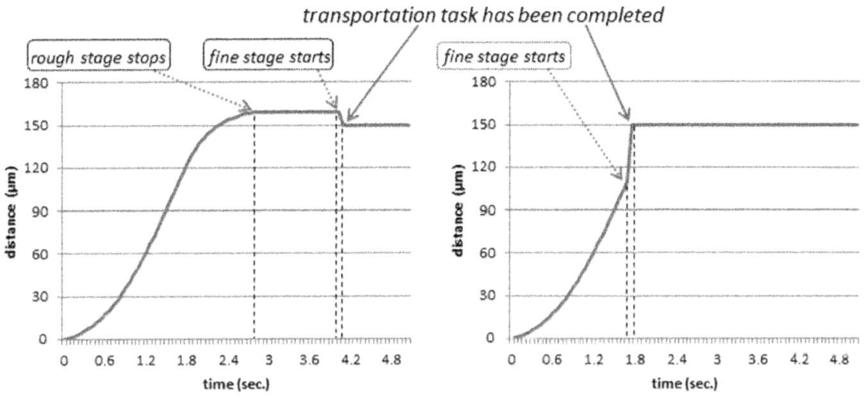

Fig. 7.10 Comparison of conventional (*left*) and proposed (*right*) rough-to-fine motion strategies

From the beginning of manipulation, the system's end-effectors and target objects are placed on the same Z focal plane manually with the help of the motorized stage. Vision feedback strategy is used for manipulator positioning in 2D. If automatic control in the Z-direction is necessary, however, we can extend our strategy to 3D by applying the All-In-Focus method [30].

7.2.7 Evaluation of Multi-Scalability: Experimental Results

The three experiments carried out with this system had the common feature that they used multi-scalability. First, various-sized microspheres were manipulated to show the ability of the system to manipulate multisized objects. Second, a cell fusion task was realized by manipulating donor and egg cells in the same scene. Third, extremely small and extremely large microbeads were manipulated in the same scene to show the precise positioning ability of the system, along with multi-sized object manipulability.

7.2.8 Evaluation of Multi-Scalability: Experimental Results: Manipulation of Various-Sized Microobjects

Microspheres, with diameters of 97, 55, 20, and 9.7 μm are manipulated to show that different-sized objects can be manipulated with this microhand. The entire process is realized in liquid on a glass and observed by 20X and 40X objective lenses. Figure 7.11 shows the manipulation process for a 20 μm microsphere. Figures 7.11a and b show the grasping task, which includes moving both fingers to bring them close enough to the target object and moving the right finger to hold the target. Figures 7.11c and d show a transportation task over a distance of 60 μm. Figures 7.11e

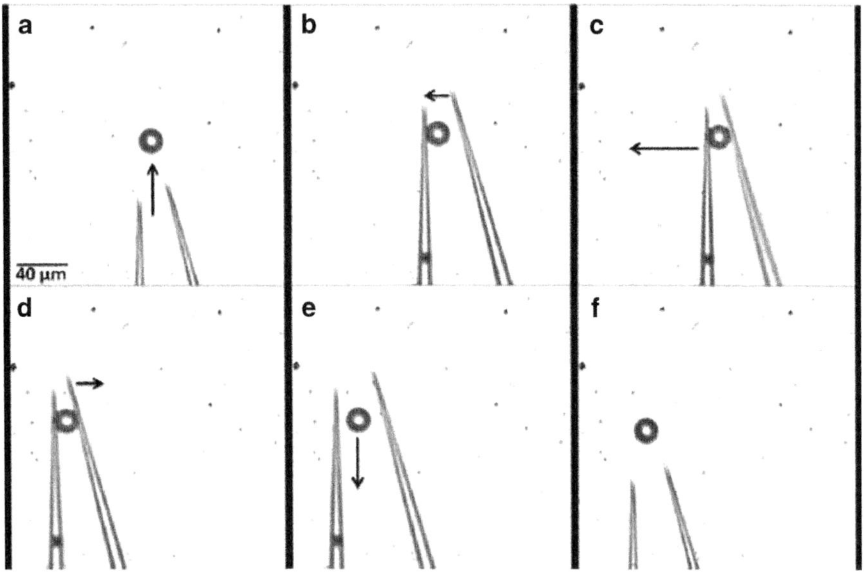

Fig. 7.11 **a-b** Grasping task for 20 µm microspheres. **c-d** Transportation task for 60 µm distance. **e-f** Release process

and f show a release task that includes moving the right finger and then moving both fingers away from the target object. This experiment was conducted as a predefined manipulation task via computer control. The entire manipulation process was the same for all target objects of different sizes. For each of the four different-sized target objects, the manipulation task was carried out 10 times to investigate the success of the process. Success of 100 % for the grasping task (Figs. 7.11a and b), 100 % success for the transportation task (Figs. 7.11c and d), and 80 % success for the release task (Figs. 7.11e and f), were achieved after 40 trials.

7.2.9 Evaluation of Multi-Scalability: Experimental Results: Cell Fusion Task

Cell fusion is one of the tasks possible for this micromanipulator in terms of multisized microobject manipulation. The procedure is as follows: a donor cell and an egg cell are placed under an optical microscope and the finger tips are moved using the stages until they are close enough to cells, then the microfingers hold, pick-up, transport, and release the different-sized cells, having them contact each other. The next stage of this task, which is the fusion of two cells with the help of an electrical current, is not performed here because the aim of this experiment is to investigate the capability of the system to be used in the manipulation of cells during a cell fusion task.

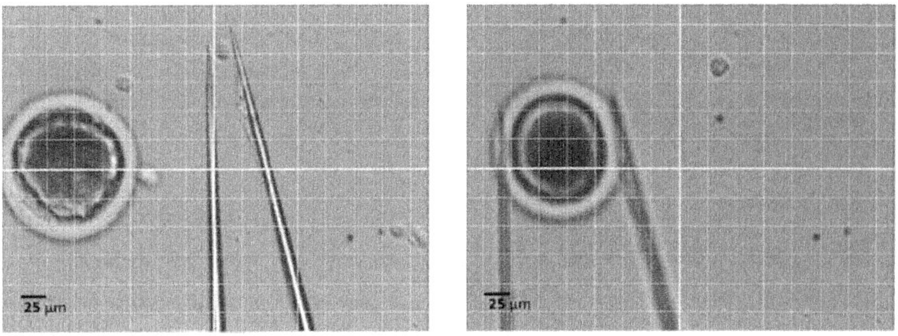

Fig. 7.12 Manipulation of donor (*left*) and egg (*right*) cells

The handling of living cells is more difficult than that of microbeads. Because the duration of life for living cells in a normal environment is very short, cells are placed in a special liquid. In addition, living cells are stickier than microbeads, which makes the release task more difficult because adhesion is more effective on a micro-scale. We therefore prefer to release the cell by moving it to the end of the end-effector, which has the smallest contact area compared to the rest of the end-effector because adhesion between the tip of the finger and the grasped cell is weakest. Furthermore, the target object is a soft living cell that poses no bending problem for the end-effector. Donor cells, which are about 16 μm, and egg cells, which are about 100 μm, were used. We were able to successfully manipulate—hold, pick-up, transport, and release—the donor and egg cells in the same scene, as shown in Fig. 7.12 and to make them contact each other with the help of teleoperation. We thus say hence that cell fusion tasks can be performed with this system.

7.2.10 Evaluation of Multi-Scalability: Experimental Results: Multi-Scalability of the System

To show the system's multi-scale manipulability, extremely small and extremely large microbeads are manipulated in the same scene and observed with a 10X objective lens as shown in Fig. 7.13.

The microhand grasps and transports a 115 μm microbead to the target position, which is the center of the screen, and releases it. Then, in the same situation, the microhand successfully manipulates an 8 μm microbead using a similar procedure. The entire process is conducted combining teleoperation for grasping and release tasks and the rough-to-fine motion strategy for the automatic transportation task. Precise positioning with 0.15 μm error is achieved with the proposed strategy.

Because the resolution of the parallel mechanism is on the sub-micron level, the manipulation of smaller objects of about 1 μm can be realized with this system. Because finding the center of a 1 μm microbead is difficult and small objects are easily affected by liquid flow, however, it is extremely difficult to accurately grasp objects a few microns in size in liquid without a special strategy.

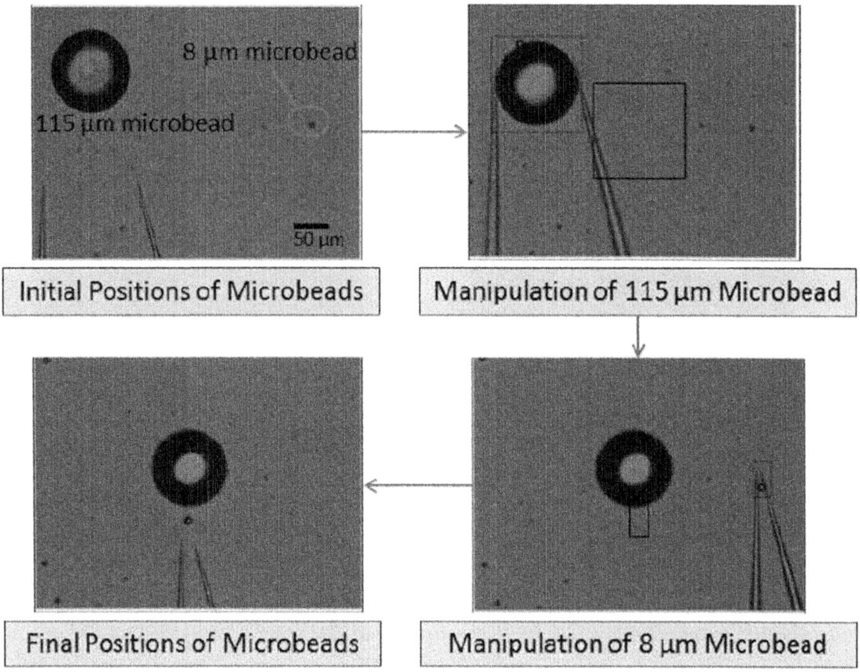

Fig. 7.13 Realization of system multi-scalability

Table 7.3 Micromanipulator performance

	Global motion	Local motion
Workspace (mm)	25.1 × 25.1 × 10.1	0.13 × 0.04 × 0.02
Max. speed (mm/sec)	1.7 (distance ≤ 0.1 mm) 0.1 (distance > 0.1 mm)	110
Observable resolution (μm)	0.1	0.1
Accuracy (μm)	0.18	2.1
Repeatability(μm)	0.15	2.5

Table 7.4 Comparison to similar manipulation systems

	mh (1)	mh (2)	mh (3)
Multisized objects manipulability (μm)	3–17	20–100	1–132

7.2.11 Evaluation of Multi-Scalability: Discussion

The performance of the developed system is shown in Table 7.3, and a comparison of multisized objects manipulability with similar studies is shown in Table 7.4. In similar studies, the manipulation of microobjects was realized by different types of microhands that we call microhand (1) [31] and microhand (2) [32]. These two similar studies were chosen for their multi-scalability and similar target objects. As shown in Table 7.4 the manipulability of multisized objects achieved by our proposed system, called microhand (3), is better than that of the studies compared.

Our micromanipulation system can be used to manipulate various-sized microobjects from 1 to 132 μm. If the target object size, e.g., tissues, is more than 132 μm, however, this system cannot be used and further improvement of workspace size is necessary. This system is by no means the final multi-scalable system and its multiscalability can be further improved to realize a generalpurpose microhand.

7.2.12 Conclusions

The system presented here allows single cell manipulation, including grasping, transporting, and releasing different-sized microobjects in the air and in aqueous environments.

The micromanipulator was designed as a generalpurpose microhand to replace task-specific systems, enabling us to use it in various microapplications. This general-purpose microhand can grasp multisized microobjects and transport them in a large workspace with precise positioning, which we call multi-scalability. To realize large workspace with precise motion for a transportation task, an automatic rough-to-fine motion strategy based on visual servo-control has been realized. The visual feedback method developed for this study is a simple robust technique well-suited for micromanipulation systems. Currently, our system can transport microobjects for up to 25 mm with sub-micron precision. The system is compact enough for the microhand to be set up on an inverted microscope.

To realize large workspace for manipulating differentsized microobjects, the parallel link mechanism has also been optimized using inverse kinematics. We can currently manipulate microobjects of sizes between 1 and 132 μm, and a comparison with similar studies indicates the superiority of the proposed system.

In future work, we will apply our proposed procedure to the manipulation of cells and tissues with the same system and will attempt to create the larger workspace necessary for completing tissue regeneration tasks.

7.3 Automated Construction System for 3D Lattice Structure Based on Alginate Gel Fiber Containing Living Cells

In this section, we propose the automated 3D tissue construction system in arbitrary forms from hydrogel fiber containing cells. Hydrogel fiber that both supplies oxygen and nutrients to the cells it contains and is maneuverable by a robot system. In this paper, we describe the effectiveness of developed construction system through the experiments of 3D lattice structure construction from hydrogel fiber.

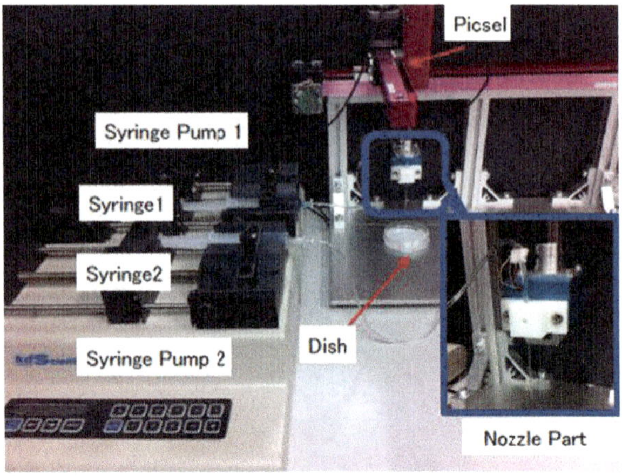

Fig. 7.14 System setup for 3D lattice construction

Table 7.5 Characteristics of robot arm

	X-axis	Y-axis	Z-axis
Stroke [mm]	300	130	100
Minimum Step [mm]	0.0125	0.0125	0.005
Maximum Speed [mm/s]	400	400	80

7.3.1 3D Lattice Structure Construction System: System Configuration

We developed a system for constructing 3D lattice structures as shown in Fig. 7.14. The system consists of a Cartesian coordinate robot arm with the specifications in Table 7.5 (Picsel, THK Co. Ltd.) and two syringe pumps (KDS-210 S, KD Scientific Inc.), both of which are controlled by the PC.

Each syringe has a Luer fitting for connecting a silicon tube with an inner diameter of 0.8 mm and an outer diameter of 2.4 mm. The syringe for syringe pump 1 uses a glass tube (G-100, Narishige) with an inner diameter of 0.8 mm and an outer diameter of 1.0 mm that has been processed with a capillary puller (PC-100, Narishige) and abrader (EG-44, Narishige) fitted at the tip. The glass tube is fixed to the robot arm by means of a fixture on the vertical axis.

A 60-mm-diameter dish is used as an area for constructing 3D structures. These system configurations enable 3D structures to be constructed in accordance with forms preset by the PC.

7.3.2 Generating Hydrogel Fiber

We picked alginate gel fiber as the scaffolding material. Alginate gel fiber, fibered gel generated from sodium alginate and calcium chloride, features high gelling reaction rates, the strength of the fiber itself, and high biocompatibility. Alginate gel fiber also has the advantage of being degradable with alginate lyase as a degrading enzyme after cells are grown in alginate gel fiber, thus enabling cells to be removed from the fiber. Taking advantage of these features, we generate 3D tissues from alginate gel fiber using the development of 3D lattice construction.

Making the best use of the features of alginate gel fiber, cell-suspension sodium alginate solution was put in syringe 1 and discharged into dish brimmed by a calcium chloride to get it gelled. After that, robot arm is moved to make a gel fiber. In this way, any structures can be sequentially constructed in arbitrary form.

Generating alginate gel fiber for constructing structures in small, arbitrary forms by using the above method requires the following:

- Small fiber dais, because this affects structure size.
- Strength, because this affects the ease with which structures are made.
- Capability to cultivate cells.

We conducted two types of experiments in searching for the conditions under which alginate gel fiber meets the above requirements:

- How the relationships between sodium alginate solution concentration and viscosity and how the generated alginate gel fiber radii are derived?
- How is alginate gel fiber generation so that cells are cultivated?

7.3.3 Generating Hydrogel Fiber: Fiber Radii Measurement

To generate alginate gel fiber with small radii, we measured radii of fiber generated from sodium alginate with different concentrations and calcium chloride with a fixed concentration 0.1 M.

Sodium alginate solution is put into a 10 ml syringe that is then fitted onto a syringe pump. The syringe has PTFE tubes with two different inner diameters, i.e., 0.5 mm and 0.3 mm, to measure alginate gel fiber radii as they are discharged from the two sizes of tubes. In this experiments, flow speed of syringe pump is set to 800 ml/min.

Measurements in Fig. 7.15 show that gelling rates increase with increasing sodium alginate concentration, making any deviations between tube diameters and generated fiber radii smaller. In other words, the higher the sodium alginate solution concentration, the thinner the alginate gel fiber produced and the closer its to tube diameters.

7.3.4 Generating Hydrogel Fiber: Evaluation of Alginate Gel Fiber Capability of Cultivating Cells

Based on experimental results in the previous section, we verified whether cells are actually cultivated in the alginate gel fiber. In these experiments, we prepared the solution by mixing 1.2 % sodium alginate solution and cellsuspension solution at a

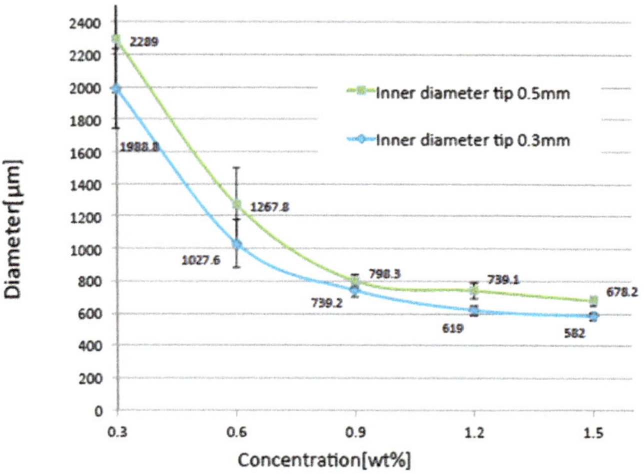

Fig. 7.15 Fiber radius measurement

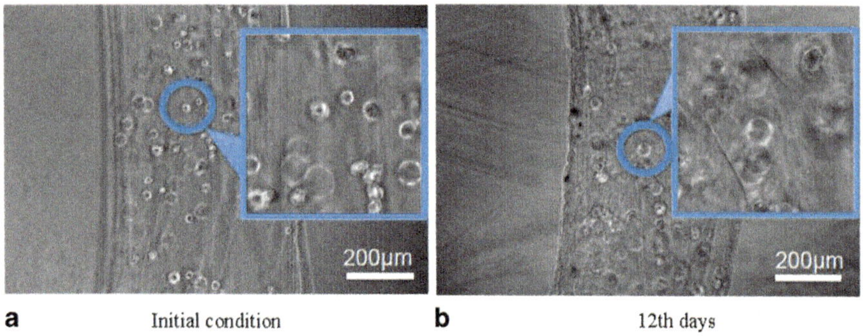

Fig. 7.16 Cultivation results for alginate gel fiber with NIH3T3 cells. **a** Initial condition, **b** 12th days

ratio of 8:2. The cell-suspension solution was prepared using a cultivation solution constituted of 89.1 % cultivation medium (Dulbecco's Modified Eagle's medium-high glucose, Sigma-Aldrich), 0.9 % penicillin (penicillin-streptomycin (× 100), Wako) and 10 % bovine serum (bovine serum, Gibco), to which the murine fibroblast cell line NIH3T3 was added to make a cellular density of 3.1×10^5 cells/ml. The final sodium alginate solution concentration was 0.96 %. This solution was put in a 10 ml syringe that was then fitted onto a syringe pump to discharge the solution from a PTFE tube 0.3 mm in inner diameter at a rate of 800 ml/min. Last, the fiber with cells was cultivated after calcium chloride solution in the dish was substituted into the above cultivation solution.

Figure 7.16 shows alginate gel fiber generated under the above conditions. From the figure, cells did not grow normally in alginate gel fiber, which seems to be attributable to the adhesiveness of alginate gel to cells.

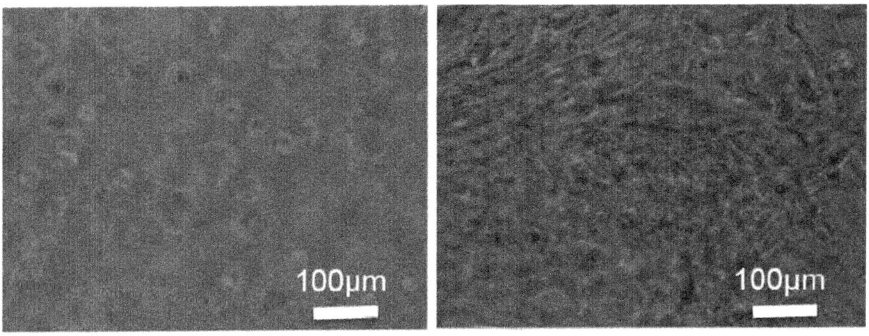

Fig. 7.17 Cultivation results for alginate-collagen gel fiber with NIH3T3 cells. **a** Initial condition, **b** 12th days

To improve cell growth, we added collagen to the sodium alginate solution as scaffolding for cultivating cells. Specifically, we prepared a solution into which we mixed 2.0 wt% sodium alginate, collagen (Cell matrix Type I-A, Nitta Gelatin Inc.), reconstitution buffer solution (NaOH: 50 mM, $NaHCO_2$: 260 mM, HEPES: 200 mM), a ten-fold concentration medium (MEM Hanks Medium) and cell-suspension solution (cell density: 1.8×10^4 cells/ml at a ratio of 40:40:5:10:5. The sodium alginate solution concentration was 0.8 %. Alginate gel fiber with cells was generated by same procedure in previous experiments and cultivated it after calcium chloride in the dish was substituted into cultivation solution.

Figure 7.17 shows the alginate gel fiber generated. The growth of cells can be seen in the fiber 12 days after cultivation.

From these results, the proposed system can generate a hydrogel fiber, which is able to cultivate cells. Thus, we prepared the solution for subsequent experiments by using the same mixing ratio as in the above experiments.

7.3.5 Experiments for Automatically Constructing 3D Structures: Procedures

This section describes the functions of the proposed automated system in experiments. The automated system has two kinds of syringe pumps. In experiments, we put sodium alginate solution in syringe 1 and calcium chloride solution in syringe 2.

Moving the tip of syringe 1 close to the liquid surface in the dish was defined as the initial condition. Paper was laid in the bottom of the dish to improve the adhesiveness of the generated gel fiber. And, a certain amount of calcium chloride solution was contained to the dish.

Under the initial condition, syringe 1 started discharging sodium alginate solution when syringe pump 1 was run. To prevent the movement of constructed structures,

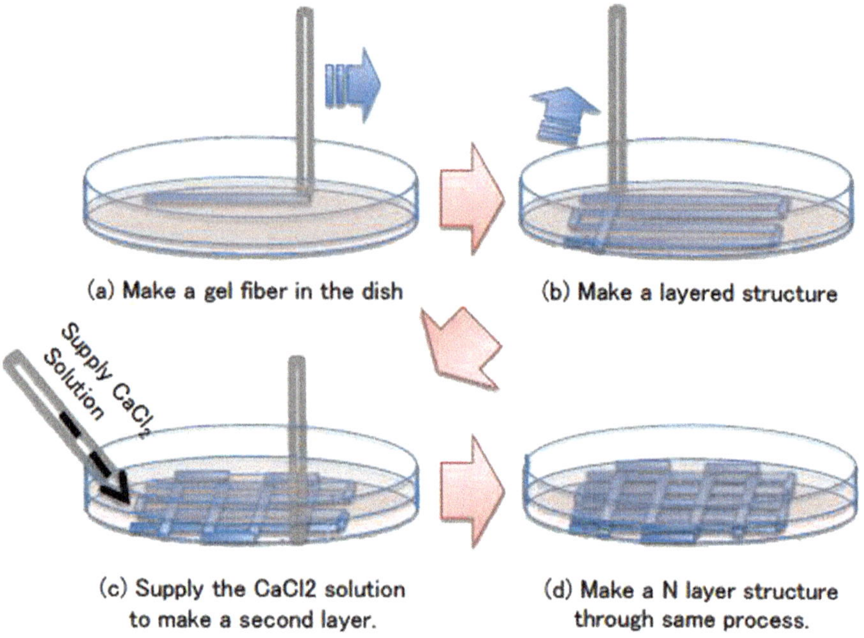

Fig. 7.18 Procedure for lattice construction. **a** Make a gel fiber in the dish, **b** Make a layered structure, **c** Supply the $CaCl_2$ solution to make a second layer, **d** Make a N layer structure through same process

anchor gel was set at a position slightly away from the area where structures were constructed. Robot arm then began to move following a preset trajectory until it reached the end of the trajectory, at which time the system completes its operation.

In the lamination process, the vertical drive axis of the robot arm was controlled to move away from the bottom of the dish. After this process, calcium chloride solution was added in sufficient amounts to generate alginate gel fiber after lamination. Structures were then constructed as described in the above paragraph. Repeating these procedures led to the construction of multilayered structures.

The whole procedure is shown in Fig. 7.18, and the flow char of system is shown in Fig. 7.19.

7.3.6 Experiments for Automatically Constructing 3D Structures: Basic Experiments in Constructing 3D Lattice Structure

We conducted basic experiments in constructing 3D lattice structures, which have possibilities to have the abilities supplied oxygen and nutrients. In experiments, the rate of movement of the robot arm was set at 75 mm/s and the solution-feeding rate of syringe 1 to 1500 ml/min. Syringe 1 has a glass tube 430 µm in diameter. The scene of 3D lattice structure construction by the proposed system is shown in Fig. 7.20.

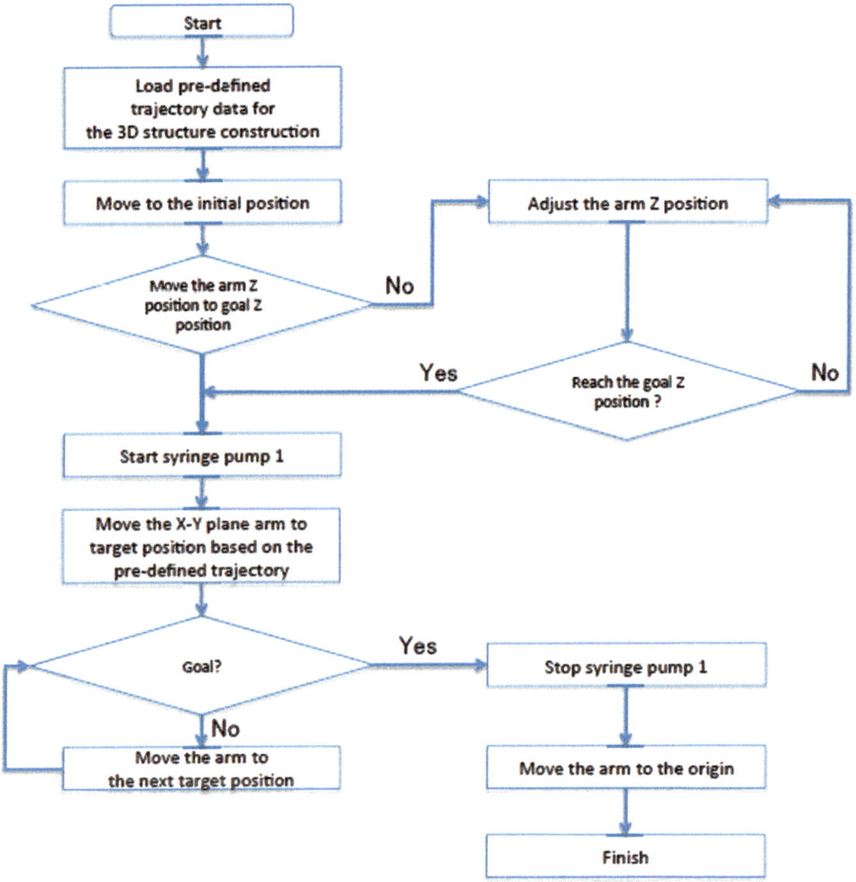

Fig. 7.19 System control flowchart

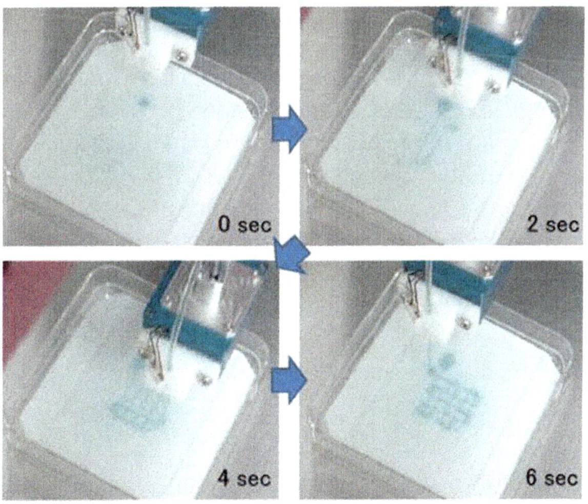

Fig. 7.20 Experiments in lattice construction

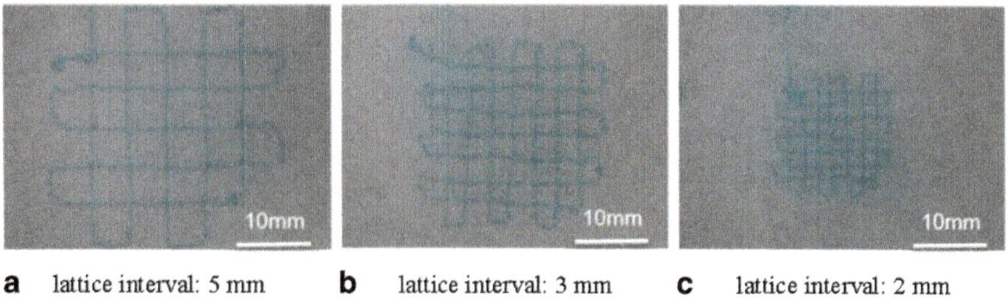

Fig. 7.21 Experimental results for 1-layer lattice construction. **a** lattice interval: 5 mm, **b** lattice interval: 3 mm, **c** lattice interval: 2 mm

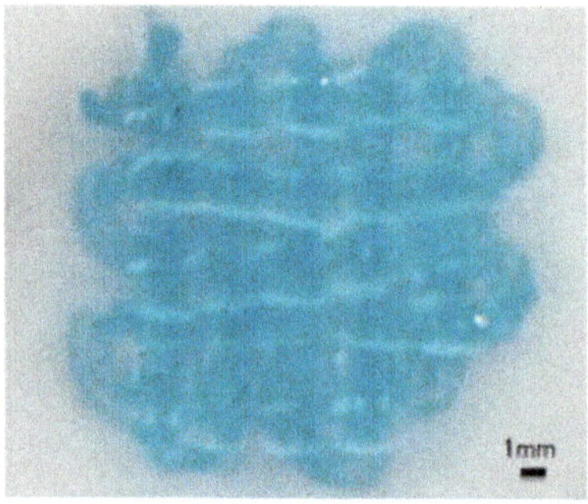

Fig. 7.22 Experimental results for 3-layer lattice construction based on alginate gel fiber without living cells

Table 7.6 Accuracy in lattice construction

	5 mm		3 mm		2 mm	
	Fiber diameter	Lattice space	Fiber diameter	Lattice space	Fiber diameter	Lattice space
Ave. [μm]	440.3	4643	440.9	2673	444.6	1813
S. D.	50.26	166.8	54.72	131.1	54.11	302.0

Figure 7.21 shows a one-layer lattice structure and Fig. 7.22 a three-layered lattice structure. The standard deviation between average fiber diameter and average lattice space in the construction of one-layer structures is given in Table 7.6. The average of fiber diameters in Figs. 7.21 and 7.22 is 450 μm.

Figures 7.21 and 7.22 show that the proposed automated system effectively constructed lattice structures as designed. Table 7.6 shows that fiber diameters were kept within a certain deviation range, irrespective of lattice intervals. Accurate lattices were constructed for 3 mm and 5 mm lattice intervals, while standard deviation in lattice space for 2 mm interval are large compared with other intervals. This suggests that alginate gel fiber did not adhere to the bottom of the dish where lattice

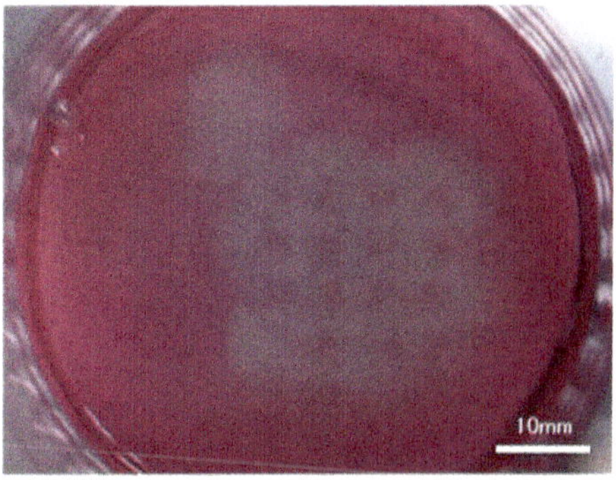

Fig. 7.23 Experimental results for 3-layer lattice construction based on alginate gel fiber with living cells

structures were constructed and that it did not exactly follow the preset discharge trajectory. To construct lattice structures with much higher accuracy, the bottom of the dish must therefore be processed so that alginate gel fiber easily adheres to it. The small mass of alginate gel fiber at the terminals of the trajectory seems to be responsible for the constant feeding of the solution from syringe 1 while the system is in operation. Thus the solution-feeding system also must have its discharge accuracy improved.

7.3.7 Experiments for Automatically Constructing 3D Structures: Experiments in Constructing 3D Lattice Structures from Alginate Gel Fiber with NIH3T3 Cells

To verify that cells are cultivatable even after lattice construction, we constructed 3D structures from alginate gel fiber containing NIH3T3 using the same procedures as in basic verification experiments., i.e., syringe 1 had a diameter of 800 mm, solution feeding to syringe pump 1 was set to 2.3 ml/min and that of syringe pump 2 to 20 ml/min, the rate of movement of robot arm was 75 mm/s, and the lattice interval was 5 mm.

Figure 7.23 shows 3D lattice structures constructed in experiments. Through the result, 3D lattice structures can be constructed from alginate gel fiber containing NIH3T3.

To check whether cells grow in these lattice structures, we cultivated cells within a CO_2 incubator after substituting calcium chloride solution in the dish into the cultivation solution. The CO_2 incubator was set to 37 °C and its CO_2 concentration to 65 %. Figure 7.24 shows results of 10-day cultivation showing that cells have grown normally in fiber.

Through the results, it can be said that the proposed system can construct 3D structures by using cell-growing hydrogel fiber.

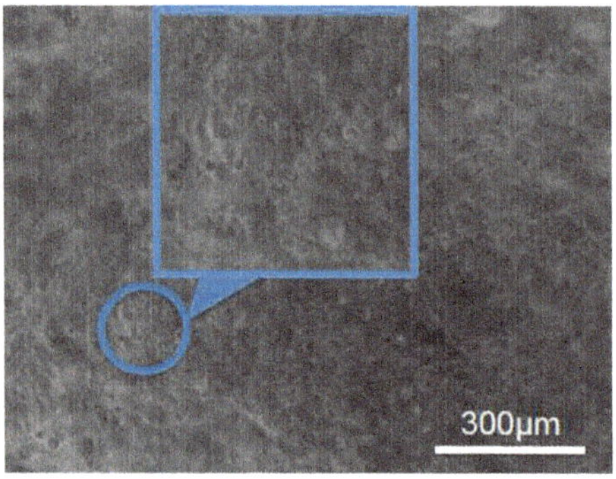

Fig. 7.24 Cell cultivation results 10 days later

7.3.8 Conclusions

In this study, we have developed an automated system for constructing 3D structures from alginate gel fiber. We found first sodium alginate solution and parameters of collagen-mixed solution suited to the developed system through a series of experiments. We then used this solution to construct lattice structures susceptible to the availability of oxygen and nutrients as an example of 3D structures. We have confirmed cell growth in constructed lattice structures, which we consider constitute basic verification of the developed system's capability to construct 3D tissues.

The status of this study is that we have just confirmed cell growth. To maintain constructed structures with cells alone, it is important to be formed by binding one cell to another in alginate gel fiber. It is also necessary to have densely, continuously arranged cells in such fiber. The current way of discharging sodium alginate makes it difficult to control processes for meeting the above conditions, however, and must be improved. Another issue to be addressed is improving the accuracy of constructed structures at terminals of lattice structures. We intend to solve this problem by installing a valve to control solution feeding.

7.4 Design and Fabrication of Changeable Cell Culture Mold

In this section, the changeable cell culture C^3 mold is proposed for engineered tissue fabrication. This mold forms a temporary microscaffold. The displacement of each actuator varies according to the pressure of the working fluid inside its corresponding microchannel. The actuator is made of polydimethylsiloxane (PDMS), a material chosen for its biocompatibility, optical transparency, high elasticity, leakage prevention, and easy modification of actuator features. In our design, the C^3 mold consists of a 3×3 microactuator array with an actuator diameter of 500 μm and spacing of 650 μm. The physical characteristics of this polymeric actuator are also discussed, and the possibility of using this device for real tissue fabrication is analyzed. Fabricated micropatterns on the agarose gel sheet are very similar to the tissue structures that would be obtained when real cells are used in future experiments.

7.4.1 Concept: C^3 Mold Design

The C^3 mold consists basically of 2 main layers: the actuator and the support layer. The elastic microactuator array is designed as an actuator layer because it offers many temporary scaffolds for the creation of many tissue structures. This is a requirement crucial to the fabrication of efficient TE scaffolds [33]. Our microactuator is a type of membrane actuator. Compared to other actuator types [34], the superiority of the membrane actuator makes it suitable for cell cultivation because there is no gap in the actuator layer, which prevents the leakage of liquids such as culture media. To manufacture the support layer, a polymer that has the ability to replicate the mold precisely on a micron scale is needed.

An actuator layer made of PDMS is used in our design, due to its many advantages, such as biocompatibility, optical transparency, high elasticity, leakage prevention, and easy modification of actuator features. PDMS has proven to be a very suitable material for cell cultivation [18, 35, 36], due to its oxygen-permeability. Hypoxia, which causes cells within thick tissues to die, is thus prevented. PDMS membrane gas-permeability [37] is expressed as

$$P = Nt / \Delta p \tag{7.5}$$

where P is the permeability coefficient, N is the gas flux through the PDMS membrane, t is the membrane thickness, and Δp is the pressure difference between the outside and inside of the evaporation chamber. Michael Johnson et al. [38] have demonstrated that a larger pressure difference causes a higher air diffusion rate through the thin PDMS membrane. It has also been proven that the air diffusion rate decreases as time progresses because the PDMS diffusion area decreases. Cells can be cultivated on PDMS, so the optical transparency of PDMS makes biological observation with a microscope easy. The high elasticity of PDMS enables the high vertical displacement of the PDMS actuator. Thus the higher displacement enables thicker micropatterned tissue to be fabricated. Another important benefit of using PDMS as an actuator layer is leakage prevention. Since there is only one component, i.e., the thin PDMS membrane, layering on the cultivation surface of the cell culture device helping prevent the leakage of culture media and cells. Instead of a flat surface like a conventional cell culture dish, the simultaneous activation of multiple actuators generates a temporary scaffold with a rough surface. As shown in Fig. 7.25, this generated scaffold structure is replicated on fabricated tissue. PDMS is made by mixing a prepolymer PDMS solution with curing agents, so the fabricated PDMS membrane is given higher elasticity by varying the mixing ratio [39]. The nanometer layer thickness of the PDMS membrane is also achieved

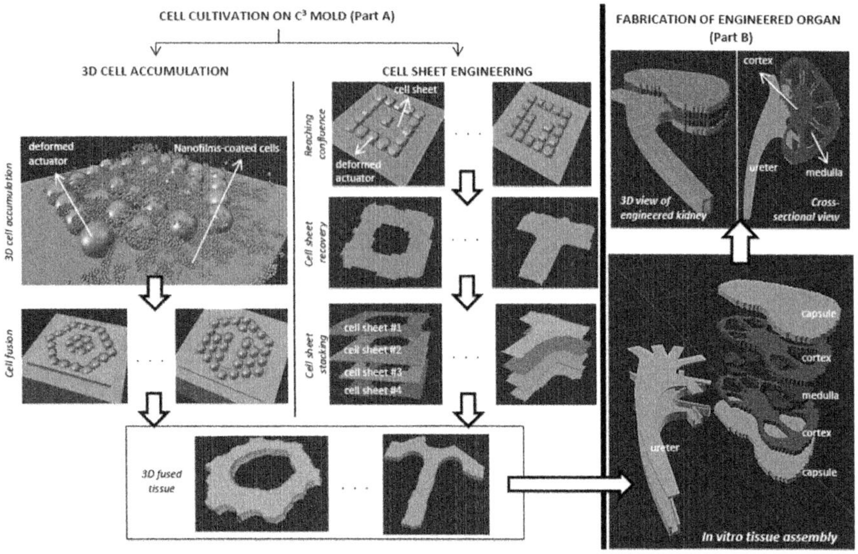

Fig. 7.25 Conceptual idea of using C^3 mold for tissue fabrication. Part A: thick tissue having micropatterns is fabricated by both 3D cell fusion of nanofilm-coated cells and stacking of multiple intact cell sheets. Part B: fabricated thick tissue is then ready for *in vitro* assembly into an engineered organ. An engineered kidney, for example, is fabricated by the assembling composite parts such as the ureter, capsule, cortex, and medulla

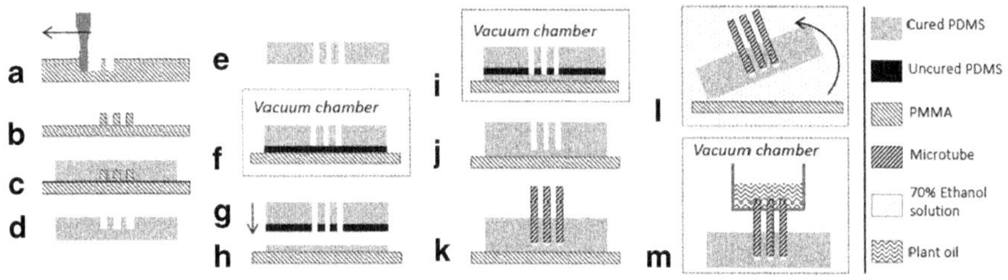

Fig. 7.26 Fabrication process of C^3 mold. **a**, **b** Micromachining of PMMA. **c**, **d** Replication of microstructure by PDMS. **e** Hole punching. **f** Coating with adhesive layer. **g**–**i** Bonding PDMS membrane and PDMS support layer together. **j** Insertion of microtubes. **k** Recovery of C^3 mold from PMMA substrate. **l** Pouring of working fluid, i.e., plant oil

by the addition of an extra chemical, such as hexane [40, 41] or toluene [42]. PDMS actuator properties are thus easily improved.

Precise microstructure replication for preparing the support layer (Figs. 7.26b–d) is done by PDMS replica molding [43]. As shown in Figs. 7.27d and e, PDMS actuator size depends on the resolution of the replicated microstructure, so the smaller microstructure likely results in a smaller actuator.

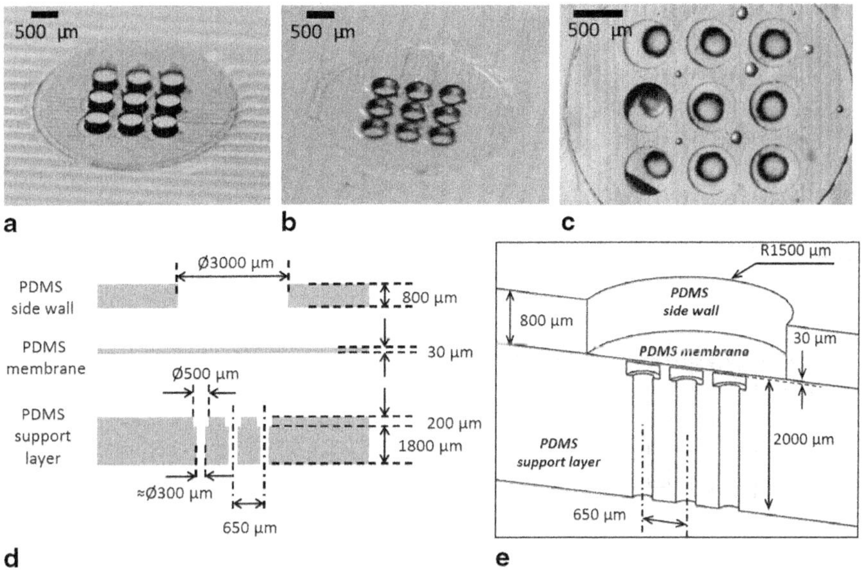

Fig. 7.27 **a** Micromachined 3 × 3 pin array PMMA-mold, with diameter of 500 µm, spacing of 650 µm, and pin height of 200 µm. **b**, **c** PDMS microchannel layer replicated from micromachined PMMA mold. **d**, **e** Layout dimensions of composite PDMS parts, including PDMS sidewall, PDMS membrane, and PDMS support layer

7.4.2 Concept: Conceptual Framework of Tissue Fabrication

As shown in part A of Fig. 7.25, the changeable scaffold concept enhances the cellular approach, e.g., 3D cell accumulation and cell sheet engineering, in dealing with the fabrication of part of a micropatterned tissue. Instead of a conventional acellular fixed scaffold being used, the changeability of the C^3 mold scaffold is used to fabricate various micropatterned tissues.

In the 3D cell accumulation in Fig. 7.25, cells are previously coated with nanofilms [7] and then seeded on the C^3 mold, where a pattern was prepared beforehand. Due to inter-cell interaction, e.g., natural ECM, multilayered tissue about 50 µm thick forms after cell fusion. Hollows are created in this tissue in areas replaced by activated PDMS actuators. After the thick formed tissue is recovered, it is ready for further *in vitro* organ assembly, as shown in part B of Fig. 7.25.

In the cell sheet engineering in Fig. 7.25, cells without any film coating are seeded on the C^3 mold. As in conventional cell cultivation, a 2D intact cell sheet is formed after confluence is reached. Theoretically, the recovered cell sheet will contain hollows, as does the tissue recovered in the 3D cell accumulation approach, but its thickness is usually less than 5 µm. Fabricating thick tissue requires the stacking of multiple cell sheets [4]. After cell fusion, fused micropatterned tissue is ready for assembly.

Miniature tissue parts having different structures are prepared individually, and then assembled into an engineered organ. As shown in part B of Fig. 7.25, tissue parts such as the ureter, capsule, cortex, and medulla are used to assemble an engineered kidney, for example. The assembled organ is then kept under conditions similar to the human body to fuse it into a single piece.

7.4.3 C^3 Mold Fabrication

The C^3 mold consists of 2 main layers—actuator and support layer—made of PDMS. Figure 7.2 illustrates the fabrication process. A negative polymethyl-methacrylate (PMMA) mold was first prepared by micromachining (Micro MC-2, PMT Corporation, Japan) plain PMMA into a pattern with a 3×3 pin array (Figs. 7.26a and b). As shown in Fig. 7.27a, this pin array has spacing of 650 μm, a diameter of 500 μm, and a height of 200 μm. Micromachinned PMMA was then submerged in a 70% ethanol solution (Kaneichi Pharmaceutical Co., Ltd., Japan) and shaken at 43 kHz for 30 min by an ultrasonic cleaner (ASU-3D, ASU Cleaner, AS ONE Co., Ltd., Japan) in order to remove residual surface dust. To fabricate the PDMS support layer, a 20:1 mixture of PDMS elastomer (Silpot 184, Dow Corning Toray, Co., Ltd., Japan) and curing agents (Silpot 184 CAT, Dow Corning Toray, Co., Ltd., Japan) was mixed thoroughly and degassed in a vacuum chamber for about 20 min to remove any air bubbles in the mixture. The prepolymer PDMS solution was then poured on micromachined PMMA and cured in an oven for 30 min at 80 °C (Fig. 7.26c). The PDMS mold was then gently peeled off micromachined PMMA (Fig. 7.26d). As shown in Fig. 7.27b, 200 μm-depth holes resulting from replica molding were used as a guide for hole punching positions. Holes were punched by a hole puncher (Harris Uni-Core, Tip ID 350 μm, Ted Pella, Inc.) based on observation through a microscope (SZX7, Olympus Co., Ltd., Japan) (Fig. 7.26e). Punched holes, which have a diameter of about 300 μm, are clearly observable in Fig. 7.27c. The top of the punched PDMS mold must be coated with an adhesive layer before it is brought into contact with the PDMS membrane (actuator layer). Since the usage of uncured PDMS as an intermediate layer in the bonding of PDMS parts has shown reliable bonding strength [42], an adhesive layer was prepared by spin-coating a mixed 30:1 PDMS solution at 4000 rpm for 100 s on the PMMA substrate. The PDMS mold was then placed gently on the adhesive layer and degassed in a vacuum chamber for 30 min to remove any air bubbles trapped between the PDMS mold and the adhesive layer (Fig. 7.26f). As a result, the entire flat PDMS mold surface is completely coated with an adhesive layer.

The PDMS mold was then brought into contact with the PDMS membrane (Fig. 7.26g). The PDMS membrane (actuator layer), which has a thickness of about 30 μm, was prepared previously by spin-coating a mixed 30:1 PDMS solution at 4000 rpm for 100 s on the PMMA substrate, and subsequently curing it at 80 °C for 15 min. The PDMS mold placed on the PDMS membrane was then degassed for 1 h (Fig. 7.26h); so that the flat surface between the PDMS mold and the PDMS membrane adhered entirely, without air bubbles. The assembled PDMS mold was then

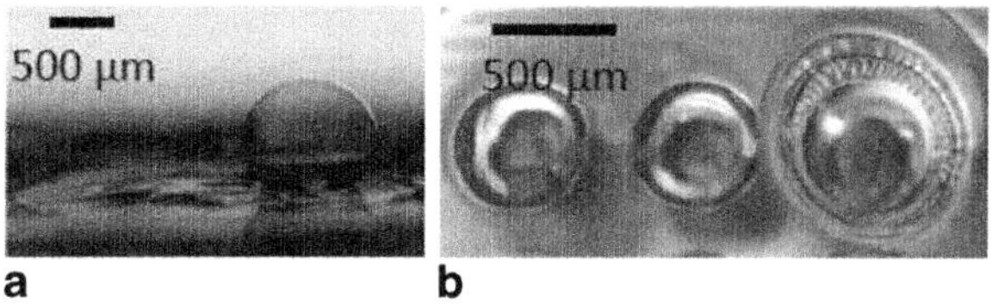

Fig. 7.28 **a** Side view and **b** top view of deformed PDMS actuator with air used as working fluid

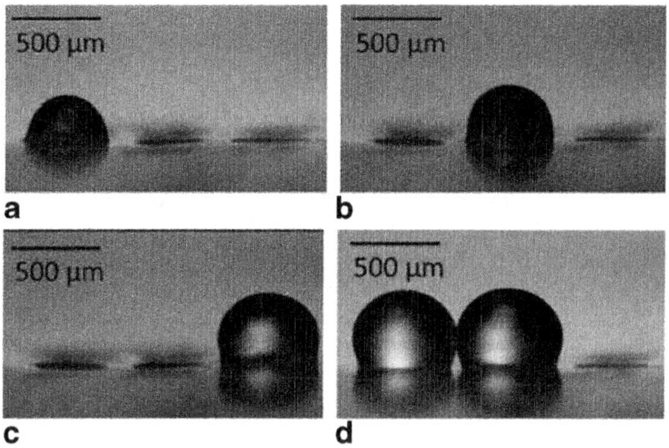

Fig. 7.29 Pattern formation performed by 3 actuators where DI water is poured in as working fluid. **a–c** Individual activation at different heights. **d** Activation of 2 adjacent actuators where actuator surface contact is observed. This behavior demonstrates formation of a temporary enclosed boundary for the creation of a cutout shape

cured at 80 °C for 15 min in an oven to allow 2 PDMS parts to bond (Fig. 7.26i). PFA microtubes (0.1 mm ID x 0.3 mm OD) were then inserted gently into punched holes from the back of the assembled PDMS mold (Fig. 7.26j), and sealed with a mixed 30:1 PDMS solution to prevent the leakage of working fluid. Although holes were previously punched by a 350 µm ID puncher, their diameters were measured to be approximately 300 µm due to the high elasticity of the 20:1 PDMS mold, so the 0.3 mm OD tubes fit well in these punched holes. The insertion of microtubes is achieved—with some effort—through observation via a microscope.

The assembled PDMS mold was then submerged in an ethanol solution to reduce adhesion between PDMS and the PMMA substrate so that the assembled PDMS mold could be peeled easily off the PMMA substrate (Fig. 7.26k). Since plant oil is used as the working fluid, air trapped inside microtubes must be replaced by connecting microtubes with a plant oil reservoir and degassing in a vacuum chamber for 2 h (Fig. 7.26l). After the PDMS sidewall mold was put on the assembled PDMS mold, the mold was ready for bioapplication.

7.4.4 Actuator Verification: Physical Analysis

We observed the displacement stability of the deformed PDMS membrane affected by different working fluids, including air (Fig. 7.28), water (Fig. 7.29), and plant oil, over time. To activate a PDMS actuator, the working fluid was first pushed by a syringe pump at a constant speed of 0.01 ml/min. Fluid pumping was active until the actuator reached a desired displacement of about 340 µm. The shape of the de-

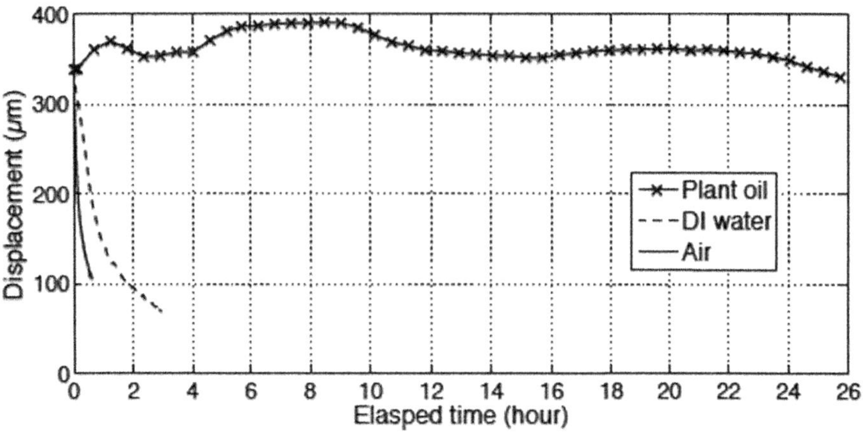

Fig. 7.30 Displacement variations of PDMS actuator over time with air, DI water, and plant oil used as working fluids

Fig. 7.31 3D micropatterned agarose gel sheet fabrication process. **a** Activate desired actuators. **b** Pour gel solution into C^3 mold. **c** Deactivate actuators. **d** Recover micropatterned gel sheet

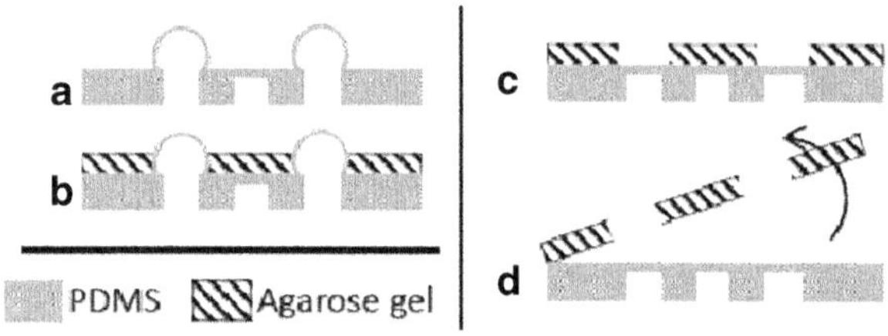

formed PDMS was observed over time while the syringe pump was deactivated. Its shape at a particular moment in time was captured by a CCD camera connected to a microscope and analyzed to find displacement at the corresponding time.

Figure 7.30 shows variations in actuator displacement by different working fluids. When air is used as the working fluid, actuator size decreases rapidly within about 20 min due to the air permeability of the PDMS membrane [37, 38]. Actuator displacement is maintained above 300 μm for 10 min when deionized water (DI water) is used. After 10 min, actuator size decreases continuously due to the water-vapor permeability of PDMS membrane [44]. The usage of liquid as working fluid slightly improves actuator stability. The 10-min stability provided by DI water, however, is not adequate for real cell applications because the minimum time required for engineered tissue fusion by nanofilm-coated cells [7] is 24 h. As a result, a liquid that has a much slower evaporation rate and does not evaporate visibly at room temperature, such as plant oil, is considered for the working fluid. As shown in Fig. 7.30, plant oil maintains a displacement higher than 300 μm for more than 24 h. Although actuator displacement fluctuates within a range of 340–390 μm, it does not significantly affect to the fabricated structure. We can thus say that the 24 h stability achieved by the usage of plant oil is indeed adequate for the fabrication of real 3D tissue. The advantages of employing the C^3 mold for tissue fabrication are demonstrated through the fabrication of various thin gel sheet micropatterns.

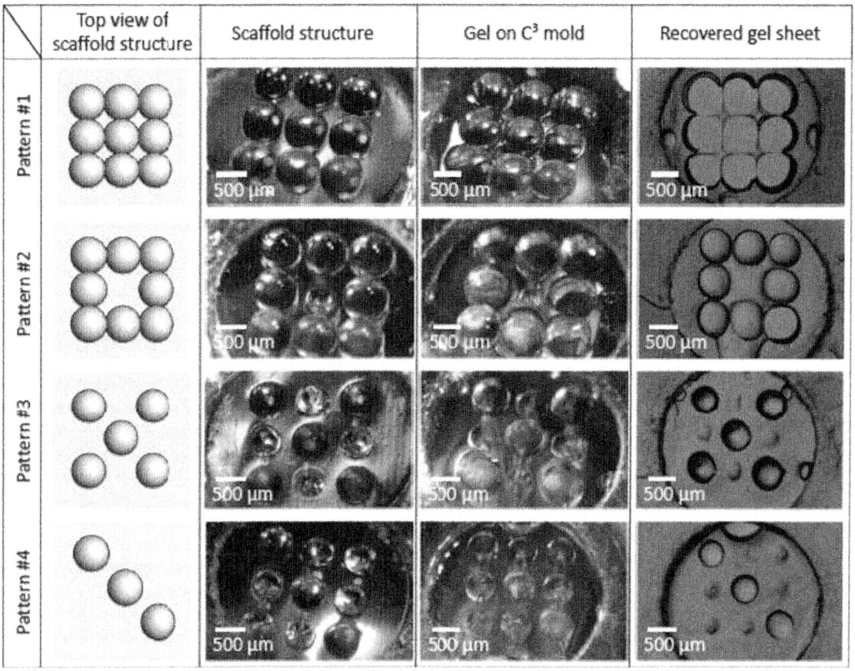

Fig. 7.32 200-µm-thick micropatterned gel sheets. Column 1: top view of prepared scaffold. Column 2: scaffold formed by PDMS actuators. Column 3: agarose gel solidified on the scaffold. Column 4: recovered thin gel sheet

7.4.5 Actuator Verification: Application Tests

In order to illustrate C^3 mold applicability for tissue fabrication, tissue fused by 3D cell accumulation according to the concept in Fig. 7.25 is represented theoretically by the formation of 200 µm-thick agarose gel sheet in the experiments below.

As shown in Fig. 7.31, a pattern was formed by the activation of multiple actuators (Fig. 7.31a). To prepare a 1 % agarose solution, a 1X TBE buffer solution (diluted 10X TBE buffer, GLP 20-131, Funakoshi Co., Ltd.) and agarose powder (Agarose 36GU, GA-001, Funakoshi Co., Ltd.) were mixed at a ratio of 10 mL TBE: 1 g agarose powder. The mixed agarose solution was then heated to 80 °C in order to fully dissolve agarose powder. Next, 5 µl of heated agarose solution was immediately aspired by pipette and seeded on the C^3 mold (Fig. 7.31b). This process represents the cell seeding step in real cell cultivation. After 10 min, the agarose solution became solid. Actuators were then deactivated (Fig. 7.31c) before the agarose gel was recovered from the mold (Fig. 7.31d). Micropatterned agarose gel was then kept in Phosphate Buffered Saline (Sigma-Aldrich) to prevent it from drying. Gel sheet fabrication illustrates the tissue fabrication process in which the solid agarose gel represents fabricated engineered tissue.

Four patterns of 200 μm-thick fabricated agarose gel sheet are shown in Fig. 7.32. Pattern #1 is created by activating all actuators simultaneously. As shown in Fig. 7.29d, the formation of a temporary enclosed boundary is observed in pattern #2. Patterns #3 and #4 display diagonal patterns.

7.4.6 Conclusions

We have developed a cell culture device called a C^3 mold in response to the need for a flexible scaffold for the fabrication of different tissue structures. The actuator layer, which has a 3×3 microactuator array, has been made of PDMS due to its biocompatibility, optical transparency, high elasticity, leakage prevention, and easy modification of actuator features. The flexible stamp resin of PDMS has made it suitable for the precise replication of microstructures for the fabrication of the support layer. Physical analysis has shown that plant oil, which has a lower evaporation rate at room temperature than DI water or air, keeps the actuator displacement higher than 300 μm for more than 24 h. This long-term stability is a significant factor that recommends the usage of the C^3 mold in the fabrication of real tissue, since the shortest time required for tissue fusion by the accumulation of nanofilm-coated cells [7] is 24 h. The achieved displacement of about 300 μm is adequate because the maximum thickness of *in vitro* 3D tissue that survives for a long time under conditions of sufficient nutrients is limited to 100 μm. To demonstrate the features of fabricated structures, 3D tissue fabricated by the usage of a C^3 mold is represented theoretically by a 200 μm-thick agarose gel sheet in our experiments. The 4 patterns replicated on the agarose gel sheet indeed prove the controllability of the formed scaffold structures and C^3 mold applicability to the fabrication of real tissue. Future work will thus include further discussion and experiments.

7.5 Summary and Perspectives

In this chapter, we have reviewed our recent progress, (1) Cell Manipulation by microhand, (2) Construction of 3D lattice by Hydrogel fiber and (3) Changeable Cell Culture mold for Advanced Cell Sheet. These proposed new technologies contribute to realize "Bio Assembler". We can say a cell manipulation technique is for a zero-dimensional structure, hydrogel fiber technique is for a one-dimensional structure, and cell sheet technique is for a two-dimensional structure. Micromanipulation technique can be applied to assemble cellular parts with different structures. Applying micro robotics to tissue fabrication is a promising way for automatic and high-speed fabrication of tissues. We expect that integrating the presented and other technologies will realize complex 3D cellular system with functions in future.

Acknowledgments This work was supported by Grant-in-Aid for Scientific Research on Innovative Areas "Bio Assembler" (23106005) from the Ministry of Education, Culture, Sports, Science and Technology of Japan.

References

1. Langer R et al (1993) Tissue engineering. Science 240:920–926
2. Griffith LG et al (2002) Tissue engineering-current challenges and expanding opportunities. Science 295:1009–1014
3. Shimizu T, Yamato M, Kikuchi A, Okano T (2003) Cell sheet engineering for myocardial tissue reconstruction. Biomaterials 24(13):2309–2316
4. Haraguchi Y, Shimizu T, Sasagawa T, Sekine H, Sakaguchi K, Kikuchi T, Sekine W, Sekiya S, Yamato M, Umezu M, Okano T (2012) Fabrication of functional three-dimensional tissues by stacking cell sheets *in vitro*. Nat Protoc 7(5):850–858
5. Sugibayashi K, Kumashiro Y, Shimizu T, Kobayashi J, Okano T (2013) A molded hyaluronic acid gel as a micro-template for blood capillaries. J Biomater Sci Polym Ed 24(2):135–147
6. Kachouie NN, Du Y, Bae H, Khabiry M, Ahari AF, Zamanian B, Fukuda J, Khademhosseini A (2011) Directed assembly of cell-laden hydrogels for engineering functional tissues. Organogenesis 6(4):234–244
7. Nishiguchi A, Yoshida H, Matsusaki M, Akashi M (2011) Rapid construction of three-dimensional multilayered tissues with endothelial tube networks by the cell-accumulation technique. Adv Mater 23:3506–3510
8. Asakawa N et al (2009) Pre-vascularization of *in vitro* threedimensional tissues created by cell sheet engineering. Biomaterials 31:3903–3909
9. Tsuda Y et al (2007) Cellular control of tissue architectures using a three-dimensional tissue fabrication technique. Biomaterials 28:4939–4946
10. Arai K et al (2011) Three-dimensional inkjet biofabrication based on designed images. Biofabrications 3. doi:10.1088/1758-5082/3/3/034113.
11. Norotte C et al (2009) Scaffold-free vascular tissue engineering using bioprinting. Biomaterials 30:5910–5917
12. Jakob K et al (2010) Tissue engineering by self-assembly and bioprinting of living cells. Biofabrication 2. doi:10.1088/1758-5082/2/2/022001.
13. Miller JS et al (2012) Rapid casting of patterned vascular networks for perusable engineered three-dimensional tissues. Nature Mater 11:768–774
14. Yamada M et al (2012) Microfluidic synthesis of chemically and physically anisotropic hydrogel microfibers for guided cell growth and networking. Soft Matter 8:3122–3130
15. Yamada M et al (2012) Controlled formation of heterotypic hepatic micro-organoids in anisotropic hydrogel microfibers for longterm preservation of liver specific functions. Biomaterials 33:8304–8315
16. Iwase M et al (2012) Fabrication of vascular tissue models by assembling multiple cell types inside hydrogel microchannels. Proceedings of 2012 International Symposium on micro-nanomechatronics and human science, pp 402–405
17. Tsang VL, Bhatia SN (2004) Three-dimensional tissue fabrication. Adv Drug Deliver Rev 1635–1647
18. Masuda T, Takei N, Nakano T, Anada T, Suzuki O, Arai F (2012) A microfabricated platform to form three-dimensional toroidal multicellular aggregate. Biomed Microdevices 14(6):1085–1093
19. Anada T, Masuda T, Hosnda Y, Fukuda J, Arai F, Fukuda T, Suzuki O (2010) Three-dimensional cell culture device utilizing thin membrane deformation by decompression. Sensor Actuat B: Chem 147:376–379
20. Gwyther TA, Hu JZ, Billiar KL, Rolle MW (2011) Directed cellular self-assembly to fabricate cell-derived tissue rings for biomechanical analysis and tissue engineering. J Vis Exp 57
21. Kawashima K, Arai T, Tadano K, Fujita T, Kagawa T (2010) Development of coarse/fine dual stage using pneumatically driven bellows actuator and cylinder with air bearings. Precis Eng 34:526–533
22. Ren L, Wang L, Mills JK, Sun D (2008) Vision-based 2-D automatic micrograsping using coarse-to-fine grasping strategy. IEEE Trans Ind Elect 55:3324–3331

23. Avci E, Ohara K, Takubo T, Mae Y, Arai T (2013) Development of Multi-Scalable Microhand System with Precise Motion Ability. J Robot Mechatron 25(1):183–191
24. Onoe H et al (2010) Core-shell gel wires for the construction of large area heterogeneous structures with biomaterials. Proceedings of 2010 IEEE 23th International Conference on Micro Electro Mechanical Systems, pp 248–251
25. Onoe H et al (2011) Living cell fabric. Proceedings of 2011 IEEE 24th International Conference on Micro Electro Mechanical Systems, pp 908–911
26. Ohara K, Kojima M, Fukushima A, Onozaki S, Horade M, Yamada M, Seki M, Mae Y, Arai T (2013) Automated Construction System for 3D Lattice Structure Based on Alginate Gel Fiber Containing Living Cells. J Robot Mechatron 25(4): 665–672
27. Chumtong P, Kojima M, Ohara K, Mae Y, Horade M, Akiyama A, Yamato M, Arai T (2013) Design and fabrication of changeable cell culture mold. J Robot Mechatron 25(4): 657–664
28. Avci E, Ohara K, Takubo T, Mae Y, Arai T (2011) High speed micromanipulation system with multi-scalability. IEEE, International Conference on Robotics and Biomimetics, pp 1212–1217
29. Avci E, Ohara K, Takubo T, Mae Y, Arai T (2010) Workspace optimization for multi-scale micromanipulation system. MHS2010 & Micro-Nano Global COE, pp 328–333
30. Nguyen C, Ohara K, Avci E, Takubo T, Mae Y, Arai T (2011) Automated micromanipulation of a microhand with all-in-focus imaging system. IEEE/RSJ International Conference on Intelligent Robots and Systems, pp 427–432
31. Chen BK, Zhang Y, Sun Y (2009) Active release of microobjects using a mems microgripper to overcome adhesion forces. J Microelectromech Syst 18:652–659
32. Chen T, Chen L, Sun L, Rong W, Yang Q (2010) Micro manipulation based on adhesion control with compound vibration. Proceedings IEEE/RSJ International Conference on intelligent robots and systems, pp 6137–6142
33. Guillotin B, Guillemot F (2011) Cell patterning technologies for organotypic tissue fabrication. Trends Biotechnol 29(4):183–190
34. Volder M de, Reynaerts D (2010) Pneumatic and hydraulic microactuators: a review. J Micromech Microeng 20(4)
35. Ryoo JH, Jeong GS, Kang E, Lee SH (2011) Ultrathin, hyperelastic PDMS nano membrane: fabrication and characterization. Int. Conf. on miniaturized systems for chemistry and life sciences, Seattle, Washington, USA, pp 686–688
36. Anada T, Fukuda J, Sai Y, Suzuki O (2012) An oxygen-permeable spheroid culture system for the prevention of central hypoxia and necrosis of spheroids. Biomaterials 33(33):8430–8441
37. Merkel TC, Bondar VI, Nagai K, Freeman BD, Pinnau I (2000) Gas sorption, diffusion, and permeation in Poly(dimethylsiloxane). J Polym Sci: Part B 38:415–434
38. Johnson M, Liddiard G, Eddings M, Gale B (2009) Bubble inclusion and removal using PDMS membrane-based gas permeation for applications in pumping, valving and mixing in microfluidic devices. J Micromech Microeng 19(9)
39. Khanafer K, Duprey A, Schlicht M, Berguer R (2009) Effects of strain rate, mixing ratio, and stress-strain definition on the mechanical behavior of the polydimethylsiloxane (PDMS) material as related to its biological applications. Biomed Microdevices 11(2):503–508
40. Sang S, Witte H (2010) Fabrication of a surface stress-based PDMS micro-membrane biosensor. Microsyst Technol 16(6):1001–1008
41. Thangawng AL, Ruoff RS, Swartz MA, Glucksberg MR (2007) An ultra-thin PDMS membrane as a bio/micro-nano interface: fabrication and characterization. Biomed Microdevices 9(4):587–595
42. Wu H, Huang B, Zare RN (2005) Construction of microfluidic chips using polydimethylsiloxane for adhesive bonding. The Roy Soc Chem 5:1393–1398
43. Jo B-H, van Lerberghe LM, Motsegood KM, Beebe DJ (2000) Three-dimensional microchannel fabrication in Polydimethylsiloxane(PDMS) Elastomer. J Microelectromech Syst 9(1):76–81
44. Zhang Y, Ishida M, Kazoe Y, Sato Y, Miki N (2009) Water-vapor permeability control of pdms by the dispersion of collagen powder. Trans Electr Electr Eng 4(3):442–449

Chapter 8
High-Throughput Cell Assembly Featuring Heterogeneous Hydrogels Produced by Using Microfluidic Devices

Masumi Yamada and Minoru Seki

Abstract We describe microfluidic devices developed for producing heterogeneous hydrogel materials including sandwich-type hydrogel fibers, hydrogel fibers having a highly complex cross-sectional morphology, stripe-patterned hydrogel sheets, and yarn-ball-shaped hydrogel beads. Cells encapsulated within these hydrogel materials exhibit behaviors that are distinct from those of cells examined using conventional cell-culture techniques. The cells are rapidly encapsulated in the hydrogel materials, and the cell-containing materials obtained could function as unit structures in constructing large tissues. Here, we also briefly discuss the use of hydrogel-based microfluidic devices in the preparation of multilayer blood-vessel models.

Keywords Microfluidics · Microfabrication · Hydrogel · Tissue engineering · Coculture

8.1 Introduction

When tissues are closely examined in vivo, most of them can be observed to be composed of a 3-dimensional (3D) hierarchical assembly of cells and extracellular matrices (ECMs). For example, muscle is composed of precisely aligned muscle-cell filaments and embedded blood vessels. The liver, the largest organ in the human body, contains a parallel arrangement of cord-like unit structures called "hepatic cords," each of which is composed of a row of parenchymal hepatocytes and surrounding sinusoidal endothelial cells. The pancreatic islet is composed of several types of major cells including α, β, and δ cells, and embedded vascular networks. To create such complex tissues in vitro, various types of tissue-engineering approaches have been developed. However, because most of the conventional platforms used

M. Yamada (✉) · M. Seki
Department of Applied Chemistry and Biotechnology,
Graduate School of Engineering, Chiba University,
1-33 Yayoi-cho, 263-8522 Inage-ku, Chiba, Japan
e-mail: m-yamada@faculty.chiba-u.jp

T. Arai et al. (eds.), *Hyper Bio Assembler for 3D Cellular Systems*,
DOI 10.1007/978-4-431-55297-0_8

for culturing animal cells are 2D, planar surfaces, critical differences exist between these in vitro environments and the tissues in vivo. Researchers have attempted to generate functional tissues or organ models that can be used in regenerative medicine, in cell-based drug-evaluation systems, and as platforms for studying cellular physiology in environments that mimic the in vivo setting. Because 3D cell cultures feature structures that are similar to in vivo tissue structures, cells cultivated in 3D can benefit from the upregulation of cellular functions and vitality, especially in the case of cells that are prone to lose their functions in vitro. New technologies are being developed to assemble multiple types of cells into large 3D tissues in which the positions of distinct types of cells and the appropriate arrangement of the ECMs can be controlled.

One of the strategies used most frequently for cultivating cells in 3D is the generation of spherical aggregates of cells, or "spheroids." When suspended cells are seeded in non-cell-adhesive dishes or into microwells, the cells typically form spherical aggregates because of the contraction force between cells. Spheroids have been used as unit structures to create large tissue models such as blood vessel-like structures [1, 2]. However, one problem encountered with these cultures is that the cells located at the center of the aggregates are not viable when the spheroid size exceeds ~200 μm [3]. This example indicates that the size of the unit structures used for constructing tissues should be less than ~200 μm if specific vascular structures are not embedded. Moreover, the spherical shape of the cell aggregates does not necessarily represent the configuration most suitable for constructing large tissues; for example, toroidal aggregates could be suitable for constructing vascular-tissue models.

Hydrogels are materials that facilitate the culturing of cells in a 3D environment. The permeable matrices of hydrogels enable oxygen and nutrients to be transported to the cells embedded within the hydrogels. Various types of hydrogel materials have been used for culturing animal cells, including collagen, gelatin, alginate, gel-forming peptides, and synthetic polyethylene glycol (PEG)-based polymers. Cells are unlikely to be viable if they are embedded in large hydrogels at a high density, much as in the case of cell aggregates. Hence, techniques have been developed for encapsulating cells in small hydrogel constructs, and this helps avoid the problem of necrosis that is associated with hypoxia. For use in producing small-sized hydrogels, microfluidic technologies and microfabrication processes have attracted considerable interest in the field of hydrogel-based cell cultivation and tissue engineering [4]. Because the inherent laminar-flow profile formed within microchannels can be controlled accurately, microfluidic devices have been employed for preparing small but highly functional objects including particles, droplets, fibers, and sheets. For example, in the case of fabricating particulate materials, in addition to homogeneous and spherical particles, functional particles such as "Janus" particles and particles featuring complicated morphologies have been produced [5]. Furthermore, hydrogel materials such as beads and fibers featuring controlled morphologies have been produced, which successfully encapsulate cells within the hydrogel matrix [6–8]. Anisotropic hydrogel materials have also been produced, and these enable the precise encapsulation of multiple types of cells [9]. These studies clearly demonstrated the potential usefulness of micrometer-sized hydrogel materials as a novel platform for cell cultivation, in which the position, density, and compositions of cells can be precisely controlled. Moreover, these hydrogel materials can potentially be used as unit structures to construct large, complex tissues that mimic the structures of organs in vivo. In this chapter, we briefly introduce recent advances in the preparation of functional hydrogel materials including anisotropic hydrogel microfibers, functional hydrogel particles used for cell cultivation, and heterogeneous hydrogel sheets. We also present a brief overview of several studies in which microfluidic devices and microchambers made of hydrogel materials were used for cultivating cells in a 3D environment.

8.2 Anisotropic Hydrogel Fibers Used for 3D Cell Cultivation and Heterotypic Coculture Preparation

8.2.1 Preparation of Anisotropic Hydrogel Microfibers by Using Microfluidic Devices

Tissues and organs displaying microscale alignments of linear unit structures of cells are commonly detected in vivo, as in the case of, for example, multilayer vascular tissues, nerve bundles, skeletal muscles, and hepatic cord structures. Although these tissues appear "linear" in a microscopic view, most of them comprise multiple types of cells that are precisely aligned in regular patterns. The engineering of such linear tissues in vitro is likely to be facilitated by the use of linear but heterogeneous 3D microenvironments that encapsulate multiple cell types and concomitantly regulate the proliferation direction of the embedded cells. One of the most common methods of assembling multiple types of cells into linear colonies involves using patterned adhesive and non-adhesive surfaces on planar cell-culture plates. Various strategies have been proposed for this purpose, including photolithography [10], microcontact printing [11, 12], and microfluidics-based patterning techniques [13]. Conversely, the preparation of fibrous hydrogels has also been reported, which enables mammalian cells to be cultivated within 3D hydrogel matrices or on the hydrogel surface and also concomitantly allows the embedded cells to be efficiently supplied with oxygen and nutrients. Several studies have reported the preparation of alginate [14, 15] and chitosan [16] hydrogel fibers by using either a micronozzle or a double capillary. Although these studies demonstrated the potential suitability of hydrogel microfibers that were typically 10–1000 μm in diameter, most of the fibers used displayed homogeneous and uniform cross-sections. Unlike in the case of these fibers, not many reports have described the preparation of complex hydrogel fibers composed of multiple regions that exhibit distinct compositions in the cross-section.

To develop anisotropic fibrous hydrogel materials whose diameters are less than a few hundred micrometers, we recently prepared complex microfibers composed of multiple regions exhibiting divergent physicochemical characteristics [9]. For example, sandwich-type solid-soft-solid hydrogel fibers were prepared using microfluidic devices in which the proliferation direction of encapsulated cells was guided and thus linear cell colonies were formed. We employed sodium alginate (NaA) as the primary hydrogel material, because it offers the following remarkable advantages. (1) Alginate is gelled in the presence of multivalent cations, and thus its gelation does not require temperature control, especially heating. (2) Alginate gelation is extremely rapid. (3) Intact cells can be embedded in alginate gels at high densities. (4) The mechanical strength of the hydrogel matrix is high, and thus the gel can be handled easily. (5) The alginate polymer does not induce inflammation when transplanted in vivo. (6) Alginate hydrogels can be removed using cation chelators such as EDTA and citrate, or by enzymatically digesting the alginate polymer. Because of these characteristics, alginate hydrogels are widely used for preparing cell-embedding hydrogels that are employed in tissue engineering and regenerative medicine.

To engineer complex hydrogel microfibers, we used microfluidic devices (Fig. 8.1a). Microchannels featuring 5–7 inlets were used, which were prepared by employing soft lithography and replica-molding techniques and using polydimethylsiloxane (PDMS) as the microchannel material [17]. Gelation solutions containing $CaCl_2$ or $BaCl_2$, buffer solutions, and NaA solutions were continuously introduced from the outer, middle, and inner inlet channels, respectively (Fig. 8.1a), by using syringe pumps. By introducing NaA solutions of distinct compositions through separate inlet ports, micrometer-scale hydrogel fibers displaying complex cross-sectional morphologies were obtained. Two points are critical for stably obtaining Ca-alginate fibers that have uniform diameters: (1) thin buffer layers must be introduced between the NaA and $CaCl_2$ solutions, because this controls the gelation speed of Ca-alginate hydrogels and prevents the microchannel from becoming clogged at the confluence point as a result of hydrogel formation; and (2) using a thickener such as dextran can help balance the viscosities of the introduced buffers/gelation solutions with the high viscosity of the NaA solution.

In our study, we first examined whether Ca-alginate hydrogel fibers displaying complex cross-sectional morphologies were obtained. We introduced NaA solutions containing fluorescent particles into a microchannel (~400 μm wide and ~100 μm deep) that had 6 inlets. Figure 8.1b shows an example of a microfiber that had an average diameter of ~15 μm. Because a stable laminar flow forms within the microchannel, the gelation agent (Ca^{2+} or Ba^{2+} ions) diffused into the central flow region of the precursor solution (NaA solution) through the buffer flows and thus continuously transformed the precursor solution into the hydrogel. Consequently, we successfully obtained microfibers that were 5–100 μm in diameter and displayed a uniform cross-sectional morphology. The diameter of the obtained fibers could be tightly controlled by changing the flow rate of the NaA solutions and the microchannel dimensions: when broad microchannels (width and depth, 800 μm) were used, we obtained fibers that had diameters up to 200 μm; moreover, lowering the

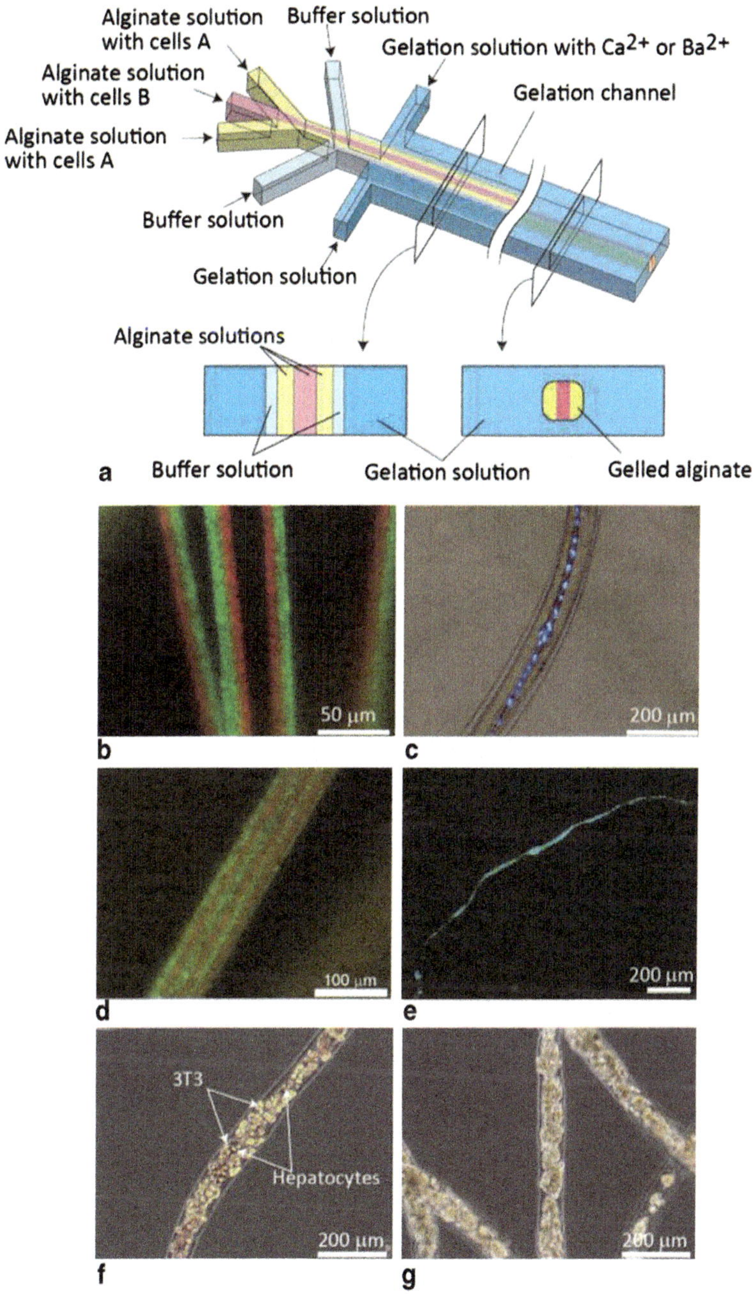

Fig. 8.1 Preparation of anisotropic hydrogel microfibers and their use in culturing cells. **a** Microfluidic system used for producing anisotropic alginate hydrogel microfibers. **b** Hydrogel microfibers embedding particles of distinct colors. **c** A linear colony of PC12 cells formed within sandwich-type solid-soft-solid hydrogel microfibers. **d** Preparation of complex hydrogel fibers containing 8 soft regions in the periphery of the fiber. **e** Millimeter-long colonies of PC12 cells formed within the hydrogel fibers containing 8 soft regions. **f, g** Primary rat hepatocytes cultivated for 7 days **f** with and **g** without 3T3 cells within the sandwich-type fibers

flow rate of the buffer solutions resulted in the formation of fibers whose widths were highly uniform.

To use the developed hydrogel fibers in cell cultures, we next prepared sandwich-type hydrogel fibers composed of 3 regions (containing distinctly colored particles) by using 7-inlet microchannels. To adjust the stiffness of the central region of the cross-section of the fibers, we used propylene glycol alginate (PGA), a non-gelling ester derivative of alginate. When an aqueous solution of NaA + PGA is used, the hydrogels obtained exhibit diminished stiffness. By introducing the NaA + PGA solution through the central inlet and NaA solutions from the neighboring inlets, we obtained sandwich-type hydrogel fibers that contained hard-soft-hard regions. Highly stiff alginate hydrogels are recognized to be unsuitable for cultivating highly proliferative cells [18]; however, we expected the cells encapsulated within the soft region of the fibers to proliferate and form intercellular networks along the fiber length. To demonstrate this, we encapsulated cells within the soft core and then observed the direction of cell growth. The soft core was made from 0.9% PGA and 0.3% NaA, whereas the solid shell was made from 2% NaA. Moreover, the surface of the obtained fibers was coated with poly-L-lysine, which prevents the hydrogel fiber from swelling in the cell culture medium and thus keeps the cells from growing out of the hydrogel matrix. We cultured NIH-3T3 cells within the hydrogel fibers for 6 days (Fig. 8.1c); the cells were initially located within the core of the sandwich-type fibers, but then gradually grew along the soft region and finally formed linear, one-cell-thick colonies. By contrast, in control experiments in which homogeneous fibers made from 2% NaA were used, spherical colonies were formed. Similar results were obtained when HeLa cells were cultivated. These results helped confirm that anisotropic hydrogel fibers serve as a suitable platform for cultivating cells in 3D environments and could potentially be applied for constructing linear tissues.

Next, we prepared hydrogel fibers whose cross-sectional morphologies were much more complex. Micronozzle array-combined microfluidic devices have been developed for generating hydrogel fibers composed of multiple soft and solid regions [19]. The patterns of the vertical micronozzles determine the cross-sectional morphology of the hydrogel microfibers. We fabricated 4-layer microfluidic devices made of poly(methyl methacrylate) employing numerical control (NC) micromachining and thermal-bonding processes. The microfibers obtained displayed 8 soft regions in the peripheral area of their cross-section, and the fiber diameter and the width of the soft regions were 60–100 and 10–20 μm, respectively (Fig. 8.1d). After confirming that 8 soft regions were successfully formed within the fibers, we encapsulated neuron-like cells (PC12 cells) in the peripheral soft regions of the fiber, and then cultured the cells under a differentiation condition in which the serum concentration in the medium was low and the medium contained nerve growth factor. Our results showed that the PC12 cells formed linear colonies and the outgrowth of their neurites was guided because of the presence of the solid-soft regions within the fiber. After cultivation for 14 days, millimeter-long cellular networks were formed (Fig. 8.1e). These heterogeneous fibers can serve as new tools for guiding cell proliferation and networking behaviors in 3D environments, and could potentially also be applied in regenerating damaged nerves and evaluating the behavior of neural cells in confined spaces.

8.2.2 Formation of Heterotypic Hepatic Microorganoids by Using Anisotropic Hydrogel Microfibers

Hepatocytes (parenchymal cells) perform major functions in the liver, including the metabolism of drugs, the production of plasma proteins, blood purification, and the storage of nutrients. Thus, engineering liver tissues by using hepatocytes can be critical for developing implantable hepatic constructs, preparing in vitro models for use in drug-metabolism and toxicity assays, and fabricating extracorporeal bioartificial liver devices for implantation in patients with liver insufficiency. Furthermore, hepatocyte-based tissue engineering could potentially be used as an alternate to whole-organ transplantation for the purpose of supporting hepatic functions. However, a major challenge faced in culturing primary hepatocytes in vitro is that hepatocytes do not proliferate and rapidly lose their functions and phenotypes, even though hepatocytes can efficiently regenerate in vitro. Therefore, new techniques designed for maintaining the functions and viability of hepatocytes in vitro are in high demand.

Various approaches have been tested to solve the problem of culturing hepatocytes in vitro. Generating spheroids by using non-adhesive surfaces/wells or patterned ECMs [20–24] is one of the most frequently used methods of upregulating hepatocyte functions, because this facilitates cell-cell interactions in a 3D environment. Hepatocytes have also been cocultured with other types of cells; for example, hepatocytes have been used in patterned cocultures prepared with fibroblasts or endothelial cells [25–31] and heterotypic cell sheets have been stacked [32, 33], and in both cases, heterotypic cellular interactions were reproduced. The human liver is composed of small unit structures called hepatic lobules, which comprise parenchymal hepatocytes and several types of non-parenchymal cells including liver sinusoidal endothelial cells (LSECs), stellate cells, and Kupffer cells. Within a lobule, hepatocytes are arranged in a cord-like linear structure, which is covered with a thin layer of LSECs. The space between the hepatocytes and LSECs is called Space of Disse, which contains various types of ECMs. This configuration allows hepatocytes to maintain close contact with the blood circulation through the fenestrations of the LSECs, and thus enables the efficient supply/exchange of nutrients, oxygen, and metabolic products. Based on considering this characteristic structure, we hypothesized that reconstructing similar microstructures composed of multiple types of cells will improve the functions and vitality of hepatocytes cultured in vitro. To fabricate such a 3D, heterotypic coculture system, we employed the anisotropic hydrogel microfibers that were produced using microfluidic devices; multiple types of cells were encapsulated close to each other within the microfibers [34].

Our strategy involved using sandwich-type, anisotropic alginate hydrogel microfibers. Hepatocytes were encapsulated in the center of the fiber, which was sandwiched by the hydrogel regions wherein non-parenchymal cells were embedded.

In the core of the hydrogel fibers, hepatocytes gradually formed linear aggregates during cultivation, which then became gradually surrounded by non-parenchymal cells; this resulted in the formation of linear but 3D microaggregates (microorganoids) that closely mimicked the hepatic cord structures observed in the liver lobules.

In our study, we used Ba^{2+} ions instead of Ca^{2+} ions as the gelation agent for the alginate hydrogel, because the Ba^{2+} ions bind the alginate polymer more strongly than do the Ca^{2+} ions, and thus the swelling of the hydrogel is suppressed. We also included Type I collagen in the NaA solutions, which helps preserve the viability and functions of hepatocytes. The hepatocytes we used were primary hepatocytes that were isolated from male F344/N rats by using a collagenase-perfusion protocol. As a model of non-parenchymal cells, we used Swiss 3T3 cells, which are frequently used in coculture experiments as feeder cells. As in the experiments described in the preceding section, we used PDMS microchannels featuring gelation channels that were 400 μm wide and 160 μm deep. Hepatocytes and 3T3 cells were suspended separately in sterile, isotonic NaA solutions containing collagen. A high density of hepatocytes was used in the precursor solution (3×10^7 cells/mL) to ensure sufficient contact between hepatocytes. After preparing and recovering the cell-encapsulating alginate hydrogel fibers, cells were cultured for up to 90 days.

First, we examined the effect of O_2 tension on the viability and functions of hepatocytes within the hydrogel fibers, because oxygen plays a critical role in hepatocyte cultures [35, 36]. We also investigated how coculturing hepatocytes with 3T3 cells affects hepatic functions and compared this with the effects measured using conventional plate cultures and hepatocyte-only cultures prepared in the hydrogel fibers. Figure 8.1f, g shows the obtained fibers containing hepatocytes with and without 3T3 cells 7 days of cultivation.

By adjusting the flow rates at which solutions were introduced, we effectively controlled the diameter of the cell-encapsulating microfibers. Hepatocytes were encapsulated in the core region that was ~50 μm wide, and 3T3 cells were embedded in the sandwiching shell regions. After cultivation for several days, we evaluated the viability of hepatocytes by using a dye-exclusion test. The hepatocytes and 3T3 cells formed linear aggregates within the hydrogel matrix (Fig. 8.1g); thus, hepatocyte viability remained above ~80 % when they were cocultured with 3T3 cells and the O_2 concentration was high (35 %). By contrast, under normal O_2 tension and in hepatocyte-only cultures, hepatocyte viability decreased rapidly, indicating that the O_2 concentration was critical for maintaining the viability of hepatocyte in this culture system.

As mentioned in the preceding section, alginate hydrogels can be readily digested by using enzymes (alginate lyase) or chelating multivalent cations. We sought to recover the linear organoids by enzymatically digesting the alginate hydrogel. By using this procedure, we recovered intact linear organoids that were 200–1000 μm long. Immunohistological examination revealed that hepatocytes located in the center of the organoids were covered with a thin layer of 3T3 cells. Furthermore, by observing the ultrastructure of hepatocytes by using transmission electron microscopy, we confirmed that hepatocytes extended microvilli in the space between adjacent 3T3 cells. These characteristic morphologies resembled those observed in tissues in vivo, indicating that the heterotypic organoids closely mimicked the hepatic cord structures observed in the liver.

Lastly, we examined several hepatic functions, and compared the results with those obtained using conventional 2D plate cultures and 3D fiber cultures in which 3T3 cells were not included. First, albumin-production rates and urea synthesis were measured directly, which revealed that (relative to controls) albumin production was enhanced substantially under coculture conditions in the hydrogel fiber, and this high level was maintained for up to ~60 days. A similar trend was observed in the case of urea synthesis. Next, we performed quantitative PCR to evaluate hepatocyte-specific functions such as the expression of cytochrome P450 and coagulation factors. These genes were expressed at high levels in the cocultures prepared within the fibers. By contrast, the expression of these genes decreased rapidly in the 3D hepatocyte-only cultures and in the plate cultures. These results clearly demonstrated that using the microscale 3D coculture system, which closely mimics the sinusoidal structures present in vivo, is a highly effective strategy for maintaining hepatic functions in vitro. The aforementioned culture technique could be applied for developing liver-tissue models designed for hepatocyte-based drug development, toxicity assays, and bioartificial liver devices. Recently, several reports have described the preparation and application of micrometer-sized hydrogel fibers designed for tissue engineering and cell culture in a confined but 3D environment, and in these studies, alginate was used as the primary hydrogel materials [37, 38]. We expect that fibrous, microengineered hydrogel materials will be widely used in constructing linear tissue models.

8.3 Preparation of Patterned Hydrogel Sheets Designed for Coculturing Cells in 3D

In Sect. 2, we introduced microfabricated hydrogel fibers and their use in generating intercellular networks and cocultures of hepatocytes. These fibers feature one of the most suitable morphologies for preparing cell cultures in terms of the diffusion-based supply of oxygen and nutrients; by comparison, however, planar hydrogel sheets might be more suitable for generating complex cell assemblies that resemble in vivo tissues. Moreover, planar structures could be more effective than fibers are in serving as building blocks used for constructing large-scale 3D tissue models by means of stacking. Furthermore, the cell-encapsulation throughput and the production speed of hydrogels could be considerably higher for sheets than for fibers. In this section, we briefly describe our recent successful production of heterogeneous, stripe-patterned hydrogel sheets, which were prepared by using micronozzle-combined planar microfluidic devices. We also describe the coculture of hepatocytes (in this case, HepG2 cells) and feeder cells (3T3 cells).

The microfluidic device used for preparing heterogeneous hydrogel sheets encapsulating distinct types of cells is shown in Fig. 8.2a, and the cell-growth behav-

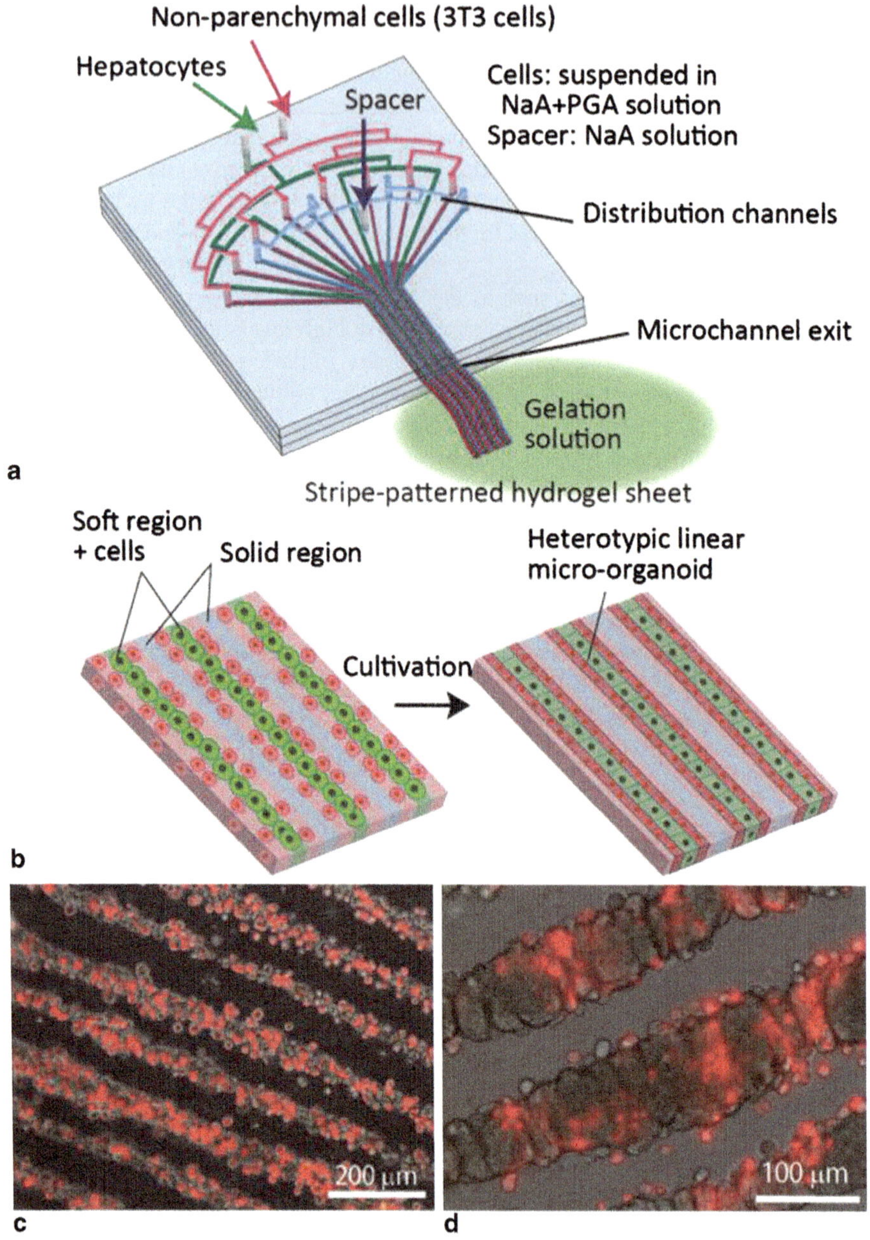

Fig. 8.2 Preparation of heterogeneous hydrogel microsheets and their use in culturing cells. **a** Microfluidic system used for preparing patterned alginate hydrogel sheets. **b** Schematic representation of cell growth in patterned hydrogel sheets. **c**, **d** Stripe-patterned hydrogel sheets encapsulating 2 types of cells (non-stained cells: HepG2; red cells: 3T3). Arrays of linear colonies were formed within the hydrogel sheet after 7 days of cultivation (**d**)

iors in the hydrogel sheet are shown in Fig. 8.2b. By introducing NaA solutions in the absence and presence of cells into the microchannel and by extruding the split/recombined streams into the gelation solution outside the microfluidic device, we can obtain hydrogel sheets incorporating arrays of parallel regions that contain distinct types of cells at high densities. By using PGA, we can control the stiffness of the hydrogel regions; cells are encapsulated in the soft regions of the sheet, and the solid regions prevent the cells from growing in the vertical direction, which results in the formation of rod-like heterotypic microorganoids. As an application, we cocultured hepatoma cells (HepG2 cells) and 3T3 cells and examined whether the functions of HepG2 cells were upregulated.

In the experiment, we fabricated 3-layer PDMS microfluidic devices. The device featured 3 inlets: Inlet 1 was used for the NaA + PGA solution containing hepatocytes; Inlet 2, for the same alginate solution but with 3T3 cells; and Inlet 3, for the NaA solution that was used for the solid spacer region. The micronozzle exit was 2.5 mm wide. The flow stream containing HepG2 cells was divided into 16 streams, each of which was sandwiched by the flows containing the 3T3 cells. The flow in the spacer region separated the flows containing 3T3 cells. The encapsulated cells could proliferate along the soft-hydrogel regions, but the solid-hydrogel regions prevented the cells from growing in the horizontal direction; this resulted in the formation of rod-like, linear organoids that were composed of multiple types of cells.

First, we examined whether the stripe-patterned heterogeneous hydrogel sheets were precisely fabricated with controlled stripe width and hydrogel thickness. When we used 2 % NaA solutions containing distinctly colored microbeads (diameter, ~1 μm) as the precursor solution, flat hydrogel fibers of a uniform thickness were obtained. The hydrogel sheet (~2.0 mm wide) was slightly narrower than the micronozzle (2.5 mm wide) because the alginate hydrogel shrank during the gelation process. By controlling the ratio of the flow rates of the introduced solutions, we were able to precisely control the width of the stripe patterns, and by adjusting the ratio of the introduced flow rates and/or the extension speed by using the roller, we could precisely control the thickness of the hydrogel.

Next, to develop cultures that mimic the sinusoid structures of the liver, we used the microchannel system described herein to prepare hydrogel sheets incorporating HepG2 and 3T3 cells; these cells served as models of hepatocytes and non-parenchymal cells, respectively. In the microchannel system, the flows of HepG2 and 3T3 cells were focused into narrow streams within the microchannel by the spacer flow, because the viscosities of the introduced solutions were dissimilar. In total, 64 streams were combined in the microchannel and had a width of 2.5 mm, and then were extruded from the microchannel into the outer gelation bath through the micronozzle. The cell-encapsulating hydrogel sheets obtained are shown in Fig. 8.2c, d. The positions of the 2 types of cells were maintained throughout the gelation process, and the heterogeneous hydrogel sheets that were obtained incorporated HepG2 and 3T3 cells (stained with a red dye) at high densities. During cultivation, HepG2 cells grew within the soft region and formed linear organoids (Fig. 8.2c), because the solid spacer regions acted as barriers. By contrast, 3T3 cells did not proliferate substantially, probably because of the contact inhibition occurring within the hydrogel matrix, but these cells adhered on the surface of the linear colonies of HepG2 cells. We observed that the HepG2 cells located at the center of the colonies were surrounded by a layer of 3T3 cells after several days of cultivation, as in the case of the cocultures prepared within the alginate fiber. Lastly, we examined whether hepatic functions were enhanced in the cocultures by using quantitative PCR analysis and ELISAs. Our results showed that albumin secretion from the cocultures was higher than that from the single culture of HepG2 cells in the hydrogel fiber, which demonstrated that hepatic functions were effectively enhanced in the 3D cocultures. Moreover, the results of quantitative PCR assays revealed that several genes associated with the differentiation of HepG2 cells were upregulated. Compared with fiber-based cultivation, this microfluidic process developed for preparing stripe-patterned alginate hydrogel sheets might enable more efficient culturing multiple types of cells at high densities, because the process can be readily applied in constructing large tissue models.

8.4 A Microfluidic Process for Producing Functional Hydrogel Beads

8.4.1 A Microfluidic System Developed for Producing Yarn-Ball-Shaped Hydrogel Microbeads

Encapsulation of biological substances into hydrogel matrices is a promising technique that can be used in diverse bioengineering and medical applications. Specifically, cell-encapsulating hydrogel microbeads could serve as a favorable unit structure for constructing large tissue models, especially when the beads are used together with hydrogel fibers or sheets, or cell aggregates. To prepare cell-encapsulating hydrogel beads that are a few hundred micrometers in diameter, various approaches have been developed that involve using microfluidic devices [6] or micronozzle structures [7]. However, most of the previous approaches are not suitable for producing hydrogel beads that exhibit diverse morphologies; mostly spherical beads have been produced because of the minimal interfacial energy of the spherical droplets of the precursor solution. Although the preparation of hydrogel beads featuring more complex morphologies has been reported, the methods used require complicated microfluidic systems that incorporate in situ photolithographic processes [39, 40]. Furthermore, most of these photolithography-based techniques could only be used to prepare simple planar structures of uniform thickness. When cells are encapsulated within hydrogel beads, morphologies that have a high surface-to-volume ratio would be highly advantageous when considering the supply of oxygen and nutrients to cells. However, hydrogel microfibers, for example, are unlikely to be

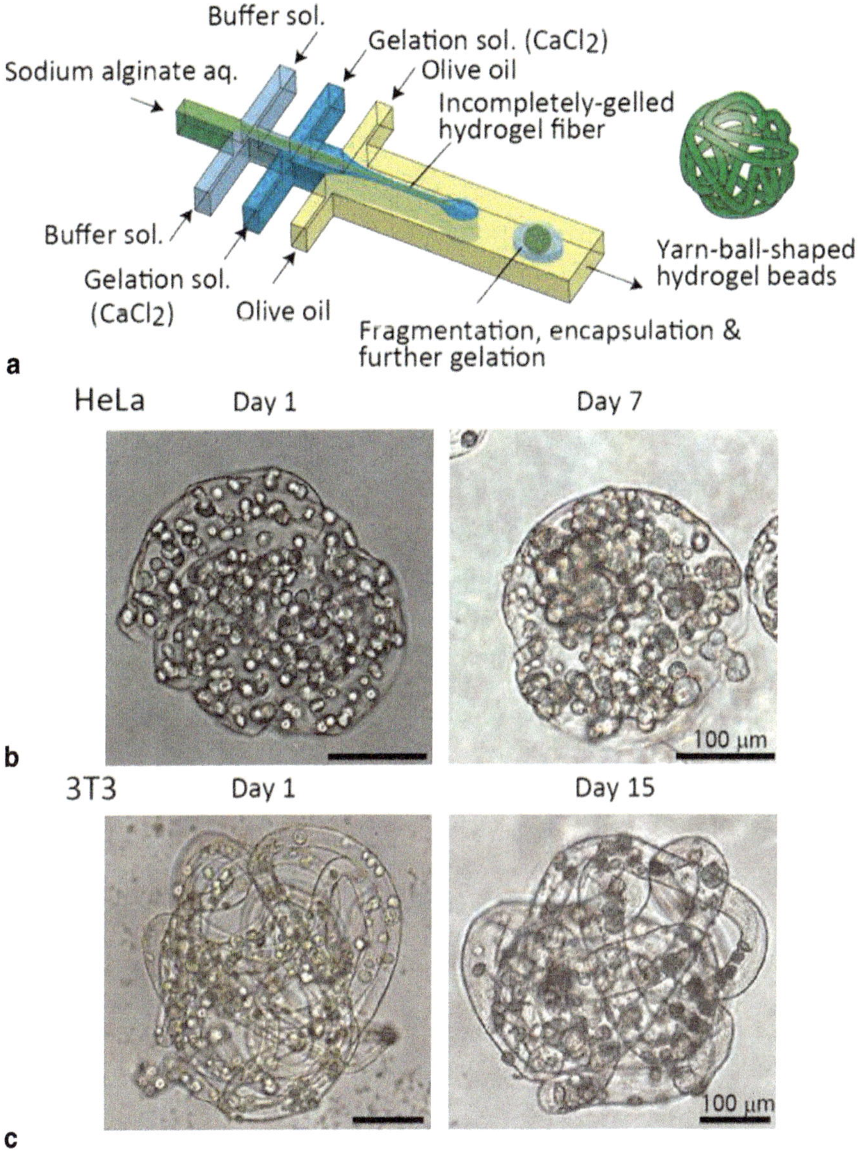

Fig. 8.3 **a** Preparation of yarn-ball-shaped hydrogel microbeads. **b** HeLa and **c** 3T3 cells cultured within the hydrogel matrices of yarn-ball-shaped hydrogel microbeads

suitable for close packing in a flow-through reactor/chamber and for intravascular injection into host organs.

Recently, we developed a simple microfluidic process designed for producing yarn-ball-shaped hydrogel microbeads [41]. In this process, we employed the procedure used for preparing hydrogel fibers [9] that is described in the preceding sections and combined this with droplet-microfluidics technology. The mechanism by which the hydrogel microbeads are formed is illustrated in Fig. 8.3a. As in the preparation of hydrogel fibers, we employed parallel flows of aqueous solutions of a hydrogel precursor (NaA), a gelation agent, and a buffer solution. Incompletely gelled alginate hydrogel microfibers were generated at the confluence, where water-in-oil (w/o) droplets formed simultaneously. The fibers were fragmented during encapsulation and they folded into a yarn-ball-like shape. This yarn-ball-shaped morphology combines several advantages associated with both spherical beads and linear fibers: this unique morphology is compact but has a high surface-to-volume ratio, and thus can potentially enable efficient supply of oxygen and nutrients to encapsulated cells; moreover, if these yarn-ball-shaped particles are used as carriers for transplantation of cells (e.g., pancreatic islets or hepatocytes) and are transplanted intravascularly, the morphology is unlikely to strongly hinder the flow of blood because of the void space.

In our experiments, we first tested the fabrication of yarn-ball-shaped hydrogel beads by using a 1.8 % NaA solution as the precursor. The concentration of the gelation agent ($CaCl_2$ or $BaCl_2$) was varied to examine their effect on the generation of the yarn-ball-shaped hydrogel beads. The experimental procedures used were similar to those described for the production of hydrogel fibers, and to generate the w/o droplets, we introduced olive oil together with lecithin into the microchannels. When we tested $CaCl_2$, we determined that at a concentration as high as 20 mM, the hydrogel fiber that was formed was not fragmented, because the gelation had already proceeded at the confluence point and the formed hydrogel fiber was highly solid. When the $CaCl_2$ concentration was too low (5 mM), nearly spherical beads were obtained because droplets were formed and the inner contents were mixed before the formation of the hydrogel fiber. When the $CaCl_2$ concentration was approximately 10 mM, the w/o droplets were generated at a point that was ~25 mm from the confluence. Incompletely gelled and fragmented hydrogel fibers were observed in the droplets, and we successfully obtained yarn-ball-shaped hydrogel beads. Based on these results, we concluded that the concentration of Ca^{2+} was critical for producing yarn-ball-shape beads. The beads were 200–300 μm in diameter, and the hydrogel fibers that formed the beads were 10–30 μm wide; these sizes could be adjusted by changing operational conditions such as the flow rates used. We were also able to produce Ba-alginate microbeads by using $BaCl_2$ as the gelation agent; however, the optimal concentration of $BaCl_2$ required for the formation of yarn-ball-shaped hydrogel beads was distinct from that of $CaCl_2$ required for generating Ca-alginate microbeads. When a surfactant (1 % bovine serum albumin) was added in the gelation solution, the beads shrank and highly became uniform in size, because the surface tension between the water and oil phases was decreased.

To apply hydrogel microbeads in high-density cell cultivation, we incorporated animal cells (NIH-3T3 or HeLa cells) into the NaA solution and prepared cell-containing hydrogel microbeads (Fig. 8.3). The initial cell densities were 4.7×10^6 and 3.0×10^6 cells/mL for HeLa and NIH-3T3 cells, respectively. The cell viabilities remained high (>80 %) even after the cells were encapsulated in the beads. The cells continuously proliferated and formed spherical colonies inside the hydrogel fibers, with the final cell densities reaching 2.4×10^7 cells/mL for HeLa cells on Day 7 and 3.8×10^7 cells/mL for NIH-3T3 cells on Day 15. This result indicated that the hydrogel beads contained voids that are suitable for high-density cell cultivation, which possibly enhanced the supply of oxygen and nutrients to the center of the beads. In summary, the fabricated hydrogel beads could be used as a new material for immobilizing and encapsulating biological materials and could potentially be applied in cell-transplantation therapies. Moreover, these hydrogel beads, which included large internal voids, could be suitable as artificial organ units that do not interrupt blood flow when injected into blood vessels. Lastly, these beads could also function as matrices for biological immobilization and cell cultivation.

8.4.2 *Production of Micrometer-Sized Hydrogel Particles*

Size is the most critical factor that determines the physicochemical characteristics of hydrogel beads. Small hydrogel beads are more suitable for biomedical application than large beads are, because chemicals are efficiently supplied and removed through their permeable matrices by means of diffusion. As mentioned in the preceding section, the production of alginate hydrogel beads by using microfluidic devices has been widely reported [6, 7]. However, the production of hydrogel beads whose diameters are in the single-micrometer size range remains challenging; this is because small droplets of highly viscous precursor solutions cannot be readily produced using a narrow microchannel or micronozzle. In this section, we describe a new approach we have developed for producing micrometer-sized hydrogel beads (diameter < 10 μm) by using a unique process of water molecule dissolution from w/o droplets into the continuous phase of a polar solvent [42]. This process enables the production of hydrogel microbeads that are considerably smaller than the droplets generated initially.

The process used for producing micrometer-sized hydrogel beads is shown in Fig. 8.4a. In this process, methyl acetate and an aqueous solution of the hydrogel precursor are introduced into the microchannels to generate w/o droplets at the first confluence. These droplets gradually shrink while flowing through the continuous phase of the polar solvent, and this occurs because of the dissolution of water molecules into the continuous phase. During this period, hydrophilic polymers are concentrated in the shrunk droplets, which are then gelled by introducing a gelation solution. Consequently, the hydrogel beads obtained are considerably smaller than the initial droplets. We prepared Ca-alginate beads and chitosan beads by using $CaCl_2$ and NaOH solutions as the gelation agents, respectively. As the polar organic solvent, we mainly used methyl acetate, which exhibits a water solubility of 8 %.

In these experiments, we again used PDMS microchannels. The microchannel depth was uniformly 55 μm, and the droplet-generation channel (the first confluence) was 50 μm wide. The water-extraction and gelation channels were 70 and 75 mm long, respectively. To produce Ca-alginate hydrogel beads, 0.025–0.15 % NaA solution was used as the precursor solution and 0.01–1 M $CaCl_2$ aq. was used

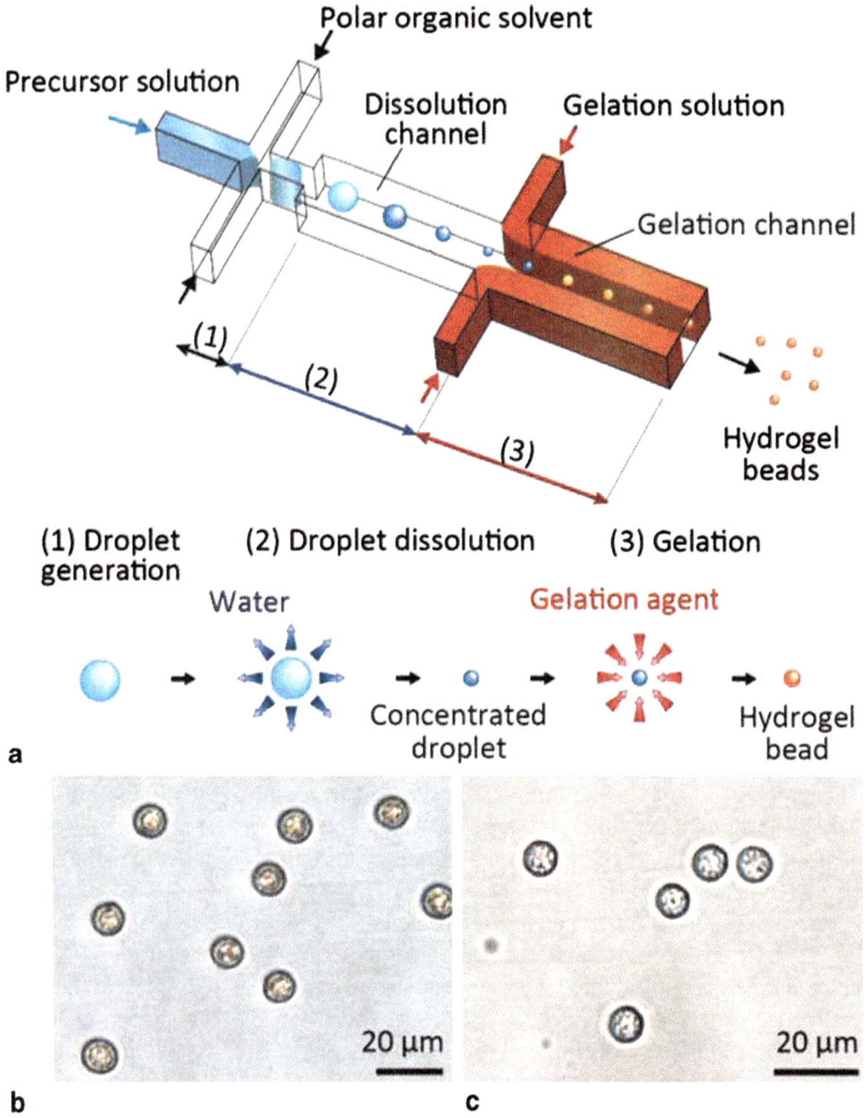

Fig. 8.4 **a** Microfluidic process used for producing micrometer-sized hydrogel beads. **b, c** Micrographs showing the prepared **b** Ca-alginate and **c** chitosan microbeads

as the gelation solution. To produce chitosan beads, 0.025 or 0.05 % chitosan solution in 0.1 M acetic acid was used as the precursor.

First, we confirmed that w/o droplets were generated at the confluence point and were in a non-equilibrium state. The droplets shrank while they flowed through the microchannel. When the flow rates of the precursor and methyl acetate were 0.9 and 30 μL/min, respectively, and the NaA concentration was 0.1 %, the initial average diameter of the droplets was ~70 μm; by contrast, the final average diameter at the time of saturation was ~9 μm (Fig. 8.4b). Thus, the volume of the droplets decreased to ~0.4 % of the volume of the initial droplets. We were able to adjust the microbead diameters by varying the initial NaA concentration: the diameter changed from 6 to 10 μm when the initial concentration of NaA was increased from 0.025 to 0.15 %, indicating that the NaA concentration in the droplet critically affects the bead size. These results demonstrated that the dissolution of water droplets in methyl acetate was possible, and this process was exploited for producing micrometer-sized hydrogel beads.

To determine whether the microbeads formed were composed of hydrogels, the microbeads were dried and re-swollen repeatedly; the beads shrank and deformed when they were dried, but recovered their initial spherical shape and volume when re-swollen by adding water, which indicated that the swollen beads contained water within their matrices. We used similar methods to prepare chitosan hydrogel beads and obtained monodispersed chitosan hydrogel beads featuring an average diameter of 10.1 μm (coefficient of variation (CV): 4.8 %) when 0.05 % chitosan solution was used (Fig. 8.4c). This result suggests that the process described here for producing micrometer-sized hydrogel beads can be applied on various types of hydrogel materials. These small hydrogel beads might be useful as tools for constructing complex tissue models and also as carriers or immobilization matrices for biological substances.

8.5 Preparation of Microfabricated Hydrogel Structures for Use in Tissue Engineering

In this section, we describe using microfluidic devices made of hydrogels to produce vascular-tissue models. Engineering vascular tissue is an approach that is essential for preparing artificial-artery grafts, fabricating large tissues embedded with capillary networks, and studying the cellular physiology associated with blood vessels such as cancer metastasis, angiogenesis, and thrombosis. Over the past decade, various strategies have been proposed for building vascular structures, including the rolling of cell sheets [43], the incorporation of cells into fibrous hydrogels [44], and the 3D deposition of cells by using inkjet printing [45]. However, preparing functional vascular tissues remains a challenge in terms of obtaining a multilayer configuration, branch structures, and thick walls; this is mainly because no techniques are available that can be used to effectively assemble multiple cell types into layered capillary structures. Specifically, vascular tissues are composed of inner endothelial cells on the lumen side, smooth muscle cells that provide physical support, and vascular tissue-specific ECM components such as elastin and collagen. Therefore, at least 2 types of cells have to be assembled into fabricated capillary to structurally mimic in vivo vascular tissues. Here, we present the approach we recently developed for assembling multiple types of cells into multilayer capillary structures [46]. In this procedure, multiple Ca-alginate hydrogel layers are deposited within hydrogel microchannels made of agarose.

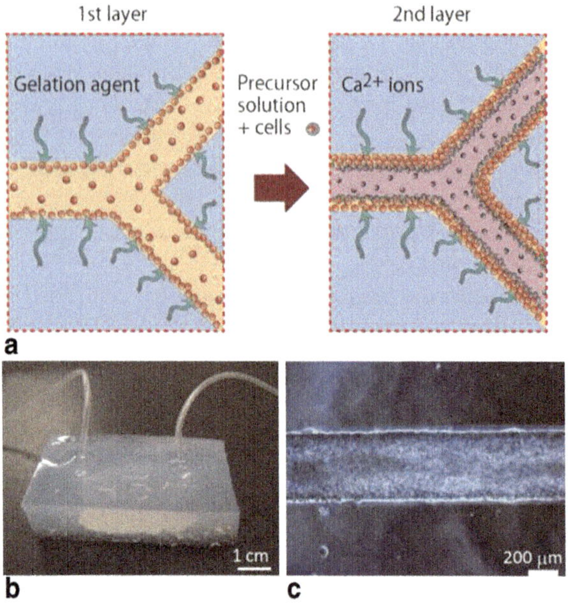

Fig. 8.5 **a** Schematic image showing the preparation of vascular tissue-like cell assemblies by using hydrogel microchannels and sodium alginate (NaA) solutions containing cells. First, the NaA solution containing Type 1 cells is introduced into the microchannel made of agarose and Ca^{2+} ions, and a Ca-alginate hydrogel layer is deposited within the microchannel. Next, the NaA solution containing cells of a distinct type (Type 2) is introduced, which results in the formation of a thick, multilayer, and heterotypic vascular-tissue model. **b** Photograph of a hydrogel microfluidic device incorporating a microchannel structure made of agarose hydrogel. **c** A multilayer vascular-tissue model prepared using 2 types of cells

The process used for preparing vascular tissue-like structures is presented schematically in Fig. 8.5a. In this approach, microchannels made of agarose hydrogel containing Ca^{2+} ions are first prepared by employing molding and physical-bonding processes. An aqueous solution containing NaA and cells is introduced into the agarose microchannel and then Ca^{2+} ions are supplied to the NaA solution through the permeable hydrogel matrix; this results in the formation of a Ca-alginate hydrogel layer on the microchannel surface. By repeating this deposition process with a solution of NaA that contains a different type of cells, multilayered Ca-alginate hydrogels incorporating distinct cell types are obtained. The tissues that are formed can be recovered from the hydrogels by detaching the bonded hydrogel plates.

We prepared agarose hydrogel microchannels containing Ca^{2+} ions in the matrix (Fig. 8.5b) by using a previously reported molding scheme [47]. An aqueous solution of 5% agarose and 10 mM $CaCl_2$ was poured onto a silicon mold with photoresist microstructures, and this was incubated at 4°C for 2 h to completely gel the agarose solution. Next, the hydrogel plate was peeled off from the mold and trimmed, a microstructured plate was physically bonded to the flat hydrogel plate to generate microchannel structures, and, lastly, inlet and outlet tubes were attached. The microchannels were uniformly 500 μm deep and wide.

Before introducing the NaA solution containing cell suspensions, the surface of the agarose microchannel was modified by introducing an aqueous solution of poly-L-lysine and 10 mM $CaCl_2$, which can improve the attachment force of Ca-alginate hydrogels. To control the high gelation speed of alginate, an aqueous solution of a chelator (trisodium citrate) was introduced, after which we introduced 0.5–1 % NaA solutions containing cells or model particles into the microchannel by means of aspiration. After the cell-incorporating hydrogel structures were prepared, the agarose hydrogel microdevices were dipped in cell-culture medium overnight. We used several types of cells including human vascular endothelial cells (HUVECs), fibroblasts (NIH-3T3), and vascular smooth muscle cells; the cells were suspended separately in isotonic solutions of NaA at high densities ($0.5–1 \times 10^8$ cells/mL). These NaA solutions containing the suspended cell were introduced stepwise into the microchannel, and the deposition of the hydrogel layers was observed using an inverted optical microscope. After the hydrogels had been deposited, complete gelation of the alginate hydrogels was further ensured by introducing a 10 mM $CaCl_2$ solution. The cells were cultured under either fluid-perfusion or static-culture conditions for up to 2 weeks.

First, using a branched microchannel, we demonstrated the stepwise deposition and formation of multilayer Ca-alginate hydrogel structures within the agarose microchannel. After the first NaA solution was introduced into the agarose microchannel for 3 min at a flow rate of 20 μL/min, an approximately 150-μm-thick Ca-alginate hydrogel layer was formed on the microchannel surface. After we introduced the second NaA solution, double-layered Ca-alginate hydrogels were formed. When a particle suspension was introduced continuously, the particles were observed to flow through the hydrogel, which indicated that a hollow capillary structure was formed in a manner that depended on the microchannel geometry. Furthermore, we could control the thickness of the deposited Ca-alginate layer by changing the introduction period and the flow rate of the NaA solution.

Next, we used the NaA solutions containing suspended cells of distinct types and successfully constructed capillary-like tissues featuring double-layered hollow structures, through which we could observe the flow of fluorescent particles. Figure 8.5c shows an example of a capillary tissue that was formed and contained 2 types of cells. Using perfusion cultures resulted in improved viability and function of the embedded cells, as compared with the use of the static cultures. Moreover, the formed tissues could be readily recovered from the hydrogel microchannels by detaching the 2 bonded agarose hydrogel plates, and the tissues were adequately strong to allow them to be handled using tweezers. These results demonstrated that the method presented here enables the assembly of cells into 3D structures and the fabrication of multiple vascular-tissue models.

In summary, we have developed a new process that allows vascular-tissue models to be prepared easily and accurately. By changing the microchannel geometry, the flow rate, and the introduction period of the NaA solution, vascular-tissue models featuring diverse configurations can potentially be prepared. We expect this method to become a highly effective approach for use in vascular tissue engineering.

8.6 Conclusions and Perspectives

In this paper, we have briefly reviewed our recent progress in producing (1) cell-encapsulating hydrogel materials by using microfluidic devices and (2) capillary-like tissue models by using microfluidic devices made of hydrogels. These materials reproduced the in vivo microenvironment and provided a 3D cell-cultivation platform that enhanced cellular functions and vitality. Furthermore, the materials and tissues presented herein can potentially be used as unit structures for constructing larger tissue models through a bottom-up approach. We expect that by combining the use of the presented hydrogel materials with robotics and automation technologies, versatile strategies can be implemented that facilitate the construction of functional tissues, which will be valuable for use in regenerative medicine, cell-based assay systems designed for drug development, and physiological investigations on cells.

Acknowledgments We would like to thank Dr. S. Sugaya, Dr. R. Utoh, Mr. M. Iwase, Mr. Y. Kitagawa, Ms. E. Yamada, Mr. Y. Nagamuna, Ms. A. Kobayashi, Ms. A. Miyama, Mr. Y. Yajima, Ms. A. Hori for their contribution in conducting experiments. This study was supported in part by Grants-in-aid for Scientific Research (KAKENHI 23106007, 23700554, and 30546784) from the Ministry of Education, Culture, Sports, Science, and Technology, Japan.

References

1. Dean DM, Napolitano AP, Youssef J, Morgan JR (2007) Rods, tori, and honeycombs: the directed self-assembly of microtissues with prescribed microscale geometries. FASEB J 21:4005
2. Norotte C, Marga FS, Niklason LE, Forgacs G (2009) Scaffold-free vascular tissue engineering using bioprinting. Biomaterials 30:5910
3. Anada T, Fukuda J, Sai Y, Suzuki O (2012) An oxygen-permeable spheroid culture system for the prevention of central hypoxia and necrosis of spheroids. Biomaterials 33:8430
4. Chung BG, Lee KH, Khademhosseini A, Lee SH (2012) Microfluidic fabrication of microengineered hydrogels and their application in tissue engineering. Lab Chip 12:45
5. Nisisako T, Torii T, Takahashi T, Takizawa Y (2006) Synthesis of monodisperse bicoloured Janus particles with electrical anisotropy using a microfluidic co-flow system. Adv Mater 18:1152
6. Tan WH, Takeuchi S (2007) Monodisperse alginate hydrogel microbeads for cell encapsulation. Adv Mater 19:2696
7. Sugiura S, Oda T, Izumida Y, Aoyagi Y, Satake M, Ochiai A, Ohkohchi N, Nakajima M (2005) Size control of calcium alginate beads containing living cells using micro-nozzle array. Biomaterials 26:3327
8. Sugiura S, Oda T, Aoyagi Y, Matsuo R, Enomoto T, Matsumoto K, Nakamura T, Satake M, Ochiai A, Ohkohchi N, Nakajima M (2007) Microfabricated airflow nozzle for microencapsulation of living cells into 150 micrometer microcapsules. Biomed Microdev 9:91
9. Yamada M, Sugaya S, Naganuma Y, Seki M (2012) Microfluidic synthesis of chemically and physically anisotropic hydrogel microfibers for guided cell growth and networking. Soft Matter 8:3122

10. He W, Halberstadt CR, Gonsalves KE (2004) Lithography application of a novel photoresist for patterning of cells. Biomaterials 25:2055
11. Zhang S, Yan L, Altman M, Lassle M, Nugent H, Frankel F, Lauffenburger DA, Whitesides GM, Rich A (1999) Biological surface engineering: a simple system for cell pattern formation. Biomaterials 20:1213
12. Kalinina S, Gliemann H, Lopez-Garcia M, Petershans A, Auernheimer J, Schimmel T, Bruns M, Schambony A, Kessler H, Wedlich D (2008) Isothiocyanate-functionalized RGD peptides for tailoring cell-adhesive surface patterns. Biomaterials 29:3004
13. Tourovskaia A, Figueroa-Masot X, Folch A (2005) Differentiation-on-a-chip: a microfluidic platform for long-term cell culture studies. Lab Chip 5:14
14. Su J, Zheng Y, Wu H (2009) Generation of alginate microfibers with a roller-assisted microfluidic system. Lab Chip 9:996
15. Sugiura S, Oda T, Aoyagi Y, Satake M, Ohkohchi N, Nakajima M (2008) Tubular gel fabrication and cell encapsulation in laminar flow stream formed by microfabricated nozzle array. Lab Chip 8:1255
16. Lee KH, Shin SJ, Kim CB, Kim JK, Cho YW, Chung BG, Lee SH (2010) Microfluidic synthesis of pure chitosan microfibers for bio-artificial liver chip. Lab Chip 10:1328
17. Duffy DC, McDonald JC, Schueller OJ, Whitesides GM (1998) Rapid prototyping of microfluidic systems in poly(dimethylsiloxane). Anal Chem 70:4974
18. Pokrywczynska M, Drewa T, Jundzill A, Lysik J (2008) Alginate is not a good material for growth of rapidly proliferating cells. Transplant Proc 40:1664
19. Kitagawa Y, Naganuma Y, Yajima Y, Yamada M, Seki M (2014) Patterned hydrogel microfibers prepared using multilayered microfluidic devices for guiding network formation of neural cells. Biofabrication 6:035011
20. Landry J, Bernier D, Ouellet C, Goyette R, Marceau N (1985) Spheroidal aggregate culture of rat liver cells: histotypic reorganization, biomatrix deposition, and maintenance of functional activities. J Cell Biol 101:914
21. Dvir-Ginzberg M, Elkayam T, Aflalo ED, Agbaria R, Cohen S (2004) Ultrastructural and functional investigations of adult hepatocyte spheroids during in vitro cultivation. Tissue Eng 10:1806
22. Fukuda J, Sakai Y, Nakazawa K (2006) Novel hepatocyte culture system developed using microfabrication and collagen/polyethylene glycol microcontact printing. Biomaterials 27:1061
23. Brophy CM, Luebke-Wheeler JL, Amiot BP, Khan H, Remmel RP, Rinaldo P, Nyberg SL (2009) Rat hepatocyte spheroids formed by rocked technique maintain differentiated hepatocyte gene expression and function. Hepatology 49:578
24. Fukuda J, Nakazawa K (2011) Hepatocyte spheroid arrays inside microwells connected with microchannels. Biomicrofluidics 5:22205
25. Bhatia SN, Yarmush ML, Toner M (1997) Controlling cell interactions by micropatterning in co-cultures: hepatocytes and 3T3 fibroblasts. J Biomed Mater Res 34:189
26. Corlu A, Ilyin G, Cariou S, Lamy I, Loyer P, Guguen-Guillouzo C (1997) The coculture: a system for studying the regulation of liver differentiation/proliferation activity and its control. Cell Biol Toxicol 13:235
27. Tsuda Y, Kikuchi A, Yamato M, Nakao A, Sakurai Y, Umezu M, Okano T (2005) The use of patterned dual thermoresponsive surfaces for the collective recovery as co-cultured cell sheets. Biomaterials 26:1885
28. Khetani SR, Bhatia SN (2007) Microscale culture of human liver cells for drug development. Nat Biotechnol 26:120
29. Hui EE, Bhatia SN (2007) Silicon microchips for manipulating cell-cell interaction. Proc Natl Acad Sci U S A 104:5722
30. Kidambi S, Sheng L, Yarmush ML, Toner M, Lee I, Chan C (2007) Patterned co-culture of primary hepatocytes and fibroblasts using polyelectrolyte multilayer templates. Macromol Biosci 7:344

31. Elloumi Hannachi I Itoga K Kumashiro Y Kobayashi J Yamato M Okano T (2009) Fabrication of transferable micropatterned-co-cultured cell sheets with microcontact printing. Biomaterials 30:5427
32. Harimoto M, Yamato M, Hirose M, Takahashi C, Isoi Y, Kikuchi A, Okano T (2002) Novel approach for achieving double-layered cell sheets co-culture: overlaying endothelial cell sheets onto monolayer hepatocytes utilizing temperature-responsive culture dishes. J Biomed Mater Res 62:464
33. Kim K, Ohashi K, Utoh R, Kano K, Okano T (2012) Preserved liver-specific functions of hepatocytes in 3D co-culture with endothelial cell sheets. Biomaterials 33:1406
34. Yamada M, Utoh R, Ohashi K, Tatsumi K, Yamatoc M, Okano T, Seki M (2012) Controlled formation of heterotypic hepatic micro-organoids in anisotropic hydrogel microfibers for long-term preservation of liver-specific functions. Biomaterials 33:8304
35. Kidambi S, Yarmush RS, Novik E, Chao P, Yarmush ML, Nahmias Y (2009) Oxygen-mediated enhancement of primary hepatocyte metabolism, functional polarization, gene expression, and drug clearance. Proc Natl Acad Sci U S A 106:15714
36. Cho CH, Park J, Nagrath D, Tilles AW, Berthiaume F, Toner M, Yarmush ML (2007) Oxygen uptake rates and liver-specific functions of hepatocyte and 3T3 fibroblast co-cultures. Biotechnol Bioeng 97:188
37. Onoe H, Okitsu T, Itou A, Kato-Negishi M, Gojo R, Kiriya D, Sato K, Miura S, Iwanaga S, Kuribayashi-Shigetomi K, Matsunaga YT, Shimoyama Y, Takeuchi S (2013) Metre-long cell-laden microfibres exhibit tissue morphologies and functions. Nat Mater 12:584
38. Kang E, Jeong GS, Choi YY, Lee KH, Khademhosseini A, Lee SH (2011) Digitally tunable physicochemical coding of material composition and topography in continuous microfibres. Nat Mater 10:877
39. Panda P, Ali S, Lo E, Chung BG, Hatton TA, Khademhosseini A, Doyle PS (2008) Stop-flow lithography to generate cell-laden microgel particles. Lab Chip 8:1056
40. Chung SE, Park W, Shin S, Lee SA, Kwon S (2008) Guided and fluidic self-assembly of microstructures using railed microfluidic channels. Nat Mater 7:581
41. Miyama A, Yamada M, Sugaya S, Seki M (2013) A droplet-based microfluidic process to produce yarn-ball-shaped hydrogel microbeads. RSC Adv 3:12299
42. Sugaya S, Yamada M, Hori A, Seki M (2013) Microfluidic production of single micrometer-sized hydrogel beads utilizing droplet dissolution in a polar solvent. Biomicrofluidics 7:54120
43. Yuan B, Jin Y, Sun Y, Wang D, Sun J, Wang Z, Zhang W, Jiang X (2012) A strategy for depositing different types of cells in three dimensions to mimic tubular structures in tissues. Adv Mater 24:890
44. Santos MI, Tuzlakoglu K, Fuchs S, Gomes ME, Peters K, Unger RE, Piskin E, Reis RL, Kirkpatrick CJ (2008) Endothelial cell colonization and angiogenic potential of combined nano- and micro-fibrous scaffolds for bone tissue engineering. Biomaterials 29:4306
45. Pataky K, Braschler T, Negro A, Renaud P, Lutolf MP, Brugger J (2012) Microdrop printing of hydrogel bioinks into 3D tissue-like geometries. Adv Mater 24:391
46. Iwase M, Yamada M, Seki M (2012) Construction of vascular tissues via multilayer cell deposition inside hydrogel microchannels. Proc MicroTAS 2012:572
47. Ling Y, Rubin J, Deng Y, Huang C, Demirci U, Karp JM, Khademhosseini A (2007) A cell-laden microfluidic hydrogel. Lab Chip 7:756

Chapter 9
On-Chip Fabrication, Manipulation and Self-Assembly for Three-Dimensional Cell Structures

Toshio Fukuda, Tao Yue, Masaru Takeuchi and Masahiro Nakajima

Abstract We present an integrated multi-functional microfluidic devices for conducting various tasks, including on-chip fabrication, on-chip manipulation and on-chip self-assembly, for building 3D cellular structures. This system is used for building target cell structures for a "factory line on a chip". This chapter gives a description of the tasks and achievements to show the procedure to our goal, which is focus on the engineered vascular-like cell embedded structures.

Keywords On-chip fabrication · Self-assembly · Photo-crosslinkable resin · Cellular vascular-like microtubes · Microfluidic device · Dielectrophoresis

9.1 Introduction

9.1.1 Motivation in Three-Dimensional Cell Structure for Tissue Engineering

Lost or damage of an organ due to disease or injury is a serious problem for human body. Transplantation of organs to replace the incurable ones is a highly successful therapy [1]. A successful transplant requires a compatible and willing donor. However, the need for donor organs far exceeds the supply [2]. This donor scarcity problem has resulted in new technique which is to build artificial organs by cells and organic materials, which are defined as a tissue engineering [3].

Hence, it will be a great approach to realize artificial organs in vitro environment. Stem cells show promising cell type to achieve the goal. It is required to be grown into a various cell types from the stem cells [4]. Living cells are genetically programmed to assemble themselves into tissues and organs. However, scientists are still in the early stages of understanding how this works [5]. Like in our bodies,

T. Fukuda (✉)
Meijo University, Nagoya University, Beijing Institute of Technology
e-mail: tofukuda@meijo-u.ac.jp

T. Yue · M. Takeuchi · M. Nakajima
Nagoya University

T. Arai et al. (eds.), *Hyper Bio Assembler for 3D Cellular Systems*,
DOI 10.1007/978-4-431-55297-0_9

the tissues and organs are with complex three-dimensional (3D) shapes and structures. Therefore, how to control the positions of cells and form certain shapes by cells as the real tissues are key issues to generate artificial cell tissues [6].

Currently, some researches have been conducted by effective approaches for building two-dimensional (2D) cell structures. The cell growth is controlled on the surface of substrate to form different 2D patterns by changing the surface adhesive properties [7]. Cell positions are controlled and immobilized to 2D shapes via micromanipulation approaches [8]. Arbitrary shaped 2D cell embedded structures are fabricated inside microfluidic devices [9]. Traditionally, many useful and important *in vitro* methods for tissue engineering are contributed by these 2D cell structures. However, numbers of researchers have questions on the validity of studying cells in an environment that is so far away from the *in vivo* state [10]. Tissues and organs are 3D arrangements of cells. 3D shaped structures are necessary for building functional artificial tissues. Besides, in 2D cell structures, the environment is constantly changing while nutrients in the media are depleted and metabolites accumulate. This is clearly different with the *in vivo* state where cells are maintained in a 3D multi-cellular environment of a constant fresh supply of nutrients and the removal of waste products via the circulatory system [11].

Assembly of cells to form 3D shapes and structures offers a practical alternative to natural tissue models. These systems provide an environment in which one or more cell types can be encouraged to form tissues-like constructs. To assemble an artificial organ with working properly, specific cells must be placed in the to specific locations [12]. This is a fundamental issue to be solved for artificial tissue and organ assembly in all living organisms. A specific 3D multi-cellular structure is often critical for proper cellular function. The ability to build such a structure would be useful for the tissue engineering [13]. One useful application would be for organ repair. Another useful purpose would be the study of diseases whose progress is impacted by multi-cellular topology [14].

3D cell models also offer numerous benefits over simple 2D cell models. In recent studies, cell viability, proliferation, differentiation, morphology, gene and protein expression, and function have been shown to exhibit significant differences in 3D compared to 2D, with 3D structures more closely mirroring what is observed *in vivo* [15]. Undoubtedly, 3D cellular structures provide a similar environment for cells as in the real tissues, which enhance the cell proliferation and cell-cell interaction during building artificial tissues. The cell-cell interactions established in 3D play a significant role in the function of both healthy and diseased tissues and can affect the cellular behavior to form tissues. The use of 3D cell structures will add benefits of improved long-term cell viability with a tissue-specific function [16].

9.1.2 Construction Methods for Artificial Tissues

Currently, there are mainly two fundamental construction methods in tissue engineering, which are top-down and bottom-up approaches. The traditional method mentioned in last section belongs to the top-down approaches [17]. In top-down approaches, cells are seeded on a biodegradable polymeric scaffold and grow according the shape of the scaffold to be a tissue for example. The cells are expected to populate the scaffold and create the appropriate microarchitecture often with the aid of perfusion, growth factors or mechanical stimulations [18]. However, even the surface patterning or more biomimetic scaffolding are used, top-down approaches often have difficulty recreating the intricate microstructural features of tissues [19], because the current biomaterials and micro fabrication methods are with certain limitation for creating intricate 3D biomimetic structures.

On the other hand, bottom-up approaches aim to address the challenge of recreating 3D biomimetic structures by designing structural features on the microscale to build modular tissues that can be used as building blocks to create larger tissues. These modules can be created in many ways, such as through self-assembled aggregation, micro fabrication of cell embedded hydrogels, creation of cell sheets or direct printing of tissues [20, 21]. Once created, these modules can be assembled into larger 3D tissues through a number of methods such as random packing, stacking of cell layers or directed assembly [22–24]. One strong biological basis for this bottom-up approach is that many tissues are also comprised of repeating functional units, such as the lobule in the liver [25]. Consequently, bottom-up approaches are promising to create more biomimetic engineered tissues.

The summary of current construction methods for engineered tissues is shown in Table 9.1. It briefly describes both top-down and bottom-up approaches. In 1997, Dr. Joseph Vacanti grew a human ear from cartilage cells on the back of a mouse, which demonstrated the scaffold based 3D cell structures and tissues [26]. Also,

Table 9.1 Summary of the construction methods for artificial tissues

Approach	Material and method	Advantages	Disadvantages	Refs.
Scaffold	Biodegradable polymer	Directly 3D shape formation	Difficult for complex multi-cellular tissues and cell differentiation	[26–29]
Cell aggregation	Cell beads, cell stimulation	Scaffold-free, multi-cellular structure	Low design flexibility and shape controllability	[30, 31]
Micro fabrication	Hydrogel, photolithography	Arbitrary shape, microscale resolution	Modular tissues need further 3D assembly	[32–35]
Cell sheets and stacking	Cell layers	Scaffold-free, direct transplantation, mass products	Low design flexibility, difficult for complex shapes	[36–38]
Cell printing	Printing, position support	High design flexibility, directly 3D shape formation	Difficult for thick structures, limited resolution	[39–42]
Random assembly	Cell encapsulated units	Microscale resolution, complex structures	Low shape controllability	[24]
Directed assembly	Cell plates	Microscale resolution, self-assembly	Special designed cell plates	[43, 44]

various bottom-up methods are under rapid development. These undergoing issues, such as fabrication of cell blocks, control of cell aggregation and assembly of engineered tissue building blocks, involve manymicrotechnologies including observation, micromanipulation, microfabrication and microfluidic system. Especially, the microfluidic devices integrated with multiple functions provide the promising ways to realize the bottom-up approach (3D cell structure assembly) of tissue engineering.

9.1.3 Research Roadmap of Multi-Functional Microfluidic System

Tissue engineering has great potential to improve the therapy for the tissue and organ damage. It may solve the problem of transplantable organ shortage and suboptimal artificial organs for repair or replacement of diseased or destroyed human organs and tissues. As mentioned above, 3D cell structure assembly will contribute greatly to construct functional tissues. Assembly of cells to form 3D structures provides a biomimetic structures for cells as the real tissue and an environment in which one or more cell types can be encouraged to form tissues-like constructs. Considering the advantages of microtechnologies, the main goal is to integrate these mentioned techniques to develop a 3D cell structure assembly method for constructing artificial tissues inside microfluidic devises. Our research roadmap is depicted in Fig. 9.1, and each part of the system is described as follows.

The first issue is to overcome the limitation of the current available micro tools inside microfluidic devices. We would like to develop the on-chip fabrication and manipulation of arbitrary shaped microstructures to construct functional micro tools directly inside microfluidic devices, in order to build the on-chip cell cultivation for obtaining cell samples with similar characteristics from a single original cell. The

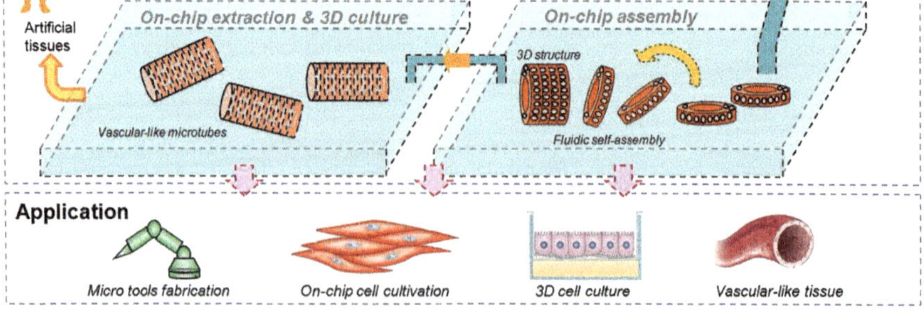

Fig. 9.1 Research roadmap of the whole 3D cell structure assembly inside microfluidic devices

optical microscope, optical tweezers and polydimethylsiloxane (PDMS) microfluidic systems are integrated in one device. Various microstructures can be fabricated and assembled to form micro tools for manipulating cells.

The second issue is to apply the cell samples for the fabrication of 2D micro cell embedded building blocks. We fabricate different shape of micro electrodes and integrate them inside the microfluidic devices. The dielectrophoresis (DEP) will be used to pattern cells. The photo-crosslinkable resin is used to encapsulate patterned cells and fabricate cell embedded components. Different shapes of 2D blocks which are movable and controllable are fabricated directly inside the device, in order to apply in different parts of the artificial tissues.

The third issue is how to make use of the fabricated cell embedded components to construct 3D structures. We develop on-chip assembly method to control the position and attitude of the 2D blocks. Fluidic assembly methods are involved inside multi-functional microfluidic devices. Our objective is to build vascular-like 3D structures based on the previous fabricated components, with high efficiency and contamination-free environment.

Finally, we extract the assembled vascular-like 3D structures outside the assembly channel for their culture. The on-chip extraction function is realized by integrating micro valve system inside the microfluidic devices. The vascular-like shape and multi-cellular condition provide the possible way to generated artificial tissues. We believe our cell assembly method will benefit both the engineering and biological fields.

9.2 On-Chip Fabrication and Manipulation of Arbitrary 2D Cell Embedded Microstructures

9.2.1 On-Chip Fabrication of Arbitrary Microstructures

To assemble 3D cell structures, the fabrication of 2D cell embedded building blocks (microstructures) is the first step. The cells inside real tissues and organs are arranged according to certain patterns and shapes, such as neural cells with line patterns, skin with reticular patterns and blood vessels with tube shapes [45]. To build artificial tissues, an important issue is how to form different patterns of cells as they are really arranged and how to immobilize these patterned cells inside certain structures for encapsulation [46].

For immobilization, the convenient approaches are based on methods such as aspiration, solution pressure and fluidic structure [47]. The advantage of aspiration and pressure is the large fixing force, while the disadvantage is the damage to the cells [48]. Using special fluidic structures, cells are immobilized inside microfluidic devices [49]. On-chip fabrication based on photo-crosslinkable resin via ultra violet (UV) illumination is a creative way of immobilizing cells [50].

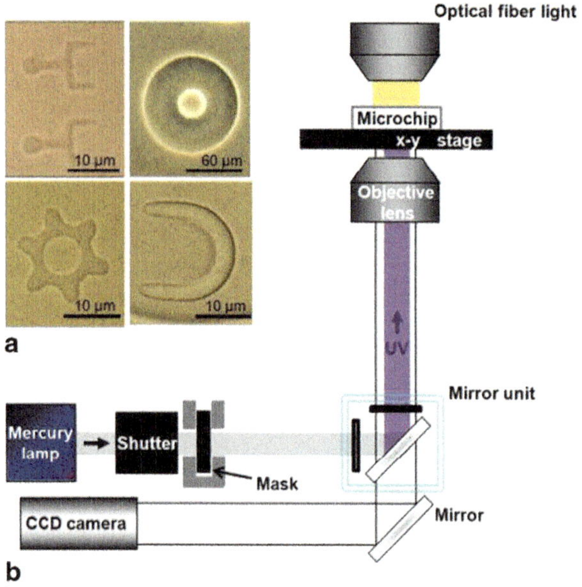

Fig. 9.2 **a** The on-chip fabricated microstructures. **b** The schematic of the on-chip UV fabrication system

We present the on-chip fabrication method for microstructures based on photo-crosslinkable resin [51]. As shown in Fig. 9.2, via the objective lens of a microscope, the patterned UV is illuminated through the mask to the photo-crosslinkable resin, which is inside the microfluidic channel made of PDMS. The microfluidic channel has both glass and PDMS surface for fabricating immovable and movable microstructures respectively. The poly(ethylene glycol) diacrylate (PEGDA) is used as a photo-crosslinkable resin. It is polymerized by the UV light which is patterned for 2D microstructures with arbitrary shapes to fabricate the structures at desired places inside a microfluidic channel [52]. PEGDA has low toxicity and good biocompatibility [53]. Besides, cell viability has been positively confirmed inside PEGDA [54]. It is not needed to prefabricate microstructures or inject the microstructures into the channel from outside. It is possible to fabricate the required microstructures at the desired places. If the cells are mixed inside the photo-crosslinkable solution, they are able to be directly immobilized inside the microstructures by UV exposure.

The fabricated microstructures are shown in Fig. 9.2a. They were fabricated by illuminating the UV-ray for 0.2 s via the objective lens of 100X and 40X. The microstructure fabricated by using 100X objective lens was about 1/100 size compared with the pattern of the mask, while it was about 1/40 size by using 40X lens.

As shown in Fig. 9.2b, the mask made by polyethylene terephthalate (PET) was set in front of the shutter. The patterns on the mask are able to be arbitrary shaped. The UV light was patterned through the mask and exposed into the PEGDA (molecular weight 700) inside the channel of PDMS microfluidic devices. The photo-crosslinkable resin was polymerized and the arbitrary shaped microstructures are fabricated inside the solution.

The position of the microstructure was controlled by the X-Y stage. When the surface of channel bottom is glass, the fabricated microstructure adheres on the glass surface. When the surface of channel bottom is PDMS, the fabricated microstructure is able to move in the non-polymerized resin freely without adhering. Because PDMS has an air permeability and it penetrates oxygen. Oxygen molecule inhibits the radical polymerization of PEGDA. The oxygen layer near the PDMS surface forms a non-polymerized PEGDA layer between the microstructure and PDMS. Thickness of non-polymerized PEGDA layer is about 2 μm. Therefore, the thickness of microstructures depends on the thickness of channel and PDMS surface. For example, inside 40 μm deep channel with both PDMS top and bottom surfaces, the thickness of fabricated microstructure is about 36 μm.

9.2.2 On-Chip Dielectrophoretic Manipulation for Cells Patterning

For manipulation and patterning, existing methods include surface adhesion, optical tweezes and DEP. By using surface adhesion, it is possible to manipulate cells with less causing damage [55]. Optical tweezers is a low-damage cell manipulation system. However, the manipulation force is weak compared with flow resistance [56]. Compared with other methods of cell manipulation, DEP is easier to control and it is non-contact [57]. Due to the low physiological stress caused by DEP, the cells remain viable after treatment and can be cultured for further purposes. By using DEP force, it is possible to manipulate particles with great selectivity. It is also possible to allow the separation of cells or the orientation and manipulation of nanoparticles and nanowires [58]. Furthermore, certain cell sheets with multiple layers are constructed by DEP. Cells aggregate and adhere with each other based on the DEP [59]. Patterned liver cell structures are also formed by DEP traps [60]. Consequently, DEP is widely used for cell manipulation, such as sorting, aggregation and patterning. For immobilization, the on-chip fabrication based on photo-crosslinkable resin via UV illumination is a creative way. Cells are directly immobilized inside the photo-crosslinkable structures. There are several advantages such as high-speed, low-cost and arbitrary shape.

For a spherical particle of radius r suspended in a medium of permittivity ε_m, the DEP force F_{DEP} is given by the expression:

$$F_{DEP} = 2\pi\varepsilon_m r^3 \,\mathrm{Re}[f_{CM}]\cdot\nabla(E_{rms}^2) \tag{9.1}$$

where E_{rms} is a root mean square value of the electric field. $\mathrm{Re}[f_{CM}]$ represents a real part of the f_{CM}, which can be represented as follows:

$$f_{CM} = \frac{\varepsilon_p^* - \varepsilon_m^*}{\varepsilon_p^* + 2\varepsilon_m^*} \tag{9.2}$$

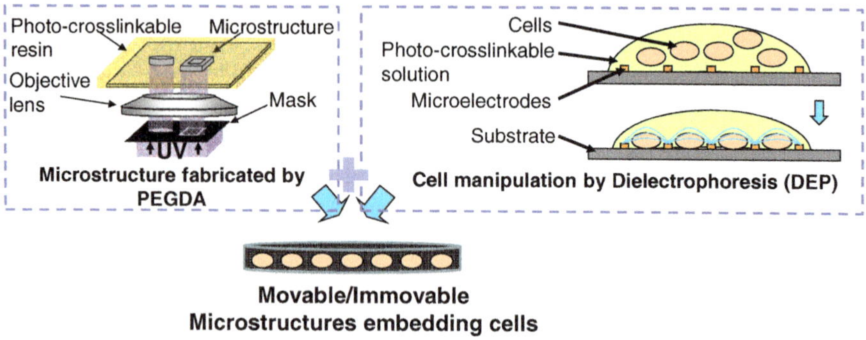

Fig. 9.3 A schematic drawing of the cell assembly method

where ε^* is a complex permittivity which is calculated by $\varepsilon^* = \varepsilon - j\sigma/\omega$, then σ is a conductivity and ω is a frequency of the electric field. Subscripts p and m represent particle and the medium, respectively. Therefore, $\text{Re}[f_{CM}] > 0$ means that the particle shows p-DEP responses so that the particle moves towards the part of the electric field with higher intensity, while $\text{Re}[f_{CM}] < 0$ represents n-DEP response and the particle moves towards the part of the electric field with lower intensity.

We present a method of manipulating particles and cells by DEP and immobilizing them using photo-crosslinkable resin inside microfluidic devices [61]. The whole research approach is shown in Fig. 9.3. The key idea is to create cell structures by combining patterning and immobilizing methods. There are mainly 3 advantages to this approach. First, the non-contact cell patterning method based on DEP is high speed and low-damage. Second, the arbitrarily shaped microstructure is effectively fabricated on-chip inside microfluidic devices. Third, the patterned particles or cells are easily immobilized inside microstructures for further assembly into complex three dimensional (3D) structures.

We developed special microelectrodes of which the size was approximately the same order as the cells. Microelectrodes are key elements in micromanipulation using DEP. Firstly, we used a Indium Tin Oxide (ITO) as a conductive material because it is transparent. The thickness of the ITO layer is 150 nm. The fabrication method was based on photolithography.

The DEP force should be generated inside the solution which means the microfluidic channel is also needed. The PDMS is one of the key components for various microfluidic applications [62]. The mold of the microfluidic channel was made with negative photoresist SU-8. The SU-8 mold was covered with PDMS liquid and cured after 24 h. After the PDMS was detached from the mold, holes were made at the ends of the channel for an inlet and an outlet. The silicone tube was put in the holes then the PDMS was attached on the glass substrate with electrodes. A solution was injected into the channel using negative pressure.

There are two DEP responses of particles: p-DEP and n-DEP. Alternating Current (AC) and NaCl water solutions were used to generate DEP force. By adjusting the frequency of the AC and the concentration of the NaCl solution (medium

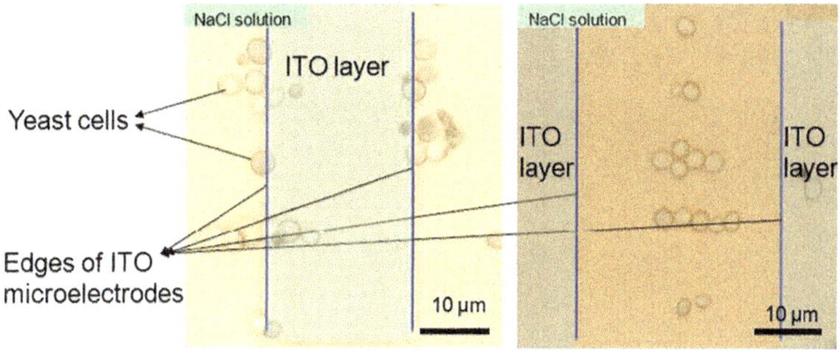

Fig. 9.4 Experiment results of positive dielectrophoresis (p-DEP) and negative dielectrophoresis (n-DEP)

Table 9.2 Different DEP responses under various medium concentrations and the AC frequencies

	100 Hz	1 kHz	30 kHz	500 kHz	1 MHz
90 mg/L	p-DEP	p-DEP	p-DEP	p-DEP	–
200 mg/L	p-DEP	n-DEP	n-DEP	n-DEP	n-DEP
400 mg/L	–	–	–	n-DEP	–

conductivity), these two DEP responses were generated. The reasons for using AC are as follows. First, operating at high frequencies eliminates any electrophoretic movement of the cell. Second, it is possible to eliminate electrochemical reactions at the electrode-electrolyte interfaces, preventing electrode corrosion and gas formation [63].

We confirmed the DEP responses of different particles experimentally such as polymer microspheres, yeast cells (*w303*) as shown in Fig. 9.4. The yeast cells were chosen as microspheres for testing the functions of our system. In order to confirm the precise experimental parameters of p-DEP and n-DEP for yeast cells, the following experimental conditions were used. To generate p-DEP, the concentration of the NaCl solution was 90 mg/L and the frequency of the AC was 1 kHz. To generate n-DEP, the concentration of the NaCl solution was 200 mg/L and the frequency of AC was 500 kHz. In both cases the Vpp of AC was 8 V. Table 9.2 shows the different DEP responses of yeast cells under various medium concentrations and the AC frequencies.

One line of 5 yeast cells was formed. As shown in Fig. 9.5, yeast cells were trapped inside the holes of which the diameter was about 10 μm. It demonstrates the capability of cell traps to manipulate and immobilize the objects in the target position without any contact and damage to the objects.

The n-DEP is often used for manipulating particles. For example, special structures of electrodes are applied to trapping cells at desired positions based on n-DEP [64]. The trapping force is calculated based on the resistance force of the cells given by the liquid flow. After one cell is trapped inside the hole, the flow velocity is

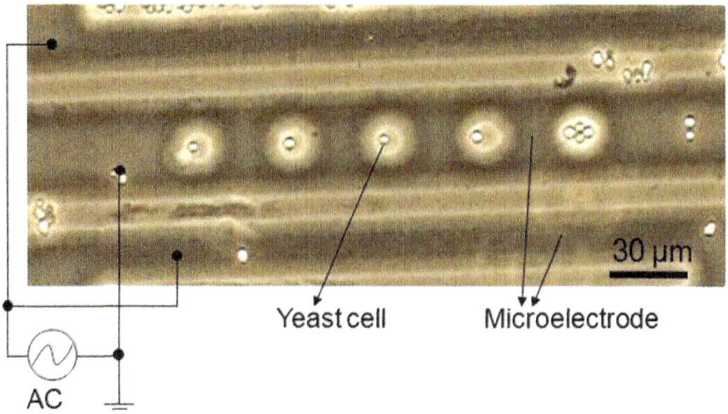

Fig. 9.5 One line of 5 cell traps and trapped yeast cells

increased continually. When the cell escapes from the trap, the trapping force is equal to the resistance force F_r which is calculated by the following expression:

$$F_r = 6\pi\eta vr \tag{9.3}$$

where η is the viscosity of solution, v is the velocity of solution and r is the radius of the cell. The trapping force by DEP in this experiment was about 6 pN.

9.2.3 On-Chip Fabrication of 2D Cell Embedded Microstructures

Based on the on-chip cell patterning method using ITO microelectrodes by DEP forces fabrication method, the fabrication of immovable microstructures embedding patterned yeast cells was conducted.

The experimental steps are follows. First, the ITO microelectrodes and the PDMS microfluidic device were prepared. The ITO microelectrodes were covered with a PDMS mold and formed a microfluidic channel. A solution containing 30% PEGDA (molecular weight 700) and 200 mg/L NaCl was prepared. The PEGDA (700) is able to dissolve in a water. Then the yeast cells (*w303*) were mixed into the solution and the solution was injected into the microfluidic channel. An AC power supply was used to generate DEP force. The frequency of the AC was 500 kHz, and the Vpp of the AC was 8 V. After the cell pattern was formed, the objective lens was adjusted to the patterned position. Then UV light pattern was exposed for 0.2 s to fabricate the microstructures at the same position as the patterned cells. The procedure is schematically shown in Fig. 9.6.

For UV light pattern, donut shaped masks were used. Therefore, after the line patterns of yeast cells were formed, donut-shaped microstructures were fabricated at the same position with UV exposure. This microstructure contains 3 lines of

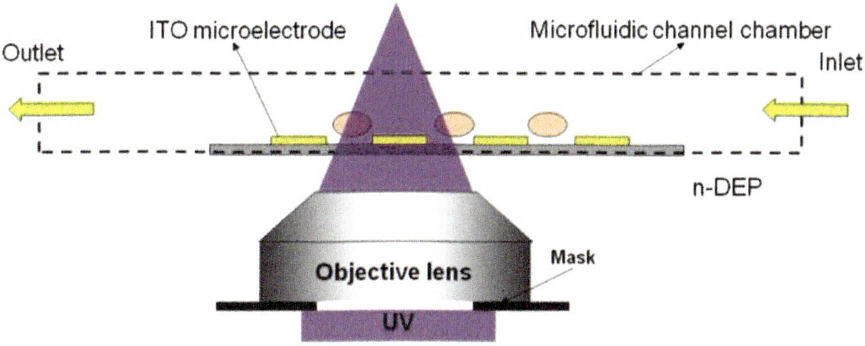

Fig. 9.6 Experimental setup for fabricating microstructures embedding patterned cells

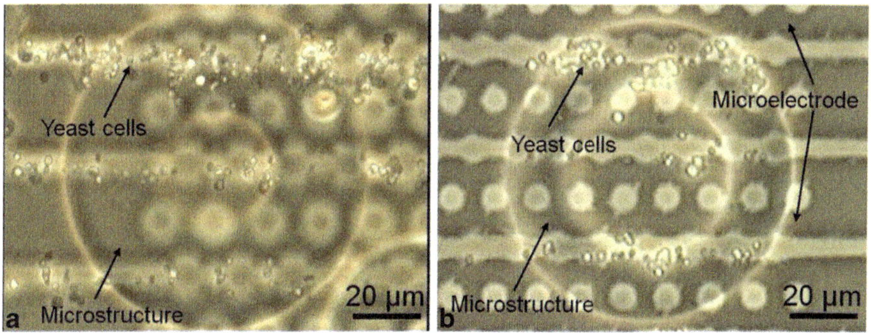

Fig. 9.7 The images of immovable microstructures embedding patterned cells

yeast cells as shown in Fig. 9.7. In Fig. 9.7a, the inner and outer diameter of the donut shapes are 35 and 100 μm respectively. There are approximately 100 yeast cells inside this microstructure. As shown in Fig. 9.7b, after cleaning cells outside the structure by a flowing solution without cells, only the immobilized cells remain clearly inside the microstructure. The pattern was generated in about 10 s and the microstructures were fabricated in 0.2 s. This high speed fabrication is easily controlled and the cells are automatically patterned. Therefore, it is possible to fabricate microstructures very quickly and immobilize large amounts of cells inside them. However, the fabricated microstructures are immovable which means it is difficult to manipulate and assemble them in order to form larger 3D structures.

Figure 9.8 shows the movable microstructures embedding beads or cells. Mammalian cell embedded movable microstructures were fabricated using mouse fibroblast NIH/3T3 cells. After NIH/3T3 cells were cultured inside an incubator, collected cells were mixed with PEGDA solution (30 % PEGDAs, 70 % Phosphate Buffered Saline (PBS)). Solution was injected inside the microfluidic chip, and then

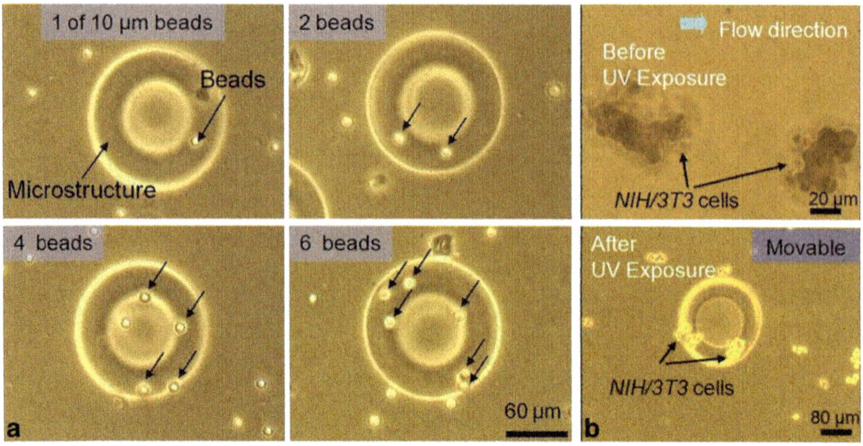

Fig. 9.8 **a** Movable microstructures embedding microbeads of which the concentration or numbers of cells are controllable. **b** Movable microstructures embedding mammalian cells (NIH/3T3)

movable 2D microstructures embedding cells (NIH/3T3) were fabricated directly inside, as shown in Fig. 9.8b.

This high speed fabrication and manipulation shows the great potential of this method for use in high-throughput cell manipulation and immobilization for tissue engineering. These movable microstructures can be further assembled to form much more complex 3D cell structures. With cell culture, these cell embedded microstructures could become functional components of artificial tissues.

9.3 On-Chip Self-Assembly of 2D Microstructures to Multilayered Microtubes

9.3.1 Fluidic Self-Assembly for Multilayered Microtubes

We present a novel method of constructing multi-layered microstructures embedding cells based on on-chip fabrication, micromanipulation system and microfluidic self-assembly inside microfluidic devices [65]. The whole procedure is with high efficiency, high accuracy and control flexibility. The concept of the self-assembly is that, the fabricated microstructures are flowed horizontally with the solution, into the micro well. Following the flow direction, the microstructures are rotated 90 degrees (axis translation from the original position to the assembly position) to be vertical inside the micro well and stopped at the end of the micro well, because they are not able to go into the micro grooves. The stopped microstructures will be assembled one by one. Finally, a tubular 3D structure embedding large amount of cells is formed inside the micro well.

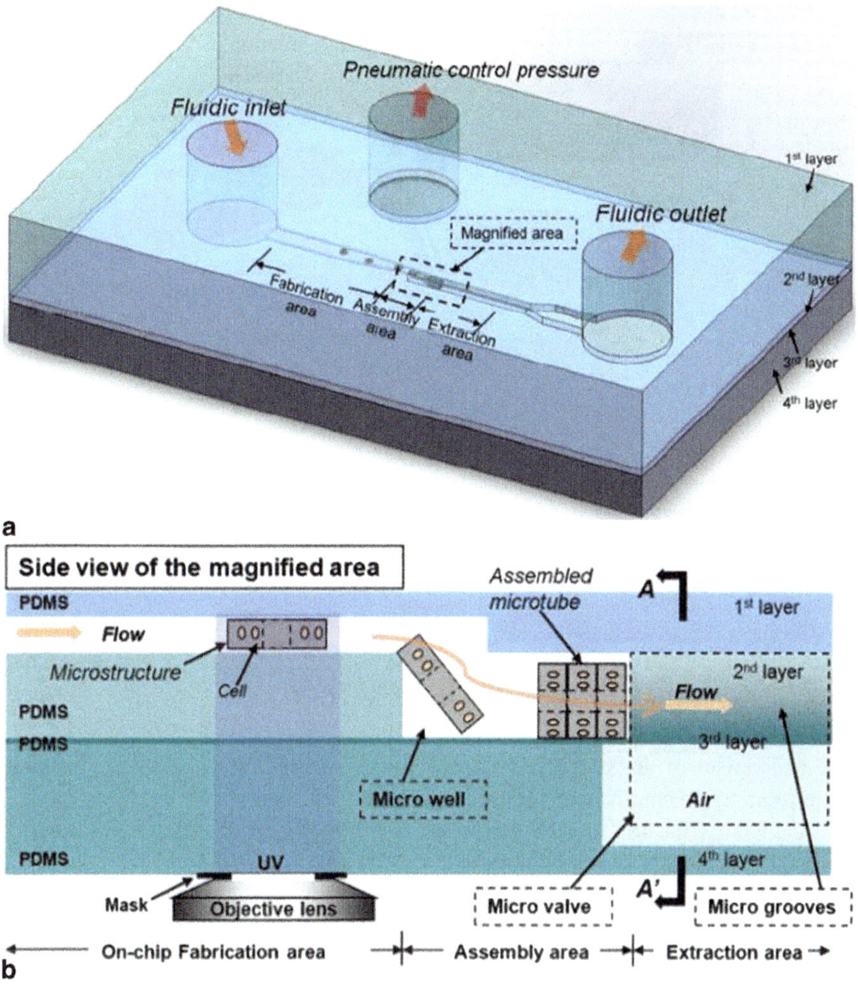

Fig. 9.9 The 4-layer PDMS microfluidic device, with fabrication, assembly and extraction areas. **a** A schematic of whole device. **b** The side view of the magnified area, the fabrication of the microstructures and the self-assembly concept inside the micro well

A microfluidic device with 4 PDMS layers is designed as shown in Fig. 9.9. The 1st layer contains a wide channel for providing a space of microstructure on-chip fabrication, as called the on-chip fabrication area. There is a micro well inside the 2nd layer for assembling the microstructures. There are 3 micro grooves inside the 3rd layer to keep the assembled microstructures in the micro well. The flow can pass through the micro grooves. This part is called the assembly area. There is a channel for acting pneumatic control in the 4th layer by applying negative pressure. The negative pressure can deform the area of micro grooves in the 3rd layer, downwards into the 4th layer. This is called a normally closed (NC) micro valve.

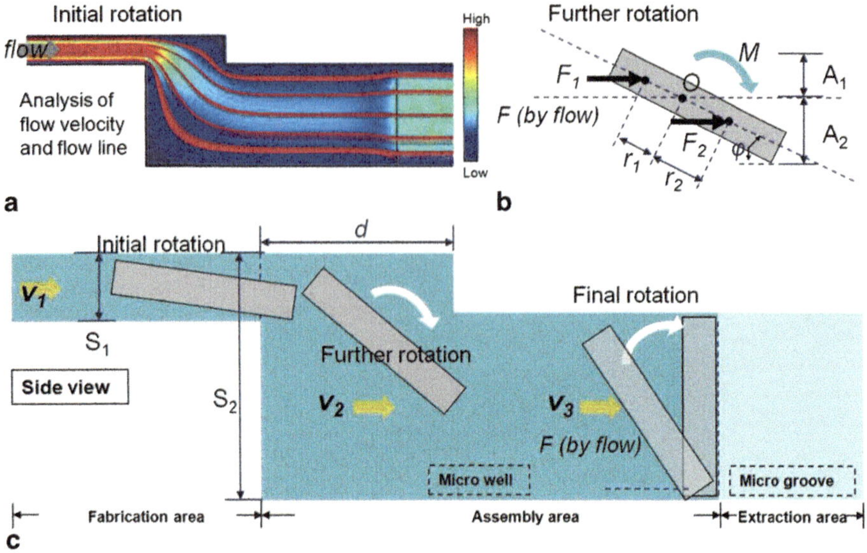

Fig. 9.10 The mechanism of the self-assembly process inside the microfluidic channel

9.3.2 Mechanism of the Self-Assembly Process

The mechanism of the self-assembly process is shown in Fig. 9.10. Its key point is the rotation mechanism of the microstructure. The fabricated microstructure is moved towards the micro well following the flow direction. The flow direction changes when it goes into the micro well as shown in Fig. 9.10a, it causes the initial rotation of the microstructure. The red lines show flow lines and the blue color area represents low flow velocity while the red area represents high. From the simulation result, the velocity difference clearly occurs at the boundary between fabrication area and assembly area. Therefore the microstructure tilts a little as Fig. 9.10b. The average flow velocity in the fabrication area is higher than the assembly area because of the different depth of the channels.

The forces which act on the microstructure in the direction of the fluid flow velocity are important. These forces are called the fluid resistance and expressed as the equation:

$$F = \frac{1}{2} C_d \rho v^2 A \tag{9.4}$$

where C_d is the drag coefficient which related with the shape of the object and the viscosity of the fluid, ρ is the density of the fluid, v is the velocity of the object relative to the fluid, and A is the cross-sectional area of the object. The fluid resistance F_1 (by flow velocity v_1) and F_2 (by flow velocity v_2) act on the different parts of the microstructure (A_1 and A_2 respectively). We also simplify the C_d as the same for these two forces. These forces generate a torque for rotating the microstructure. The rotation center is shown as point O. The torque M is calculated as the equation:

$$\begin{aligned} M &= r_1 F_1 \sin\phi - r_2 F_2 \sin\phi \\ &= \frac{1}{2} C_d \rho \sin\phi (v_1^2 r_1 A_1 - v_2^2 r_2 A_2) \end{aligned} \tag{9.5}$$

where r is the distance between the point where force F is applied and the rotation center O, and the length r is related with the acting force area A. φ is the angle between the microstructure and the horizontal plane. Obviously v_1 is larger than v_2. The r_1 and r_2 have the positive correlation with the A_1 and A_2 respectively. However, with a smaller A_1, the torque may become weaker and it will fail to rotate the microstructure. The A_1 is determined by the Depth of the fabrication channel, and the A_2 is determined by the dimension of the microstructure.

This is the secondary stage of the rotation process for the microstructure, called as a further rotation, shown in Fig. 9.10b. We consider that the further rotation is mainly generated in the place between the fabrication area and the assembly area, which is the distance d as shown in Fig. 9.10c. The cross-sectional areas S_1 and S_2 represent the relationship between v_1 and v_2 ($S_1 v_1$ equals $S_2 v_2$) under constant flow rate in the channel. The simplified different flow velocities v_1, v_2 and v_3 in the channel are shown in Fig. 9.10c. The further rotation takes the time t and the further rotation angle α can be calculated as:

$$\begin{aligned} \alpha &= \frac{1}{2}\frac{M}{I}t^2 \\ &= \frac{1}{2}\frac{\frac{1}{2}C_d \rho \sin\phi\left(v_1^2 r_1 A_1 - v_2^2 r_2 A_2\right)}{I}\left(\frac{d}{v_2}\right)^2 \\ &= \frac{1}{4}\frac{C_d \rho \sin\phi\left(S_2^2 r_1 A_1 - S_1^2 r_2 A_2\right) d^2}{I S_1^2} \end{aligned} \tag{9.6}$$

where I is the rotational inertia of the microstructure. It shows that the flow velocities v_1 and v_2 do not influence the further rotation angle α.

After the further rotation, the micro structure continues to rotate for a while because of the inertia. The microstructure is expected to be rotated less than 90 degrees before the final rotation at the end of the micro well as shown in Fig. 9.10c. With a longer d, the further rotation angle will be larger and it may be over 90 degrees. This may cause a failure of the final rotation, especially in the case of 180 degrees. As the rotation dynamic analysis is complicated, the above theoretical analysis is a simplified model for the qualitative analysis. It contributes to the following experiments with different parameters which influence the assembly results.

After a whole successful rotation, the horizontal microstructure becomes vertical and it is assembled at the end of the micro well. As this assembly process is performed by the fluid inside the microchannel, we call it the microfluidic self-assembly.

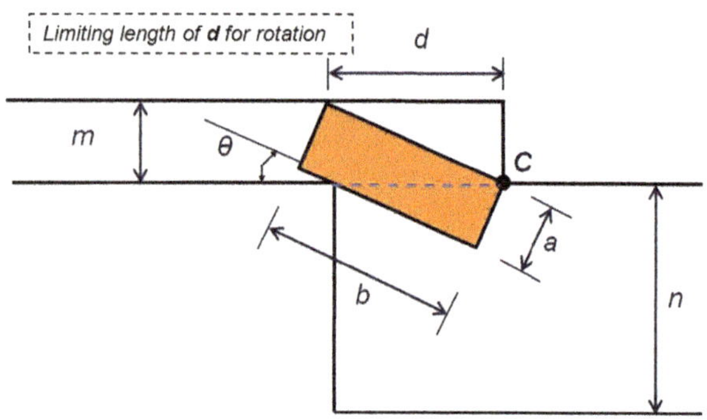

Fig. 9.11 The calculation of the opening length between fabrication area and assembly area, and the minimum length of *d* for microstructures to completely flow into the micro well (*n* is designed according *b*)

In the 2-layer microfluidic channels, a key parameter is the opening length *d* between the fabrication area and the assembly area as shown in Fig. 9.11. The rotation of the microstructures when they flow into the micro well is influenced by the length *d*. There is a minimum length of *d* for the microstructures to completely flow into the micro well. When the corner of microstructure overlaps the corner of channel at the point *C*, and the other corner of microstructure contacts the top of the channel simultaneously, this is a limiting position of the rotation which represents the minimum length of *d*, which is d_{min}. Hence, the d_{min} is calculated by the equation:

$$d_{\min} = ab / m \tag{9.7}$$

The thickness of the microstructure is controlled by the depth of the fabrication channel *m*.

$$a = m - 6 \tag{9.8}$$

It means that the d_{min} is related with the size of microstructure.

A microstructure was fabricated for testing this d_{min}. The 3 types of channels with different opening length *d* were used in the testing experiments. The flow velocity v_2 of these 3 cases was controlled as the same value which was about 25 μm/s. In 120 μm length case which is shorter than minimum length 169 μm, the microstructure did not flow into the micro well. In 170 μm case which is almost the same length as 169 μm, the microstructure flowed into the micro well and rotated about 90 degrees. In 220 μm case which is longer than 169 μm, microstructure rotated more than 90 degrees inside the micro well. A reason of the over rotation is that the *d* is too long. The microstructure rotated too much during going into the micro well. Based on these theoretical calculation and experimental results, the opening length of about 170 μm was characterized as an acceptable parameter for the experimental devices.

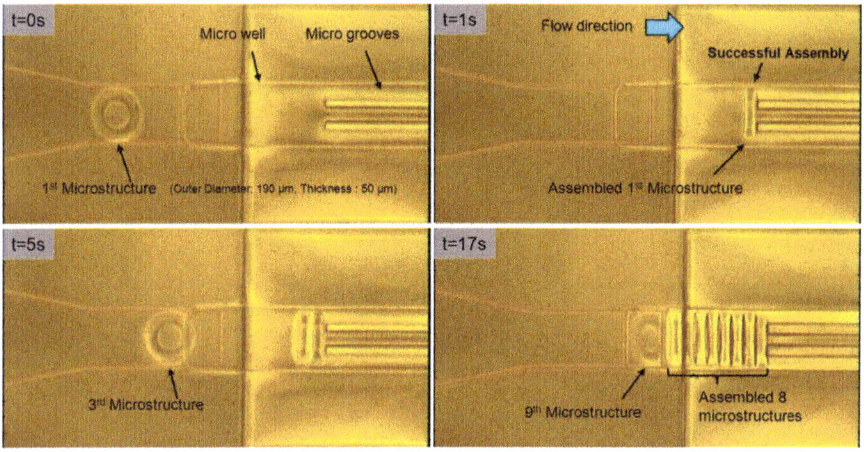

Fig. 9.12 The successful assembly results of 190 μm diameter microstructures with high-efficiency

9.3.3 Evaluation of the Microfluidic Self-Assembly

The on-chip fabrication and microfluidic self-assembly were conducted inside the device. The experimental results of self-assembly are shown in Fig. 9.12. Movable microstructures were continually fabricated inside upper part channel, then they were moved with solution flow into the micro well of bottom part. The inner hollow of this tube-shaped structure is shown clearly in these micrographs. The great advantage of this method is that the whole process is done inside microfluidic channel without any manual manipulation.

Microstructures were continuously fabricated in the fabrication area and then moved with flow into the assembly area. As shown in Fig. 9.12, the fabricated first microstructure successfully went into the micro well and was assembled within 1 s. After 18 s, 9 microstructures were assembled as a tubular structure.

Self-assembly experiments under various flow velocities were conducted to test the effects of the flow velocity. The assembly success rate is defined as the percentage of successful cases in the total assembly cases. By using the microstructures with same size, the velocity difference showed no obvious influence to the assembly success rate. The results are shown in Fig. 9.13a. Considering Eq. (9.6), the experimental results demonstrated that the rotation angle almost did not change under different flow velocities.

Based on Eqs. (9.5) and (9.6), these assembly failures are mainly influenced by the A_1 and A_2, which are the force acting areas related with the dimensions of the microstructure. For the dimensions of microstructures, there are 3 parameters which are the outer diameter (OD.), the inner diameter (ID.) and the thickness. The thickness is controlled by the depth of the fabrication channel which determines the

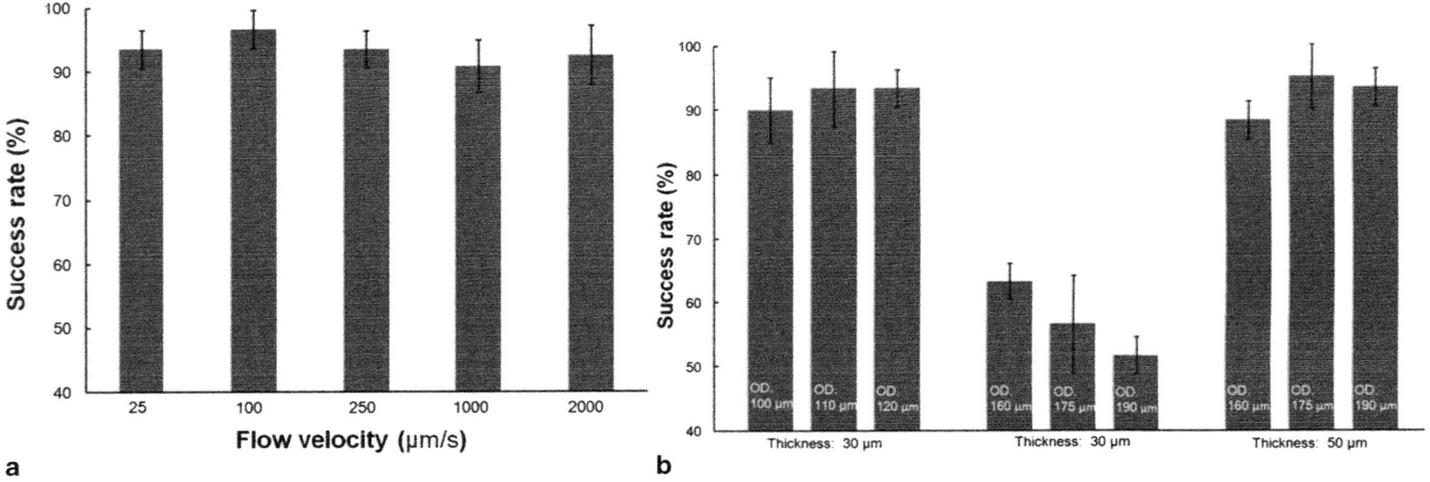

Fig. 9.13 Evaluation results of on-chip self-assembly **a** The evaluation of the relationship between the assembly success rate and the flow velocity. Microstructures with 190 μm in outer diameter were used. The results demonstrated that the rotation angle almost did not change under different flow velocities. **b**The evaluation of the relationship between the assembly success rate and the dimensions of the microstructure. When the ratio of outer diameter to thickness is larger than 5, the success rate decreases tremendously

A_1. The A_2 is mainly influenced by the diameters of the microstructures. However, we design that the inner diameter is always the half size of the outer diameter. It is considered that the outer diameter determines the A_2. Therefore, outer diameter and thickness were evaluated in our experiments.

An evaluation of the relationship between the assembly success rate and the dimensions of the microstructure was conducted as shown in Fig.9.13b. The results are shown in 3groups including 2 different thickness. There were 3 different outer diameters in one group of thickness. All of these cases were conducted under the flow velocity about 250 μm/s. When the thickness was the same, the smaller outer diameter ones showed higher success rate. When the diameter was the same, the larger thickness ones showed higher success rate. The results indicate that when the ratio of outer diameter to thickness is larger than 5, the success rate decreases tremendously.

As shown in Eqs. (9.5) and (9.6), one explanation is that the small thickness represents the small A_1, while the large outer diameter represents the large A_2. It means that the rotation torque M may become smaller and then the rotation of microstructure fails. This evaluation result contributes to determine the better size of microstructures for assembling the microtubes.

9.4 On-Chip Construction of Cellular Vascular-like Microtubes

9.4.1 4-Layer Multi-functional PDMS Microfluidic Device

We demonstrated the on-chip fabrication with extraction of assembled structures from the microfulidic channel. This novel microfluidic device was fabricated with 4 PDMS layers. The fabrication, self-assembly of cell embedded microstructures, and the extraction of assembled tubular structures are all conducted inside this device. It is a reusable microfluidic device and the production of cell embedded microtubes can be conducted continuously.

We proposed microgrooves at NC valve just after the assembly area. This area is named as extraction area, as shown in Fig. 9.14. On-chip extraction was conducted by opening the NC valve. Under the closing condition, it is possible to stop the fabricated microstructures, but release the solution. With the negative pressure, the micro grooves are deformed downwards and the assembled microtube is extracted from the assembly area. The micro valve is commonly used in microfluidic devices for controlling flow [66]. The NC valve is suitable for generating continuously closing of the channel, and also the temporary opening of the channel under the pressure control [67, 68]. Via the opening of the NC valve, the assembled vascular-like microtubes are able to be extracted from the device and transported to the culture

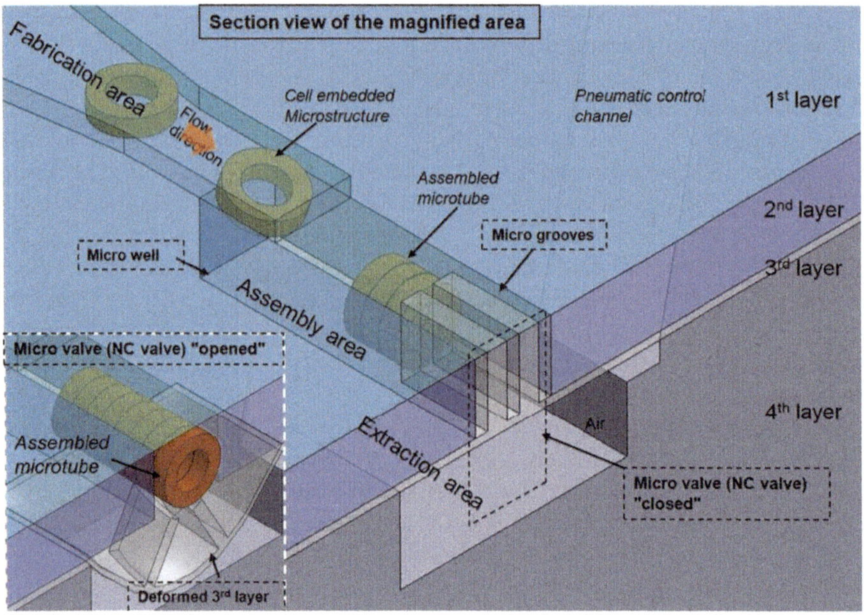

Fig. 9.14 Section view of the magnified area cut at A-A'. When the NC valve is closed without the negative pressure in 4th layer, the self-assembly was conducted in this status. When the NC valve is opened with the negative pressure in 4th layer, the extraction of the assembled microtube was conducted in this status

dishes for further treatments and applications. After extracting this microtube via the micro valve, the negative pressure is tuned off to close the valve, and the new microstructures are able to be fabricated and assembled to construct a new microtube structures.

The on-chip extraction for the microtube was conducted, as shown in Fig. 9.15a. The microtube was stopped by the micro grooves. Then the negative pressure was applied through the pneumatic control channel in the 4th layer to the NC valve area. The micro grooves above the pneumatic control channel of 3rd layer were deformed downwards. After the vertical deformation of 3rd layer was larger than 240 μm, the 3rd layer contacted with 4th layer. By enhancing the negative pressure, the elliptic contact area expanded. The length of the major axis of the elliptic contact area was called the valve opening length. When the valve opening length was longer than the length of the micro grooves, the micro grooves were fully sank in the 4th layer and the valve was fully opened for extracting the microtube. The extraction took about 40 s and the whole microtube was successfully extracted. Figure 9.15b shows the relationship between the applied negative pressure and the valve opening percentage. As the length of micro grooves is 1200 μm, the valve opening percentage is

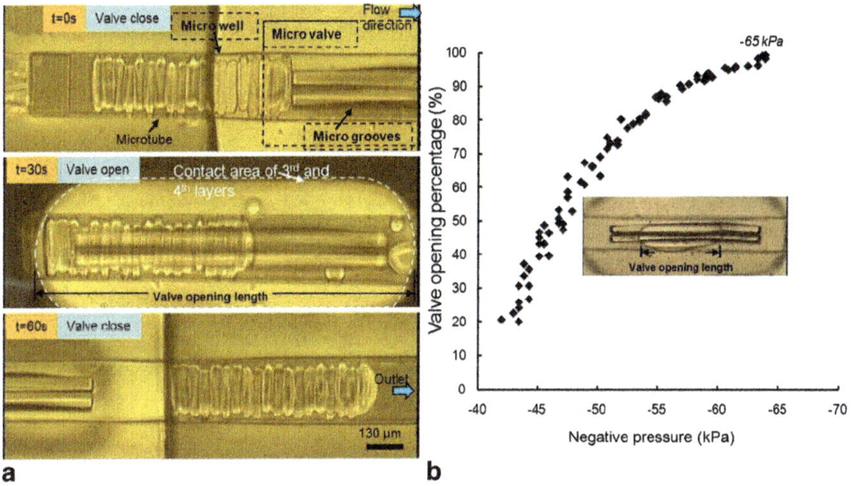

Fig. 9.15 **a** The NC valve was closed for assembly and opened for extraction of the assembled microtube. **b** The relationship between the applied negative pressure and the valve opening percentage. When the negative pressure was lower than −65 kPa, the NC valve was able to be fully opened

defined as the percentage of the valve opening length on 1200 μm. For this device, the negative pressure 65 kPa is needed for fully opening the NC valve and conducting the extraction process.

9.4.2 Assembly of Cell Embedded Microstructures to Microtubes

In order to confirm that the presented method was able to be used in biological applications, cell viability tests were conducted after embedding cells inside the donut-shaped microstructures. The cell embedded microstructures using NIH/3T3 cells were washed by Phosphate Buffered Saline (PBS) then they were immersed in the viability test solution which was the mixture of 10 μL Calcein AM (1 mg/mL), 15 μL Propidium iodide (PI, 1 mg/mL) and 5 mL PBS. The test solution with the microstructures was put inside an incubator (37 °C, 5 % CO_2) for 30 min. After that, the test solution was removed and the microstructures were washed by PBS again. The fluorescent images of the testing results are shown as Fig. 9.16a. The green represents the live cells while the red means the dead cells. Different dimensions of donut-shaped microstructures were used. Figure 9.16b shows that the relative cell viability of all the samples is higher than 80 %, which confirms that they are to be used in biological applications.

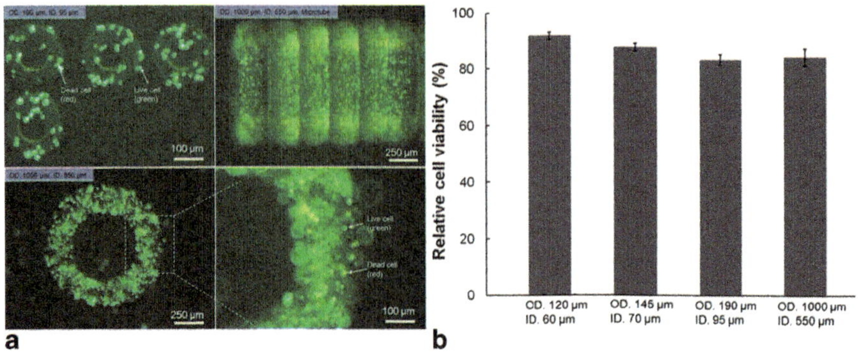

Fig. 9.16 The viability tests for the cells embedded inside microstructures. **a** The green represents the live cells while the red means the dead ones. **b**The relative cell viability of the cells embedded inside microstructures

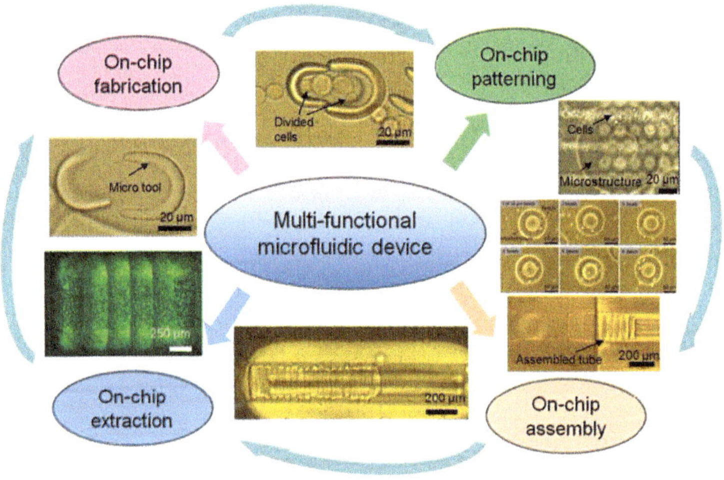

Fig. 9.17 The integrated multi-functional microfluidic system, with the on-chip fabrication, on-chip patterning, on-chip extraction and on-chip assembly

9.5 Conclusion

In this chapter, we focused on building 3D cell structures inside multifunctional microfluidic devices based on a bottom-up method, as shown in Fig. 9.17. It is used as a "factory line on a chip", including on-chip fabrication, on-chip patterning, on-chip assembly and on-chip extraction.

A novel 4-layer PDMS microfluidic device is fabricated, including fabrication area, self-assembly area and extraction area inside one channel. The on-chip fabrication and assembly of microstructures based on photo-crosslinkable resin is presented. The movable microstructures embedding patterned particles and cells are fabricated inside the microfluidic channel, as the 2D blocks for building 3D structures. The novel fluidic self-assembly of 2D microstructures based on axis-translation process is presented and experimentally demonstrated. The NIH/3T3 cells embedded microtubes are self-assembled and directly extracted outside the channel by the NC micro valve system. The 3D vascular-like microtubes are constructed by our system, with high efficiency and a contamination-free environment. This technology will contribute the construction of functional 3D cell tissues.

The fabricated structures will contribute the construction of functional 3D tissues.

References

1. Langer R, Vacanti JP (1993) Tissue engineering. Science 260:920–926
2. Stock UA, Vacanti JP (2001) Tissue engineering: current state and prospects. Annu Rev Med 52:443–451
3. Vacanti CA (2006) The history of tissue engineering. J Cell Mol Med 10:569–576
4. Emmert MY, Wolint P, Wickboldt N, Gemayel G, Weber B, Brokopp CE, Boni A, Falk V, Bosman A, Jaconi ME, Hoerstrup SP (2013) Human stem cell-based three-dimensional microtissues for advanced cardiac cell therapie. Biomaterials 34:6339–6354
5. Karamichos D, Rich CB, Hutcheon AEK, Ren R, Saitta B, Trinkaus-Randall V, Zieske JD (2011) Self-Assembled matrix by umbilical cord stem cells. J Funct Biomater 2:213–229
6. Mironov V, Boland T, Trusk T, Forgacs G, Markwald RR (2003) Organ printing: computer-aided jet-based 3D tissue engineering. Trends Biotechnol 21:157–161
7. Ishizaki T, Saito N, Takai O (2010) Correlation of cell adhesive behaviors on superhydrophobic, superhydrophilic, and micropatterned superhydrophobic/superhydrophilic surfaces to their surface chemistry. Langmuir 26:8147–8154
8. Lewpiriyawong N, Yang C, Lam YC (2008) Dielectrophoretic manipulation of particles in a modified microfluidic H filter with multi-insulating blocks. Biomicrofluid 2
9. Dendukuri D, Pregibon DC, Collins J, Hatton TA, Doyle PS (2006) Continuous-flow lithography for high-throughput microparticle synthesis. Nat Mater 5:365–369
10. Elliott NT, Yuan F (2011) A review of three-dimensional in vitro tissue models for drug discovery and transport studies. J Pharm Sci 100:59–74
11. Pampaloni F, Reynaud EG, Stelzer EHK (2007) The third dimension bridges the gap between cell culture and live tissue. Nature Rev Mol Cell Biol 8:839–845
12. Mazzoleni G, Lorenzo D, Steimberg N (2009) Modelling tissues in 3D: the next future of pharmaco-toxicology and food research? Genes Nutr 4:13–22
13. Gartner ZJ, Bertozzi CR (2009) Programmed assembly of 3-dimensional microtissues with defined cellular connectivity. Proc Natl Acad Sci 106:4606–4610
14. Huh D, Hamilton GA, Ingber DE (2011) From 3D cell culture to organs-on-chips. Trends Cell Biol 21:745–754
15. Derda R, Laromaine A, Mammoto A, Tang SKY, Mammoto T, Ingber DE, Whitesides GM (2009) Paper-supported 3D cell culture for tissue-based bioassays. Proc Natl Acad Sci 106:18457–18462
16. Moon S, Hasan SK, Song YS, Xu F, Keles HO, Manzur F, Mikkilineni S, Hong JW, Nagatomi J, Haeggstrom E, Khademhosseini A, Demirci U (2010) Layer by layer three-dimensional tissue epitaxy by cell-laden hydrogel droplets. Tissue Eng Part C-Methods 16:157–166
17. Nichol JW, Khademhosseini A (2009) Modular tissue engineering: engineering biological tissues from the bottom up. Soft Matter 5:1312–1319

18. Niklason LE, Gao J, Abbott WM, Hirschi KK, Houser S, Marini R, Langer R (1999) Functional arteries grown in vitro. Science 284:489–493
19. Hahn MS, Miller JS, West JL (2006) Three-dimensional biochemical and biomechanical patterning of hydrogels for guiding cell behavior. Adv Mater 18:2679–2684
20. L'Heureux N, Paquet S, Labbe R, Germain L, Auger FA (1998) A completely biological tissue-engineered human blood vessel. Faseb J 12:47–56
21. Yeh J, Ling Y, Karp JM, Gantz J, Chandawarkar A, Eng G, Blumling III, J, Langer R, Khademhosseini A (2006) Micromolding of shape-controlled, harvestable cell-laden hydrogels. Biomaterials 27:5391–5398
22. Koh WG, Pishko MV (2006) Fabrication of cell-containing hydrogel microstructures inside microfluidic devices that can be used as cell-based biosensors. Anal Bioanal Chem 385:1389–1397
23. L'Heureux N, McAllister TN, de la Fuente LM (2007) Tissue-engineered blood vessel for adult arterial revascularization. New Engl J Med 357:1451–1453
24. Du Y, Lo E, Ali S, Khademhosseini A (2008) Directed assembly of cell-laden microgels for fabrication of 3D tissue constructs. Proc Natl Acad Sci 105:9522–9527
25. Liu Tsang V, Chen AA, Cho LM, Jadin KD, Sah RL, DeLong S, West JL, Bhatia SN (2007) Fabrication of 3D hepatic tissues by additive photopatterning of cellular hydrogels. FASEB J 21:790–801
26. Cao YL, Vacanti JP, Paige KT, Upton J, Vacanti CA (1997) Transplantation of chondrocytes utilizing a polymer-cell construct to produce tissue-engineered cartilage in the shape of a human ear. Plast Reconstr Surg 100:297–302
27. Roh JD, Sawh-Martinez R, Brennan MP, Jay SM, Devine L, Rao DA, Yi T, Mirensky TL, Nalbandian A, Udelsman B, Hibino N, Shinoka T, Saltzman WM, Snyder E, Kyriakides TR, Pober JS, Breuer CK (2010) Tissue-engineered vascular grafts transform into mature blood vessels via an inflammation-mediated process of vascular remodeling. Proc Natl Acad Sci 107:4669–4674
28. Krawiec JT, Vorp DA (2012) Adult stem cell-based tissue engineered blood vessels: a review. Biomaterials 33:3388–3400
29. Bomhard AV, Veit J, Bermueller C, Rotter N, Staudenmaier R, Storck K, The HN (2013) Prefabrication of 3D cartilage contructs: towards a tissue engineered auricle—a model tested in Rabbits. Plos One 8:e71667
30. Matsunaga YT, Morimoto Y, Takeuchi S (2011) Molding cell beads for rapid construction of macroscopic 3D tissue architecture. Adv Mater 23:H90–H94
31. Sakar MS, Neal D, Boudou T, Borochin MA, Li Y, Weiss R, Kamm RD, Chen, Asada, H (2012) H.: Formation and optogenetic control of engineered 3D skeletal muscle bioactuators. Lab Chip 12:4976–4985
32. Chung SE, Park W, Shin S, Lee SA, Kwon S (2008) Guided and fluidic self-assembly of microstructures using railed microfluidic channels. Nat Mater 7:581–587
33. Xu F, Finley TD, Turkaydin M, Sung Y, Gurkan UA, Yavuz AS, Guldiken RO, Demirci U (2011) The assembly of cell-encapsulating microscale hydrogels using acoustic waves. Biomaterials 32:7847–7855
34. Billiet T, Vandenhaute M, Schelfhout J, Van Vlierberghe S, Dubruel P (2012) A review of trends and limitations in hydrogel-rapid prototyping for tissue engineering. Biomaterials 33:6020–6041
35. Lu T, Li Y, Chen T (2013) Techniques for fabrication and construction of three-dimensional scaffolds for tissue engineering. Intl J Nanomed 8:337–350
36. Yamato M, Okano T (2004) Cell sheet engineering. Mater Today 7:42–47
37. Yang J, Yamato M, Kohno C, Nishimoto A, Sekine H, Fukai F, Okano T (2005) Cell sheet engineering: Recreating tissues without biodegradable scaffolds. Biomaterials 26:6415–6422
38. Ohashi K, Yokoyama T, Yamato M, Kuge H, Kanehiro H, Tsutsumi M, Amanuma T, Iwata H, Yang J, Okano T, Nakajima Y (2007) Engineering functional two- and three-dimensional liver systems in vivo using hepatic tissue sheets. Nat Med 13:880–885

39. Jakab K, Norotte C, Marga F, Murphy K, Vunjak-Novakovic G, Forgacs G (2010) Tissue engineering by self-assembly and bio-printing of living cells. Biofabrication 2(2):022001
40. Shimoto T, Nakayama K, Matsuda S, Iwamoto Y (2012) Building of HD MACs using cell processing robot for cartilage regeneration. J Robot Mechatron 24:347–353
41. Forgacs G, Sun W (2013) Biofabrication: micro-and nano-fabrication, printing, patterning and assemblies. Elsevier Inc.
42. Shimoto T, Hidaka N, Sasaki H, Nakayama K, Akieda S, Matsuda S, Miura H, Iwamoto Y (2013) Bio rapid prototyping project: development of spheroid formation system for regenerative medicine. Information Technology Convergence, vol 253. Information Technology Convergence, Springer Netherlands, pp 855–862
43. Kuribayashi-Shigetomi K, Onoe H, Takeuchi S (2012) Cell origami: self-folding of three-dimensional cell-laden microstructures driven by cell traction force. Plos One 7(12):e51085
44. Kato-Negishi M, Morimoto Y, Onoe H, Takeuchi S (2013) Millimeter-sized neural building blocks for 3D heterogeneous neural network assembly. Adv Health Mater 2:1564–1570
45. Suuronen EJ, Sheardown H, Newman KD, McLaughlin CR, Griffith M (2005) Building in vitro models of organs. Intl Rev Cytol 244:137–173. (Kwang WJ ed, Academic Press)
46. Khetani SR, Bhatia SN (2006) Engineering tissues for in vitro applications. Curr Opin Biotech 17:524–531
47. Tsutsui H, Yu E, Marquina S, Valamehr B, Wong I, Wu H, Ho CM (2010) Efficient dielectrophoretic patterning of embryonic stem cells. Ann Biomed Eng 38:3777–3788
48. Tixier-Mita A, Jun J, Ostrovidov S, Chiral M, Frenea M, Le Pioufle B, Fujita H (2004) A silicon micro-system for parallel gene transfection into arrayed cells. Proc. of 8th Intl. Conf. Miniaturized System for Chem. and Life Sciences (uTAS 2004), pp 180–182
49. Di Carlo D, Aghdam N, Lee LP (2006) Single-cell enzyme concentrations, kinetics, and inhibition analysis using high-density hydrodynamic cell isolation arrays. Anal Chem 78:4925–4930
50. Maruyama H, Arai F, Fukuda T (2007) Gel-tool sensor positioned by optical tweezers for local pH measurement in a microchip. Proc. of 2007 IEEE Intl. Conf. Robot. Autom., (ICRA 2007) pp 806–811
51. Yue T, Nakajima M, Takeuchi M, Fukuda T (2014) Improved laser manipulation for on-chip fabricated microstructures based on solution replacement and its application in single cell analysis. Int J Adv Robot Syst 11:11
52. Ito M, Nakajima M, Maruyama H, Fukuda T (2009) On-chip fabrication and assembly of rotational microstructures. Proc. of 2009 IEEE/RSJ Intl. Conf. Intell. Robot. Syst., (IROS 2009) pp 1849–1854
53. Panda P, Ali S, Lo E, Chung BG, Hatton TA, Khademhosseini A, Doyle PS (2008) Stop-flow lithography to generate cell-laden microgel particles. Lab Chip 8:1056–1061
54. Chan V, Zorlutuna P, Jeong JH, Kong H, Bashir R (2010) Three-dimensional photopatterning of hydrogels using stereolithography for long-term cell encapsulation. Lab Chip 10:2062–2070
55. Bernstein RW, Scott M, Solgaard O (2004) BioMEMS for high-throughput handling and microinjection of embryos. Mems/Moems Technol Appl 5641:67–73. Ma Z et al., eds, Bellingham: Spie-Int Soc Optical Engineering
56. Hu S, Sun D (2011) Automatic transportation of biological cells with a robot-tweezer manipulation system. Int J Rob Res 30:1681–1694
57. Jaeger MS, Uhlig K, Schnelle T, Mueller T (2008) Contact-free single-cell cultivation by negative dielectrophoresis. J Phys D-Appl Phys 41:175502
58. Zhang C, Khoshmanesh K, Mitchell A, Kalantar-zadeh K (2010) Dielectrophoresis for manipulation of micro/nano particles in microfluidic systems. Anal Bioanal Chem 396:401–420
59. Sebastian A, Buckle AM, Markx GH (2007) Tissue engineering with electric fields: immobilization of mammalian cells in multilayer aggregates using dielectrophoresis. Biotechnol Bioeng 98:694–700
60. Ho CT, Lin RZ, Chang WY, Chang HY, Liu CH (2006) Rapid heterogeneous liver-cell on-chip patterning via the enhanced field-induced dielectrophoresis trap. Lab Chip 6:724–734

61. Yue T, Nakajima M, Tajima H, Fukuda T (2013) Fabrication of microstructures embedding controllable particles inside dielectrophoretic microfluidic devices. Int J Adv Robot Syst 10:132
62. Scharnweber T, Truckenmuller R, Schneider AM, Welle A, Reinhardt M, Giselbrecht S (2011) Rapid prototyping of microstructures in polydimethylsiloxane (PDMS) by direct UV-lithography. Lab Chip 11:1368–1371
63. Voldman J, Gray ML, Toner M, Schmidt MA (2002) A microfabrication-based dynamic array cytometer. Anal Chem 74:3984–3990
64. Mittal N, Rosenthal A, Voldman J (2007) NDEP microwells for single-cell patterning in physiological media. Lab Chip 7:1146–1153
65. Yue T, Nakajima M, Takeuchi M, Hu C, Huang Q, Fukuda T (2014) On-chip self-assembly of cell embedded microstructures to vascular-like microtubes. Lab Chip 14:1151–1161
66. Unger MA, Chou HP, Thorsen T, Scherer A, Quake SR (2000) Monolithic microfabricated valves and pumps by multilayer soft lithography. Science 288:113–116
67. Schudel BR, Choi CJ, Cunningham BT, Kenis PJA (2009) Microfluidic chip for combinatorial mixing and screening of assays. Lab Chip 9:1676–1680
68. Mohan R, Schudel BR, Desai AV, Yearsley JD, Apblett CA, Kenis PJ (2011) Design considerations for elastomeric normally closed microfluidic valves. Sensor Actuat B: Chem 160:1216–1223

Chapter 10
Fabrication of 3D Cellular Tissue Utilizing MEMS Technologies

Shotaro Yoshida, Daniela Serien, Fumiaki Tomoike, Hiroaki Onoe and Shoji Takeuchi

Abstract This chapter focuses on the use of micro-sized cell culture substrates fabricated by micro electro mechanical systems (MEMS) technologies in the field of three-dimensional (3D) cellular tissue constructions. First, we provide a brief overview of MEMS-based tissue engineering methods, and explain why the micro-sized cell-laden substrates are important. Second, we explain the fabrication processes of the micro-sized cell-laden substrates. The fabrication of micro-sized substrates is described by focusing on their materials, structures, cell culture processes, and handling. Third, their applications for single cell handling is described by providing information on observation, collection, and arrangement. Finally, assembly of 3D cellular constructs is explained by pick-and-place assembly, self-assembly, and self-folding methods. Key advantages of the micro-sized substrates are the ability of manipulating various types of adherent cells and also spatial control of cells in 3D cellular tissues. The single cell handling ability facilitates the cell-cell interaction analysis in the field of pharmacology, and 3D cellular assembly ability will open the route for fabrication of spatially-controlled heterogeneous 3D cellular tissues in the regenerative medicine field.

Keywords MEMS · 3D structure · Microplate · Self assembly · Cell-origami · Single cell observation · Cell handling

10.1 Introduction

Micro electro mechanical systems (MEMS) are the small-sized (~ 100 μm) machinery systems made of electrical and mechanical components. MEMS technologies have advanced with the intensively developed microfabrication techniques in the field of semiconductor integrated circuits. Main applications of MEMS technologies are micro-sized sensors and actuators such as pressure sensors, inkjet heads, and accelerometers [1]. In addition to these physical sensors and actuators, the

S. Takeuchi (✉) · S. Yoshida · D. Serien · F. Tomoike · H. Onoe
Institute of Industrial Science, The University of Tokyo, Tokyo, Japan
e-mail: takeuchi@iis.u-tokyo.ac.jp

T. Arai et al. (eds.), *Hyper Bio Assembler for 3D Cellular Systems*,
DOI 10.1007/978-4-431-55297-0_10

MEMS technologies have broadened their applications to chemical and biological research fields [2]. MEMS technology-based micro-sized systems are suitable for handling cells due to the similarity in size. Fabrication of biosensors or medical devices on a chip, so-called micro total analysis systems (μ-TAS) or lab on a chip, has emerged. In particular, handling of cells leads to three-dimensional (3D) cell culture and fabrication of cellular tissues, which is eagerly required in the field of tissue engineering for in vitro pharmacologic models and transplant therapy [3].

The fabrication of 3D cellular tissue was conventionally pursued by using 3D scaffold-based "top-down" methods [4, 5]. 3D scaffolds allow cells to adhere on their surfaces, which serve as engineered tissues. However, spatial control of cells is difficult since cells are randomly seeded on the scaffolds. More precise handling of cells is needed to create spatially controlled cellular tissues for mimicking native tissues. In this respect, microfabrication of cell-laden microparts (e.g., blocks [6], sheets [7], fibers [8, 9], spheroids [10, 11]) are promising methods for their ability in handling single to several hundreds of cells. 3D assembly of the micro-sized cellular building blocks, recognized as "bottom-up" methods, enables spatially heterogeneous 3D cellular constructs.

Among various types of microparts, cell-laden planar microparts [12–35] have specific advantages. The planar microparts allow culturing adherent cells on their surfaces, and handling by micromanipulation techniques. The advantages of the planar microparts are handleability of adherent cells maintaining cellular adhesions and morphology, and applicability for the assembly of spatially-controlled 3D structures. Conventional micro-object handling techniques are applicable such as microfluidic techniques [36–41], micromanipulation techniques [42–44], chemical/physical cell-patterning techniques [45–48], and self assembly techniques [13, 49–51]. These advantages open the path for the fabrication of spatially-controlled heterogeneous 3D cellular tissue models.

In this chapter, we provide the overview of fabrication techniques of cell-laden planar microparts and their applications from single cell handling to 3D cellular assembly.

10.2 Fabrication of Microsized Cell Culture Substrates

Unlike common cell culture substrates such as cell culture dishes and flasks, the microsized cell culture substrates enable controlled culture and handling of specific numbers of adherent cells. Fabrication of the micro-sized substrate requires microfabrication techniques established in the MEMS field. In this section, we explain the various designs and structures of the planar microparts, their material compositions, microfabrication processes, cell culture methods, and micropart handling methods.

10.2.1 Design

The planar microparts must meet the following requirements. First, the microparts need to be processed into micro-sized structure. Second, materials of the microparts must be biocompatible and transparent to facilitate cell viability and their observation, respectively. Third, top surface of the microparts must have cell-adhesive properties to enhance cellular adhesion. Finally, the microparts should be easy to release from base substrates for handling. We describe dimensions and layered structures of the microparts to meet the above mentioned requirements.

Dimensions Typical dimensions of the planar microparts are depicted in Fig. 10.1. The sizes of the planar microparts range from single cell size (10–50 μm) to multiple cell size (10–1000 μm). The typical thickness is 2–50 μm. The shapes can be changed on demand. For stable observation, the microparts are arrayed with 10–100 μm gap on a base substrate. The gap distance should be long enough to prevent cells to bridge between two microparts [15].

Layers Some planar microparts have layered structures: cell adhesion layer, core layer, and sacrificial layer from top to bottom (Fig. 10.2). The cell-adhesion layer is made of extracellular matrix (ECM) proteins or surface structured to have hydrophilic/hydrophobic properties for the promotion of cellular adhesions. The core layer is the main structure of micropart and determines its shape and functions. The sacrificial layer combines the core layer with the base substrate during fabrication. The sacrificial layer is made of material that can be dissolved after fabrication for detachment.

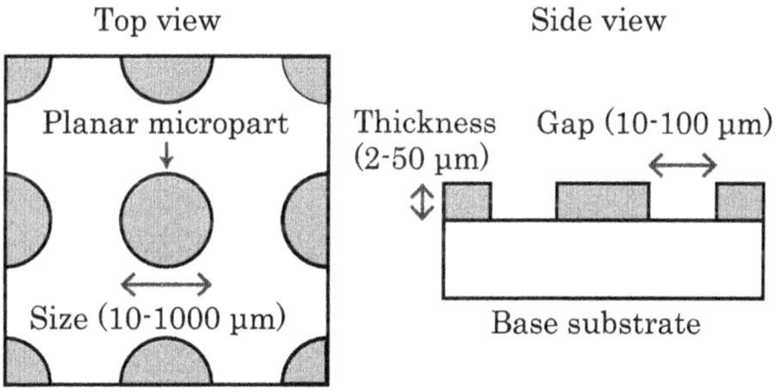

Fig. 10.1 Dimensions of planar microparts

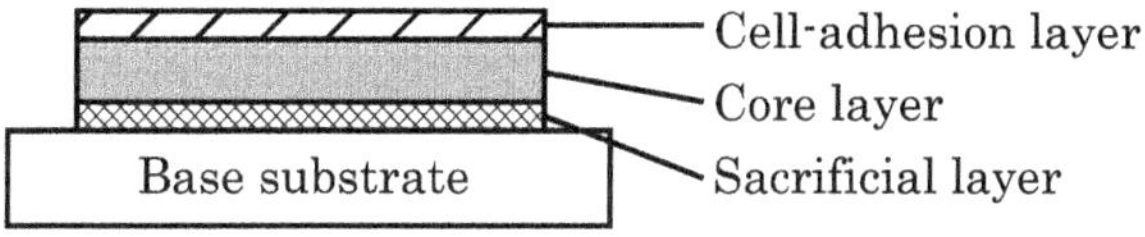

Fig. 10.2 Layered structure of planar microparts

10.2.2 Materials

We describe the materials of the previously reported planar microparts. Table 1 shows the materials used for the cell-adhesion layer, core layer, and sacrificial layer of the microparts.

Cell-adhesion Layer Cell-adhesion layers are achieved by coating the microparts surface with cell-adhesive materials such as (i) ECM proteins which bind with cell adhesion molecules, (ii) positively charged hydrophilic molecules which promote

Table 10.1 Materials used for planar microparts

Layer	Materials	Notable Characteristics	Ref
Cell-adhesion layer	Collagen	ECM protein	[12–21, 26, 27]
	Fibronectin	ECM protein	[13, 15, 16, 18–20, 30, 34, 35, 52]
	Matrigel	Mixture of ECM proteins	[22, 27]
	Gelatin	Derived from collagen	[27, 30]
	Poly-D-lysine	Hydrophilic molecule	[18–20]
	Anti-EpCAM	Antibody	[35]
Core layer	SiO_2	Low autofluorescence	[12]
	Parylene	Flexiblity	[13–16]
	SU-8	Photoresist	[18–23, 50, 53]
	Polystyrene	Low autofluorescence	[22, 25–27]
	1002F	Photoresist	[22, 24, 27, 28, 30, 34, 35]
	1009F	Epoxy resin	[22]
	PLGA	Biodegradable polymer	[32]
	PDMS	Flexibility	[52]
	PEG hydrogel	Photocurable hydrogel	[54]
	Iron oxide nanoparticle	Solution-dispersible magnetic material	[24, 26]
	Cu	Non-magnetic metal	[49, 55]
	Ni	Magnetic metal	[49, 55]
	Au	Biocompatible metal	[49, 50, 55]
Sacrificial layer	Gelatin	Dissolved by heat (37 °C)	[13, 14]
	Calcium Alginate	Dissolved by enzyme	[15, 16]
	PIPAAm	Thermally sensitive (22 °C), reversible polymer	[52]
	PVA	Dissolved by water	[53]
	PMMA	Dissolved by organic solvent	[55]
	Cu	Dissolved by iron chloride etchant	[50]

protein adhesion, and (iii) antibodies which link with specific cell adhesion molecule. Common ECM proteins are collagen [12–21, 26, 27] and fibronectin [13, 15, 16, 18–20, 30, 34, 35, 52]. Based on the ECM proteins, matrigel [22, 27] and gelatin [27, 30] are also used. Cells have their specific cell adhesion molecules on their membrane that bind with specific ECM protein. Appropriate ECM protein is needed due to the cell-type specific expression of different cell adhesion molecules (e.g., some neuronal cells adhere more frequently on collagen than fibronectin [56]). Note that some types of cells such as fibroblast cells secrete ECM proteins by themselves during culture. Positively charged hydrophilic molecules such as poly-*D*-lysine [18–20] and poly-*L*-lysine also promote cellular adhesion. Antibody to specific cell adhesion molecules such as anti-EpCAM captures target cells [35]. Surface roughness, nano-sized topography and hydrophilicity are also important for cell adhesion. In addition, cell non-adhesive coating on base substrate are often provided to promote selective cell adhesion on the microparts. Poly(dimethylsiloxane) (PDMS) [12], 2-methacryloyloxyethyl phosphorylcholine (MPC) polymers [13–16, 57], and Pluronic F127 [52] are known for cell non-adhesive materials.

Core Layer The core layer serves as supporting structure of cellular structures [13]. The layer is often designed to be enduring enough for handling by externals forces, such as fluidic forces [12], and micromanipulators [15]. Typical materials selected for the core layers are glass [12], parylene [13–16], polystyrene [22, 25–27], photoresist such as SU-8 [18–23, 50, 58] and 1002F [22, 24, 27, 28, 30, 34, 35], 1009F epoxy resin [22], poly(lactic-co-glycolic acid) (PLGA) [32], PDMS [52], poly(ethylene glycol) (PEG) hydrogel [54], iron oxide nanoparticles [24, 26], and metals such as copper (Cu), nickel (Ni) [49, 55], and gold (Au) [49, 50, 55]. The materials are selected depending on the purposes. To be specific, glass and polystyrene have an advantage in low auto-fluoresce and are suitable for fluorescent measurement of cells. Glass, parylene, polystylene, SU-8, and PDMS have optical transparency which is useful in cell observation. Thin Au (~ 20 nm) are also transparent. Parylene and PDMS have flexibility that enable folding. PLGA and PEG hydrogel are highly biocompatible and biodegradable. Iron oxide nanoparticles and Ni layers can be used for magnetic functionalization which allows the microparts to be actuated by external magnetic fields. Cu and Au layers are useful to engineer inert stress in the bilayer structure of micropart.

Sacrificial Layer Some types of planar microparts have the sacrificial layer for controlled detachment from the base substrate. The materials such as gelatin [13, 14], calcium alginate [15, 16], poly(*N*-isopropylacrylamide) (PIPAAm) [52], poly(vinyl alcohol) (PVA) [53], poly(methyl methacrylate) (PMMA) [55], and Cu [50] are reported for the sacrificial layer that can be detached. The key features of the sacrificial layers are the stablity during fabrication process and the removability by stimulus. The use of sacrificial layers provides selective handling of the microparts [15] and also their batch release [13].

10.2.3 Fabrication Processes

To fabricate the planar microparts, fundamental MEMS fabrication processes such as photolithography [11–16, 18–24, 27, 28, 30, 34, 35, 50, 53] and soft lithography [22, 25–27, 32, 52, 54] are employed. The microparts with different types of designs and materials require different fabrication processes. Photolithography technique is suitable for patterning metals and polymer with the aid of photoresist and chemical or physical etching process. Although it enables precisely-patterned multiple layered structures on a base substrate, it is unfitted to pattern soft materials such as hydrogels. On the other hand, soft lithography process can be employed to pattern curable solutions (e.g., hydrogels) without using photoresists and etching process. The soft lithography process, however, is not suitable for creating complex layered structures. In this chapter, we describe the use of the two types of major fabrication processes in micropart fabrication.

Photolithography Process Photolithographic patterning consists of (i) thin-films formation of target materials and photoresists, (ii) lithographic exposure to the photoresists, and (iii) etching of photoresists and the target materials. Figure 10.3a shows the photolithographic fabrication process of the microparts made of parylene [15]. Briefly, on a glass substrate, calcium alginate (sacrificial layer), parylene (core layer) are first coated. Alminum (Al) and photoresist thin-layers are formed and patterned by photolithography to create etching masks for the calcium alginate and the parylene. The calcium alginate and parylene layers are etched by oxygen plasma using the Al etching mask. The cell non-adhesive layer such as MPC polymer [57] is formed and patterned on the base substrate by lift-off process that is removal of Al mask structure. Fibronectin (cell-adhesion layer) is then coated only on the surface of parylene layer as a result of their repulsion by the MPC polymers-coated base substrate. The pattern of cell-adhesive fibronectin and non-adhesive MPC polymers promotes selective cellular adhesion onto the microparts. Here, the use of calcium alginate is a significant factor compared to common MEMS process that has not employed soft materials. Figure 10.3b shows the fabricated microparts.

Soft Lithography Process Soft lithographic patterning is molding of curable solutions inside a template pattern. Figure 10.4a shows the soft lithographic fabrication process of the planar microparts made of magnetic polystyrene [26]. Here, template patterns are created, and PDMS microwell arrays are fabricated by molding the templates. The PDMS microwell arrays also serve as a template of other curable solutions. The microwells are coated with polystyrene solution containing magnetic iron oxide nanoparticles. The magnetic polystyrene solutions form convex shapes in the microwell spontaneously and gradually transformed into concave-shaped solids by heating process. Other concave-shaped polystyrene solids without magnetic particles are formed in the same way. The fabricated magnetic polystyrene microparts are shown in Fig. 10.4b, c and d. The soft lithography process enables non-photoresist, etching-free fabrication of planar microparts that consist of curable liquids.

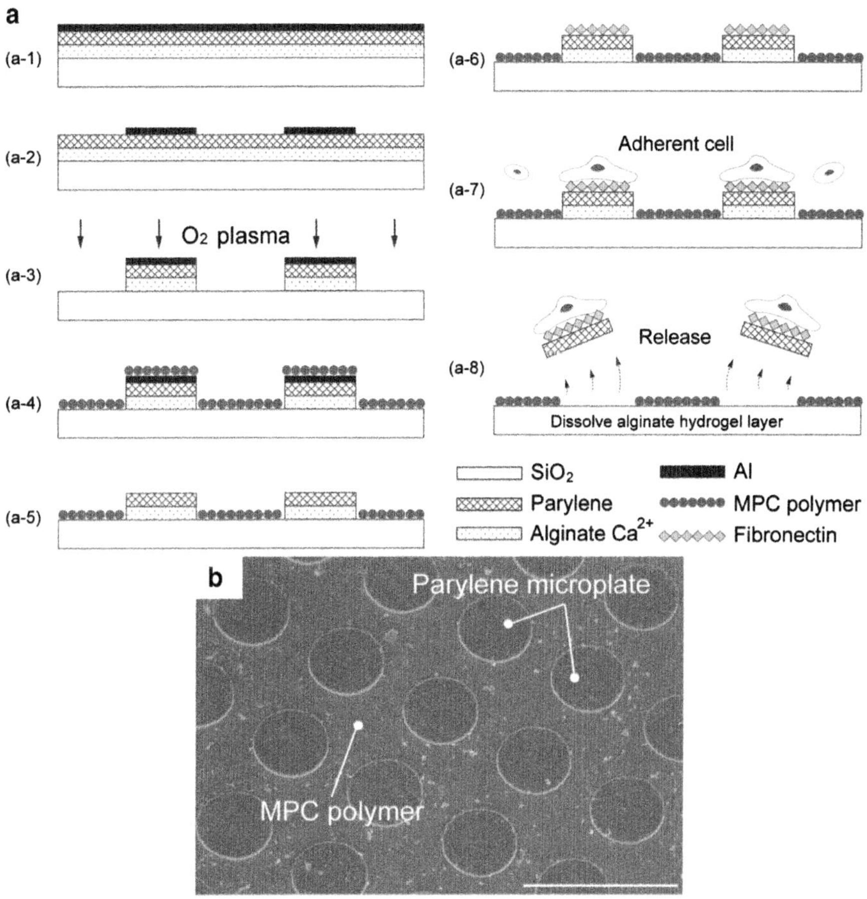

Fig. 10.3 Fabrication of the parylene microparts [15]. **a** Fabrication process using photolithography techniques. **b** An SEM image of the fabricated microparts. The scale bar indicates 100 μm. **a**, **b** Copyright 2014 John Wiley and Sons

10.2.4 Cell Culture

Cells can be cultured on the planar microparts; they are randomly seeded and selectively adhered on cell-adhesive surfaces of the microparts. Small sized (30–50 μm) microparts enable isolation and culture of single cells [12, 14–16, 18–21], and large sized (50–1000 μm) microparts enable multiple cell culture [13, 14]. Various types of adherent cells can be cultured on the planar microparts: fibroblast cells [12, 13, 15, 16, 28, 49, 50, 55, 59], hepatic cells [12, 15], muscle cells [13, 27, 52], neural cells [14, 15], cervical cancer cells [15, 18–21, 28, 32, 34], and embryonic stem cells [23, 30].

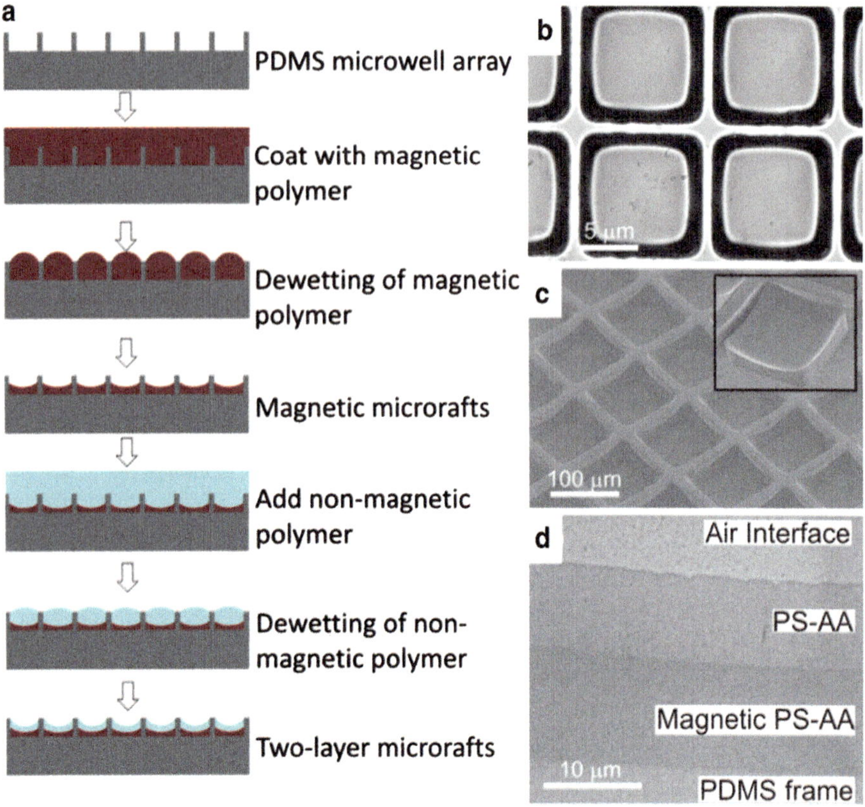

Fig. 10.4 Fabrication of the magnetic polystyrene microparts [26]. **a** Fabrication process using soft lithography techniques. **b** Bright field, c SEM, and d TEM images of the fabricated microparts. PS-AA: Poly(styrene-co-acrylic acid). **a**–**d** Copyright 2011 American Institute of Physics

Figure 10.5 shows cell culture on the planar microparts [14, 15]. Cells selectively adhere on the microparts by cell adhesive/non-adhesive surface modification (Fig. 10.5a) [15]. Cells are cultured for several hours to days for their attachment and proliferation. Yield of single cell on a micropart depends on both initial cell-seeding density and size of the microparts (Fig. 10.5b) [14]. As the size of the micropart increases, the number of cells adhered on them also increases.

10.2.5 Release and Handling Methods

The planar microparts can be released from the base substrate for handling. The release of microparts is conducted before or after cells are seeded on their surfaces. The release methods are basically categorized as ultrasound release [29], laser release [30, 34], mechanical release [32], and sacrificial layer-aided release [13–16].

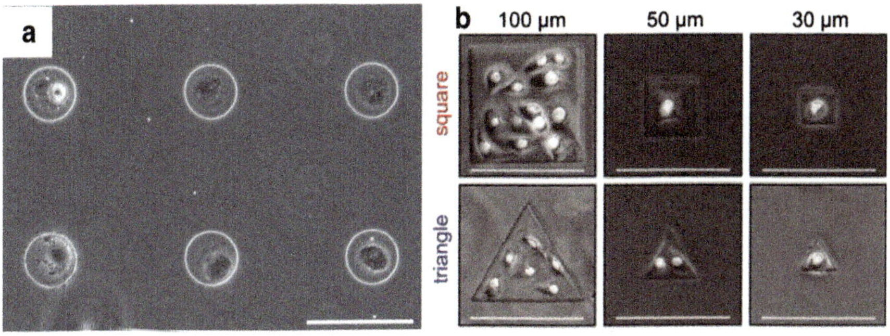

Fig. 10.5 Cell culture on the planar microparts [14, 15]. **a** Single fibroblast cell culture on circlular microparts [15]. **b** Single to multiple neural cells cultured on square- and triangle-shaped microparts [14]. White dots are fluorescently stained cell nucleus. Scale bars represent 100 μm in **a**, **b**. **a** Copyright 2014 John Wiley and Sons, **b** Copyright 2014 Fuji Technology Press

The ultrasound release of microparts is suitable for collecting large microparts (~ 500 μm) although the ultrasound may damage the cells [29]. Laser-based release enable serial release of individual cell-laden micorparts, however it is unable to detach large microparts (larger than 500 μm) due to the limitation in laser energy that is needed to be lowered enough to maintain cellular viability [29, 30]. Mechanical release using micromanipulator such as needle [32] provides precise detachment of a micropart, though on the other hand it is time-consuming. Sacrificial layers provides batch release of multiple microparts, however selectivity of microparts is relatively low compared to other serial release methods. To handle cell-laden planar microparts, fluid flow ejected from a glass capillary [12], contact manipulation by a glass capillary and a microtweezer [12, 14, 15, 32], by a pipette [18], and magnetic actuation [24] are employed.

10.3 Single Cell Handling

The planar microparts are applicable to single cell handling for their ability to culture single cells. Single cell handling, observation, and collection are demanded for in vitro pharmacologic tests such as target cell separation and single cell analysis of phenotypic or genotypic events.

Conventional single cell handling methods [36–48] have advantages in observation, collection, and handling of single cells at high accuracy and high yields. However, morphology and cell surface markers of adherent cells are changed in the cell handling process [15]. Those properties must play fundamental roles in cellular behavior, thus, technologies to handle single adherent cells are needed. In the present section, we describe the application of cell-laden planar microparts to single cell handling. The single cell handling procedure is composed of observation, collection/handling, and arrangement/immobilization.

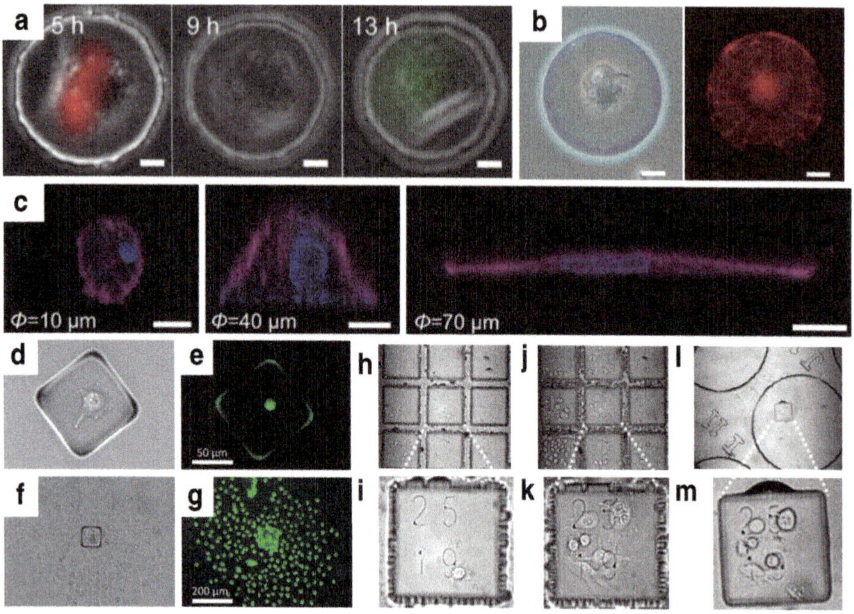

Fig. 10.6 Single cell observation on the planar micropart [15, 19, 26]. **a** Cell cycle observation by fluorescent measurement of cell cycle indicator-transfected single cell's nuclei. **b** Cell cytoskeleton observation of single cell [15]. **c** *x–z* cross sectional observation of single cell by confocal microscopy [15]. **d–g** Observation of proliferation from single cell to multiple cell clones [19]. **d** A bright field image and **e** a fluorescent image of a cell adhered on a micropart. **f** A bright filed image and **g** a fluorescent image of proliferated cells. **h–m** Tracking of single cell before and after collection by coded microparts [26]. **h**, **i** Single cell adhered on a coded micropart. **j**, **k** proliferated cells. **l**, **m** Collected cells. Scale bars represent 10 μm in **a–c**. **a–c** Copyright 2014 John Wiley and Sons, **d–g** Copyright 2007 American Chemical Society, **h–m** Copyright 2011 American Institute of Physics

10.3.1 Single Cell Observation

Figure 10.6 shows single cell observation on the planar microparts [15, 19, 26]. Optical transparency of the microparts enables fluorescent microscopy of the cells. Cell cycle of single cell can be observed by fluorescent microscopy (Fig. 10.6a) [15]. Cell nuclei containing a cell cycle indicator show cell-cycle-dependent fluorescent expression. Cellular cytoskeletons such as actin fibers are visible on the planar microparts (Fig. 10.6b) [15]. The micropart regulates the cell adhesion area and morphology. Thus, intracellular architectures of shape-defined cells can be observed. Cross-section of single cell on the micropart is measureable by confocal microscopy (Fig. 10.6c) [15]. By changing the size of the micropart, different morphologies of single cells can be observed. Proliferation of isolated single cell can be observed (Fig. 10.6d–g) [26], and tracked before and after cell handling by encoding tags on the microparts (Fig. h–l and m) [19]. The tracking of single cell is useful for the establishment of clones from targeted single cell.

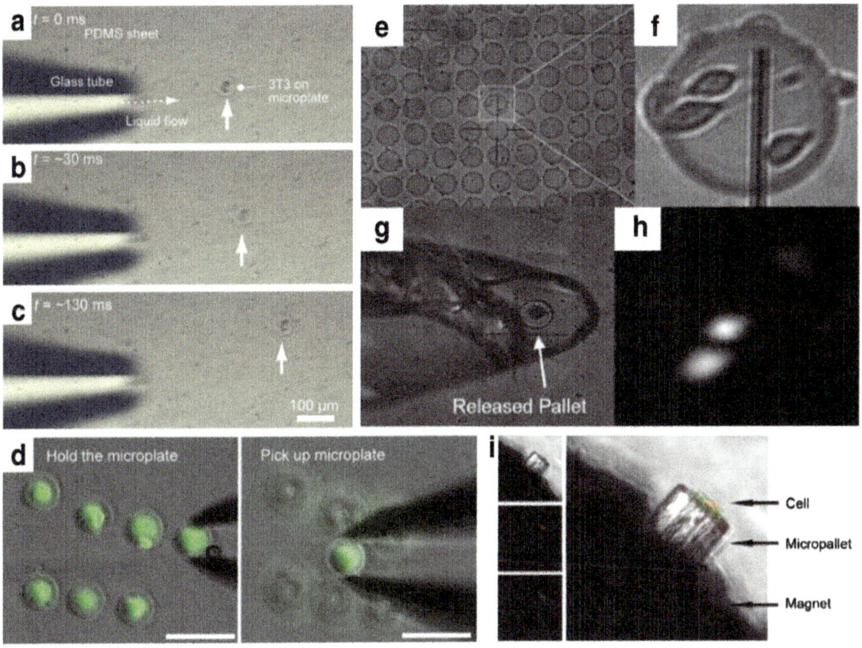

Fig. 10.7 Collection and handling of cell-laden planar microparts [12, 15, 18, 24]. **a–c** Handling by liquid flow applied from a glass tube [12]. **d** Holding and pick-up motion by a microtweezer [15]. **e–h** Laser release and handling by a pipette [18]. **i** Handling by magnet [24]. Scale bars represent 100 μm in **d**. **a–c** Copyright 2008 Institute of Physics, **d** Copyright 2014 John Wiley and Sons, **e–h** Copyright 2007 American Chemical Society, **i** Copyright 2010 American Chemical Society

10.3.2 Handling

Handling of cell-laden planar microparts is performed by indirect micromanipulation or direct micromanipulation (Fig. 10.7) [12, 15, 18]. Figure 10.7a–c shows indirect, fluid flow-based collection and handling of single cell-laden micropart [12]. The cell-laden microparts are handleable freely with maintaining cellular adhesions. Figure 10.7b shows contact micromanipulation by microtweezer [15]. By grabbing the edge of the micropart, the cell can be handled three-dimensionally without direct physical contact. Figure 10.7e–h shows laser-induced release of cell-laden microparts, and handling by pipette suction [18]. The laser is useful for releasing target micropart due to its high spatial resolution. Figure 10.7i shows magnetic collection of laser released micropart [24]. The micropart is magnetized by iron oxide nanoparticles embedded in the core layer.

10.3.3 Arrangement and Immobilization

For cellular analyses after collection and handling, the cell-laden microparts are needed to be arranged and immobilized again to a certain location (Fig. 10.8). As

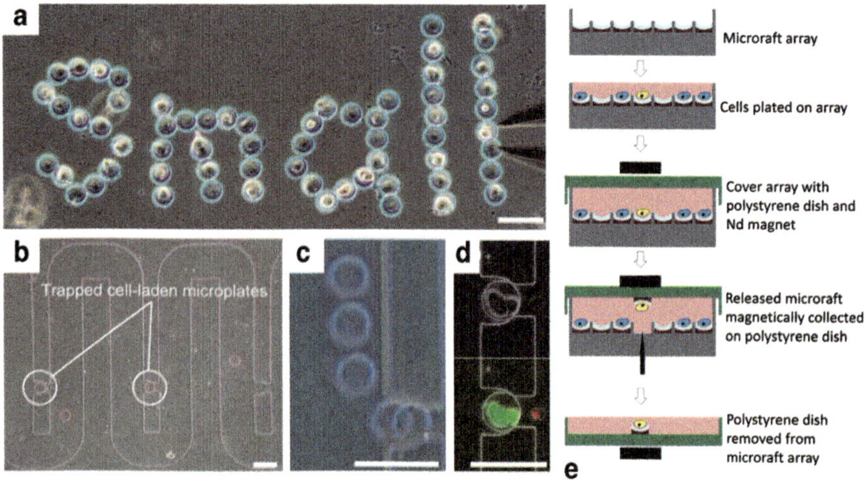

Fig. 10.8 Arrangement and immobilization of cell-laden planar microparts [15, 26]. **a** Arrangement by microtweezer-based handling [15]. **b–d** Microfluidic arrangement and immobilization [15]. **e** Magnetic arrangement [26]. Scale bars represent 100 μm in **a–d**. **a–d** Copyright 2014 John Wiley and Sons, **e** Copyright 2011 American Institute of Physics

shown in Fig. 10.8a, cell-laden microparts can be arranged in arbitrary 2D geometry by micromanipulation technique [15]. However, since the microparts are easily moved again by external forces such as fluid flow or vibration, they needed to be immobilized for stable observation and analysis. Figure 10.8b–d shows arrangement and immobilization of cell-laden microparts in microfluidic channel [15]. Laminar flow in the microfludic channels enables conveying the microparts to a certain point accurately. Hydrodynamic force and physical well immobilize the arranged microparts. Figure 10.8e shows magnetic arrangement and immobilization of magnetically-functionalized microparts [26]. Magnetic forces acting on the microparts, which is controlled by external magnet, enable transferring and immobilizing the micropart.

10.4 D Assembly of Cellular Constructs

The fabrication of tissue-like structures and 3D cellular constructs requires mimicking structural hierarchy, heterogeneity and variety of tissues and organs [60–63]. One of the most critical aspects to engineering cellular constructs is multi-cell co-culture with 3D spatial control [60]. 3D spatial control is necessary to precisely recreate the corresponding microenvironment of the desired tissue such a tumor-specific niche [61]. The microenvironment provides physical and chemical cues to the embedded cell [60–62]. These cues affect cellular behavior and cellular development [60–62].

3D cellular constructs provide a complete 3D microenvironment to cells in contrast to conventional 2D cell culture substrates [60–62]. Additionally to the creation of the appropriate microenvironment, other challenges of tissue engineering are sufficient delivery of nutrients, regulation of cell density and long term sustainability of the construct [61, 64]. When tissue engineering constructs are supposed to be used as medical treatment, further major challenges of implantation and integration are required to be solved [64, 65].

Bottom-up tissue engineering is approaching the challenges of tissue engineering by assembling cells and cell-laden microparts in 3D (3D assembly). Bottom-up microfabriaction technologies such as hydrogel casting [62] and cell spheroid molding [10] generate structures with randomized features. For precise control of a heterogeneous microenvironment however, pick-and-place, self-assembly and self-folding are approaches to direct the position of every individual building block. In the following, we introduce and discuss these bottom-up technologies for spatially controlled heterogeneous tissue-like structure formation.

10.4.1 Pick-and-Place Assembly

Pick-and-place assembly is one of the key techniques developed in the MEMS field for construction of 3D microstructures [60–65]. In the meter-scale, 3D objects are often assembled by pick-and-place procedure of their components, however, in the micro-scale, this building process is highly difficult due to the scale effects and low Reynolds number. Although the pick-and-place assembly has limitation in throughput due to its serial processing, it has advantages in high spatial precision and heterogeneity of components in 3D assembled objects. Therefore, efforts were made in the MEMS fields to develop micro-scale pick-and-place methods for construction of precisely organized microsystems.

The use of pick-and-place method in assembly of 3D cellular constructs may be important to achieve highly-precise positioning of multiple types of cells. The pick-and-place 3D cellular assembly is demonstrated in combination with micromanipulation and *in situ* gluing method (Fig. 10.9) [14]. At first, cell-laden microparts are picked up by microtweezer and placed perpendicular to each other inside a three-dimensional microgroove. Then, prepolymer solutions of ultra-violet (UV)-curable and biocompatible PEG hydrogels are applied onto the microparts from a glass capillary. Finally, UV light was exposed to polymerize the PEG hydrogel, which glues the microparts together (Fig. 10.9a). Throughout the procedure, 3D cellular constructs are built from cell-laden planar microparts. The method enables composition of cell-laden 3D microstructures with selected cells, and also in desired geometry. Figure 10.9b shows pick-and-place process of differently labeled cell-laden microparts. The green cell-laden microparts and red cell-laden microparts are transferred from different regions. Figure 10.9c shows 3D assembled cellular construct by this method. By using pick-and-place method, it is possible to assemble heterogeneous types of cell-laden microparts in 3D configuration.

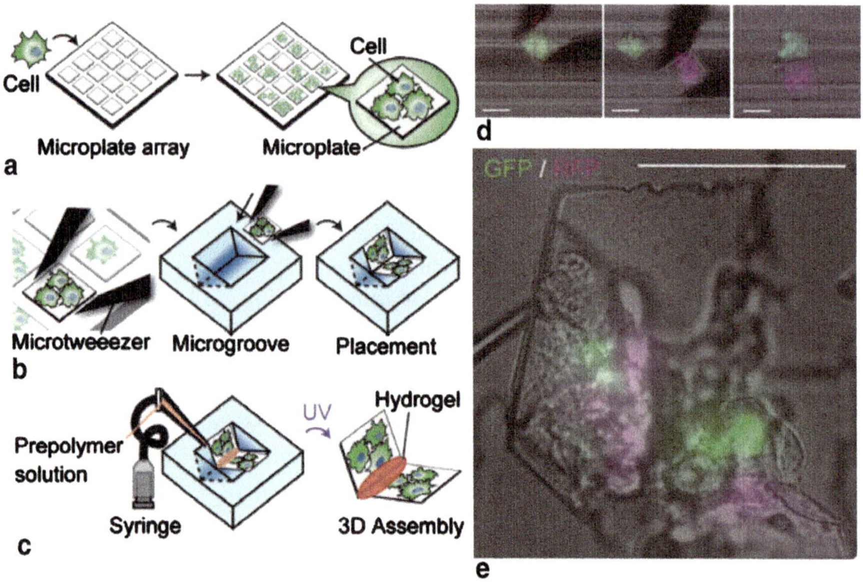

Fig. 10.9 Pick-and-place assembly of cell-laden planar microparts [14]. **a–c** Cells on the microparts were picked up and placed on a microgroove. The cell-laden microparts were assembled by a photocurable hydrogel glue. **d** Picked up and placed cell-laden microparts. The cells were labeled with fluorescent dyes. **e** Assembled cell-laden 3D structure. The cells were labeled with fluorescent proteins. Scale bars indicate 50 μm in **d**, and 100 μm in **e**. **a–e** Copyright 2014 Fuji Technology Press

10.4.2 Self-assembly of Cell-Laden Hydrogel Building Blocks

The term self-assembly has been used to describe various formation processes; in 2002 Whitesides and Grzybowski provided a strict definition: Self-assembly is the autonomous organization of components into patterns or structures without human intervention [66].

Components for self-assembly are either separated parts or distinct parts of one structure with sizes ranging from nanoscale to macroscopic scale [66–68]. It has been demonstrated that the combination of multiple self-assembly concepts enables upscaling to millimeter scale [69, 70]. Since cells are themselves the major components for cellular tissues, self-assembly is an attractive approach to tissue engineering [66, 71, 72]. In contrast to pick-and-place methods (Sect. 10.4.1), self-assembly methods are parallel independent, unsupervised mass fabrication processes with high redundancy and robustness [66, 73]. However, the fabrication yield decreases with a higher number of components due to the increase of undesired by-products (Fig. 10.10, [74]).

Among various strategies of self-assembly methods based on molecular interactions [6, 63, 66–71, 73–90], only few strategies can be utilized for tissue engineering; aqueous condition, salt and pH concentration as well as cell culture condition

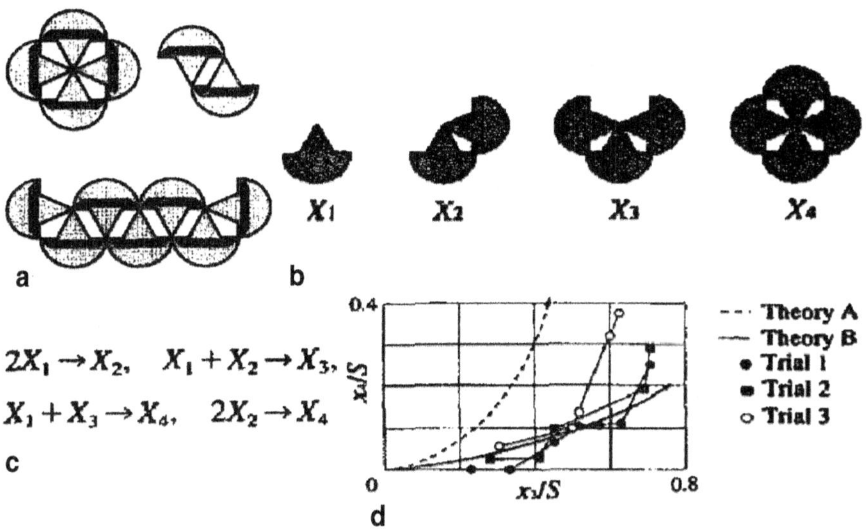

Fig. 10.10 Surface tension based self-assembly of four identically shaped microparts [74], **a** desired clover-like state and examples for undesired states, **b** representation of X1-X4, **c** rate equation, **d** self-assembling behavior, theory and experiment. Copyright 1996 Elsevier

in temperature and gas concentration need to be maintained, external fields and forces cannot interfere with life functions of cells. Due to the strict requirements, biocompatible self-assembly methods strongly rely on hydrogel technologies [5, 6, 58, 63, 90–93]. Utilization of hydrogel additionally enables encapsulation of cells at various concentrations and combination of different cell types [58, 90–93]. Further, it is compatible with microfluidic technologies [63, 77]. Current sizes of hydrogel microparts range from mm to 400 μm. Millimeter-scaled microparts are advantageous for the fabrication of large tissue-like structures. The spatial control over the cell environment is too low to recreate highly complex tissue such as a cellular niche [90].

The following technologies are the state of the art of self-assembly for tissue engineering. They are rapid and non-invasive [90, 92]. There are three common points to note about the following methods: (i) cell-laden hydrogel microparts are prepared prior to the self-assembly process, (ii) hydrogel microparts are differently shaped to enable a lock-and-key mechanism and enhance the desired assembly outcome (Fig. 10.11), and (iii) a secondary crosslinking of bulk hydrogel is implemented into the process to fixate the assembly. The crosslinking by UV light is short enough to be considered biocompatible [6, 58, 63, 90–93]. However, for medical applications possible DNA damage and its effect on changes in protein expression and cellular behavior might be limitations. Ideally, secondary arrangement fixation should be entirely without impact on cells. Soon, peptide hydrogels [84] and functionalized hydrogels [86] might become valuable additions to the current technologies.

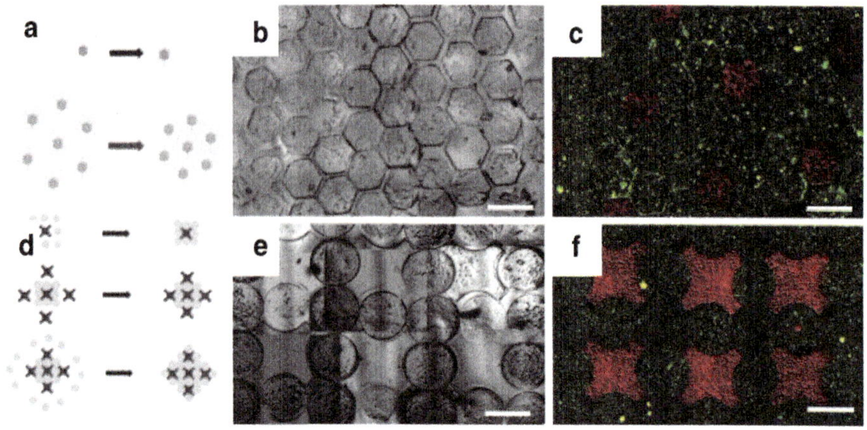

Fig. 10.11 Concept and experimental result for different shape-matching designs for lock-and-key mechanisms for controlled co-culture of HepG2 cells (*red*) and surrounding 3T3 cells (*green*) [58]. **a–c** Assembly of cell-laden hexagonally shaped hydrogel microparts. **d–f** Assembly of cell-laden lock-and-key hydrogel microparts, scale bars represent 1 mm. Copyright 2010 John Wiley & Sons

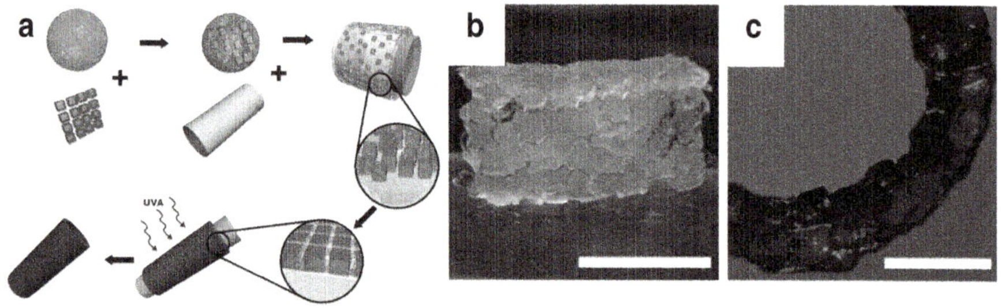

Fig. 10.12 **a** Concept of micro-masonry for a tubular structure [92], **b, c** experimental result with lock-and-key hydgrogel microparts, scale bars represent **b** 5 mm and **c** 2.5 mm. Copyright 2010 John Wiley & Sons

Interface-Directed Self-Assembly Interface-directed self-assembly is a technology that combines shape-matching surface energies and capillary forces [58, 91, 92]. The principle of interface-directed self-assembly has been varied to achieve different results: (i) Solid microstructures of different shapes have been achieved by dropping a mixture of hydrogel blocks and UV curable polymer into mineral oil [58]. Hydrophobic mineral oil might remain [6]. (ii) Utilizing the air-liquid interface, hydrogel blocks are placed in a hydrophobic solution and secondarily cross-linked in a one-layer arrangement (Fig. 10.11, [58]). (iii) Hydrogel and pre-polymer are applied to a curved PDMS shape such that the air-liquid interface is curved (Fig. 10.12, [92]). By removing the excess polymer, capillary forces enhance the formation process. Upon curing, hollow structures with a one-layer shell such as tubes and semi-spheres were created [92]. This 3D structure formation is also called micro-masonry [92]. This term refers to bricklaying technologies where bricks are an analogy for cell-laden hydrogel microparts.

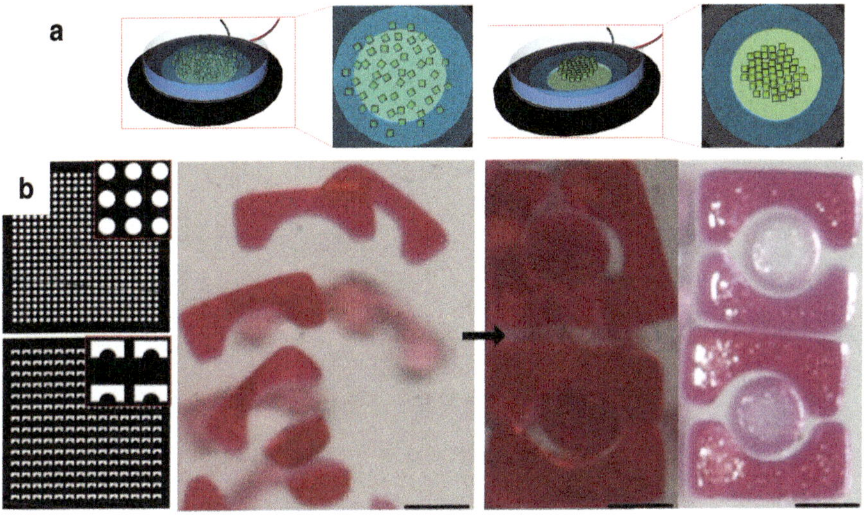

Fig. 10.13 Self-assembly by acoustic waves [90], **a** concept for assembly, **left** before acoustic excitation, *right* after acoustic excitation. **b** Hydrogel micropart lock-and-key mechanism before and after. Scale bars represent 200 micrometer. Copyright 2011 Elsevier B.V.

Acoustically Guided Self-Assembly Instead of capillary forces, acoustic waves are used to drive self-assembly (Fig. 10.13, [90]). This is not only non-invasive, but also decouples the fabrication from two-phase interfaces. Thereby, it enables self-assembly in entirely aqueous conditions and easy multi-layer fabrication [90]. Among current hydrogel self-assembly technologies, acoustically guided self-assembly has been reported with the smallest individual microparts. However, due to identic conceptual key features, the expected minimal resolution of micropart size is similar to the other technologies.

Hydrophobicity Guided Self-Assembly Similar to the interface-directed self-assembly methods, template hydrophobicity guided self-assembly uses templates with hydrophobic surfaces patterned partially hydrophilic or vice-versa exploit surface energy driven self-assembly processes (Fig. 10.14, [93]). Co-culture can be achieved by various approaches: (i) Differently cell-loaded hydrogel microparts self-assemble, (ii) hydrogel microparts and hydrogel bulk are differently loaded with cells, and (iii) a combination of the first two approaches seems very feasible to co-culture 3 different types of cells.

10.4.3 Self-Folding Origami Assembly for Tissue Engineering

Self-folding origami assembly is a self-assembly process where planar components that are tethered in 2D networks, fold or bend out-of-plane into 3D contructs [13, 54, 89, 94–97]. Self-folding provides utilization of the advantages of the advanced 2D lithographic patterning technologies with high throughput and high feature

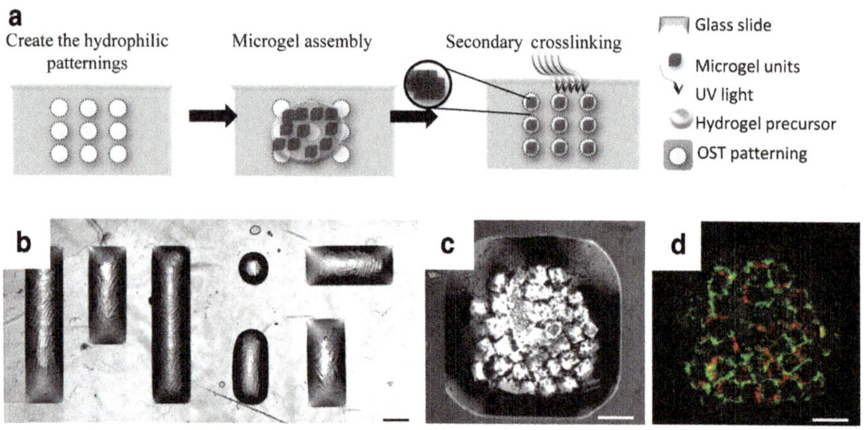

Fig. 10.14 Hydrophobic-hydrophilic template based self-assembly [93], **a** concept for self-assembly of cell-laden hydrogel microparts in cell-laden bulk hydrogel. **b** Shape demonstration of letters '**m**', '**i**' and '**t**', **c, d** hydrogel microparts encapsulating red-labeled cells in bulk hydrogel encapsulating green-labeled cells. Scale bars represent 400 mm. Copyright 2010 John Wiley & Sons

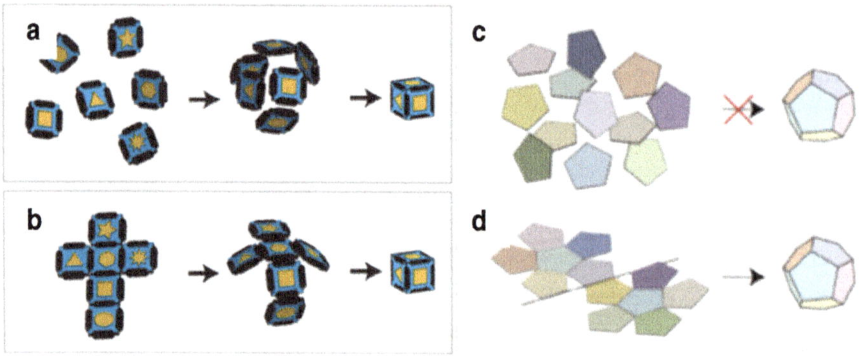

Fig. 10.15 Conceptial illustration of self-assembly and self-folding for **a, b** 6-face [96] and **c, d** 12-face [94] structures. Self-assembly is expected to have a low yield, when self-folding has sufficient yield due to the deterministic nature of the tethered net. **a, b** Copyright 2009 PLOS and **c, d** Copyright 2010 John Wileys & Sons

resolution for high control of micropart size, porosity and additional surface topography [49, 94, 95]. Simultaneously, out-of-plane folding enables the fabrication of curved, complex 3D cell scaffolds and tissue-like structures [50, 51].

Due to the given connectivity of components, self-folding origami has higher deterministic nature than free self-assembly in addition to possible cooperativity between components (Fig. 10.15, [94]). Furthermore, self-correction, i.e. automatic correction of component position, is enhanced by locking mechanisms and polymeric material choice [51, 53]. Self-correction and cooperativity between tethered components enable development of complex structures with higher fabrication yield than it was possible with classical self-assembly [51].

The mechanisms to realize self-folding range from the usage of molten solder or polymer via surface tension or liquid traction force [53, 59, 77, 78], over magnetic stimulation [80, 81] to the integration of stimuli-responsive materials at the hinge site. The release from the substrate prior or during folding-stimulus is essential when the folding is actuated by shrinkage [98], swelling [89], bilayer tension [49, 50, 54, 99, 100] or conducting polymers [101]. Simultaneous self-folding is achieved by introducing a stimuli-responsive trigger [51, 97]: thermo-responsive triggers such as heat-induced softening of hinges [53, 55] and heat-induced multi-layer stress [49], or salt or pH-responsive by induced swelling [89]. Laser light was also demonstrated to actuate polymer on a pre-stressed bilayer allowing selective and sequential folding in addition [102].

Self-folded origami structures are equally promising as drug delivery systems with self-regulated release [88, 89], as cell container as well as cell culture scaffolds [49–51, 53]. By varying the complexity of the hinge pattern, the fabrication of relevant cell culture structures such as coils, cylindrical tubes and cubicles is pursued [13, 50, 54]. The current self-folding based technologies still face various difficulties: (i) cell density can be controlled by loading multiple times or at high cellular density but as a trade-off to precise spatial control [13, 50, 54], (ii) the folding process itself is often not biocompatible and cell seeding is performed after folding. It is important to understand that heat- or pH-induced triggering might be considered biocompatible since cells are vital, however their protein expression, behavior and possible fate might be altered entirely [103], and (iii) for application as medical implants, engineering tissue-like structures requires not only biocompatible substrate materials, but biodegradable materials.

From Cell Culture Scaffolds to Bioartificial Organs Utilizing metal-polymeric bilayer tension, self-folding has been extensively used for cell culture scaffold fabrication (Fig. 10.16a, c and d [50]). By varying the complexity of the hinge pattern, the fabrication of three relevant cell culture structures - coils, cylindrical tubes and cubicles - has been achieved [50]. Cell seeding was performed consecutively due to the lack of biocompatibility of the folding process. Precise positioning of cells by prior seeding to the 2D net structure remains challenging. Cell loading after folding and coating or sputtering with biocompatible materials such as inert gold enabled the fabrication of artificial organs (Fig. 10.16i, [104]). The loading and culturing process is limited to robust cells. Due to the homogeneity of cell-loading through pores and slits, a goal structure of rather homogeneous tissue was achieved [104]. Co-culture of multiple cells and hierarchical complexity remain future challenges.

Hydrogel Origami It is envisioned that cell assemblies such as cell-sheets can be utilized in the future to position cells in a 3D structure. Current research utilizes hydrogel to pre-structure and maintain cell-laden structures [54, 105]. In this approach bilayer tension is created inside of cell-laden hydrogels [54]. Tension is introduced by spatially controlled different cross-linking or creating multiple-layer

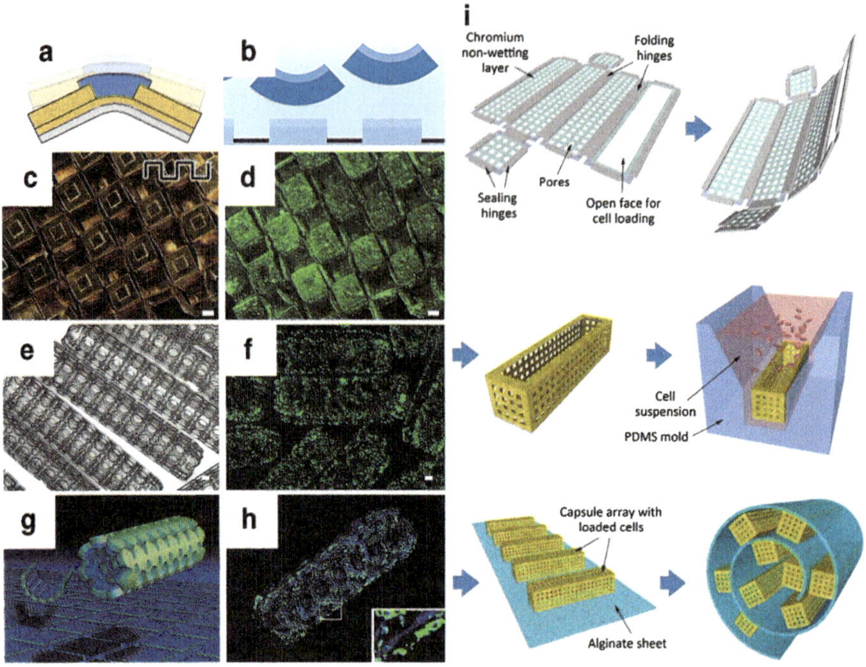

Fig. 10.16 Examples of origami assembly. **a, b** Strategies for self-folding: **a** chrom-gold- photoresist microparts [50] and **b** differently cross-linked hydrogel microparts [54]. **c, d** bidirectionally folded 3D cell culture scaffold, scale bars represent 160 mm [50], **e, f** Self-folded hydrogel cylindrical structures loaded with cells, scale bars represent 200 mm, **g, h** co-culture with two-layered hydrogel [54], **i** self-folded microchamber loaded with pancreatic cells and further assembled into a bioartifical endocrine pancreas [104]. **a, c, d** 2010 Elsevier B.V. **b, e–i** 2013 John Wiley & Sons

structures (Fig. 10.16b, e– and h). Co-culture can be achieved by loading different cells per hydrogel layer (Fig. 10.16g and h, [54]). By making appropriate material choices and aiming for guided co-culture, tissue-like structures have become feasible.

Cell Origami Cell origami is self-folding of 3D cell-laden microstructures driven by cell traction force [13]. This technology harnesses the intrinsic cell traction force (CTF) of cells for the folding process (Fig. 10.17a). Therefore, no additional folding mechanism is required to be employed and cells are cultured already in advance of the self-assembly process. However, the strength of CTF is cell-type dependent and especially soft tissue cells such as stem cells are expected to generate weak CTFs [106]. Usage of fibroblast or muscle cells enabled the fabrication of cubicle and cylindrical structures (Fig. 10.17b, c, f and g). The angle of folding was engineered by changing the geometrical dimensions of the hinge between microplates (Fig. 10.17d and e). Co-culture has been demonstrated with dodecahedronal structures (Fig. 10.17b and f). Co-culture on cylindrical structures might lead to

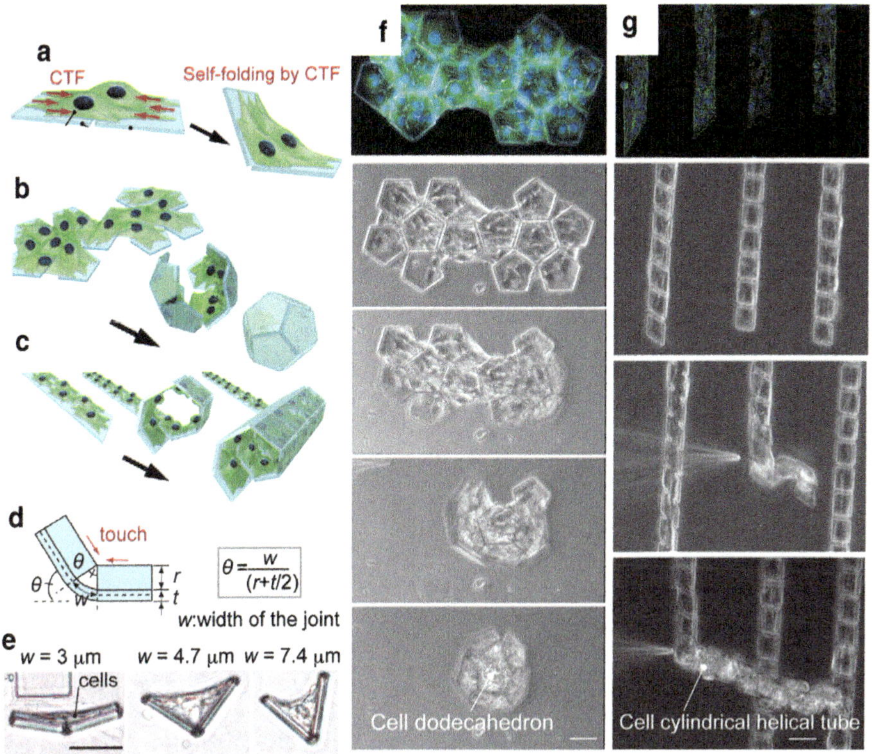

Fig. 10.17 Cell origami [13]. **a** Principle: Self-folding is driven by cell traction force (CTF). **b, c** Schematic illustration for dodecahedron and helical tube formation by cell origami. **d** Folding angle θ is dermined by hinge and plate dimensions, **e** microscope images of cells folding at different angles θ. **f, g** Time lapse observation of folding for dodecahedron and cylinder structure. Scale bars represent 50 μm. Copyright 2012 PLOS

tissue-like tubular structures such as blood vessels. For the full potential of cell origami, understanding cell-substrate interactions and prediction of the folding results will become necessary to advance to structures of higher complexity.

10.5 Conclusion

We reviewed fabrication of micro-sized cell culture substrates by MEMS technologies, and their use in 3D cellular tissue fabrications. There is growing demand for 3D tissue fabrication in regenerative medicine field, and the micro-sized planar microparts are suitable for the purpose since they enable handling adherent cells which are the main components of native tissues. The materials, structures, fabrication processes, cell culture, and handling of micro-sized substrates are explained.

Fundamental MEMS technologies such as photolithography and soft lithography enable fabrication of microsized planar substrates that can culture specific numbers of adherent cells. The application of the planar microparts for single cell observation, collection, and handling is summarized. Small sized (30–50 μm) microparts isolate and culture single cells. The single-cell-laden planar microparts are suitable for observation of intracellular events, further tracking, in vitro experiments, and tissue engineering. Large sized (50–1000 μm) microparts are capable of culturing multiple adherent cells. 3D cellular tissues are constructed by pick-and-place assembly, self-assembly, and self-folding methods. In future works, the use of various cell types and fabrication of functional tissues are required. The MEMS-based microsized cell culture substrates will be the key technology for fabrication of spatially-controlled heterogeneous 3D cellular tissues that are intensively required in regenerative medicine field.

References

1. Tadigadapa SA et al (2003) Developments in Microelectromechanical Systems (MEMS): a manufacturing perspective. J Manuf Sci Eng 125:816–823
2. Grayson ACR et al (2004) A BioMEMS review: MEMS technology for physiologically integrated devices. Proc IEEE 92(1):6–21
3. Huh D et al (2011) From 3D cell culture to organs-on-chips. Trends Cell Biol 21(12):745–754
4. Langer R, Vacanti JP (1993) Tissue engineering. Science 260:920–926
5. Khademhosseini A, Langer R (2007) Microengineered hydrogels for tissue engineering. Biomaterials 28:5087–5092
6. Gurkan UA et al (2012) Emerging technologies for assembly of microscale hydrogels. Adv Healthc Mater 1(2):149–158
7. Yang J et al (2009) Tissue engineering using laminar cellular assemblies. Adv Mater 21(32–33):3404–3409
8. Tamayol A et al (2013) Fiber-based tissue engineering: progress, challenges, and opportunities. Biotechnol Adv 31(5):669–687
9. Onoe H et al (2013) Metre-long cell-laden microfibres exhibit tissue morphologies and functions. Nat Mater 12:584–590
10. Mironov V et al (2009) Organ printing: tissue spheroids as building blocks. Biomaterials 30(12):2164–2174
11. Fennema E et al (2013) Spheroid culture as a tool for creating 3D complex tissues. Trends Biotechnol 31(2):108–115
12. Onoe H, Takeuchi S (2008) Microfabricated mobile microplates for handling single adherent cells. J Micromech Microeng 18(9):095003
13. Kuribayashi-Shigetomi K et al (2012) Cell origami: self-folding of three-dimensional cell-laden microstructures driven by cell traction force. PLoS One 7(12):e51085
14. Yoshida S et al (2014) Three-dimensional microassembly of cell-laden microplates by in situ gluing with photocurable hydrogels. Int J Auto Technol 8(1):95–101
15. Teshima T et al (2014) Parylene mobile microplates integrated with an enzymatic release and handling of single adherent cells. Small doi:10.1002/smll.201301993
16. Teshima T et al (2014) Magnetically responsive microflaps reveal cell membrane boundaries from multiple angles. Adv Mater doi:10.1002/adma.201305494

17. Wang Y et al (2006) Micropatterning of living cells on a heterogeneously wetted surface. Langmuir 22:8257–8262
18. Salazar GT et al (2007) Micropallet arrays for the separation of single, adherent cells. Anal Chem 79:682–687
19. Wang YL et al (2007) Collection and expansion of single cells and colonies released from a micropallet array. Anal Chem 79:2359–2366
20. Wang Y et al (2007) Broadening cell selection criteria with micropallet arrays of adherent cells. Cytometry A 71 A(10):866–874
21. Wang YL et al (2008) Micropallet arrays with poly(ethylene glycol) walls. Lab Chip 8:734–740
22. Wang YL et al (2010) Micromolded arrays for separation of adherent cells. Lab Chip 10:2917–2924
23. Xu W et al (2010) Microcup arrays for the efficient isolation and cloning of cells. Anal Chem 82:3161–3167
24. Gunn NM et al (2010) Ferromagnetic micropallets for magnetic capture of single adherent cells. Langmuir 26(22):17703–17711
25. Xu W et al (2010) Contact printing of arrayed microstructures. Anal Bioanal Chem 397:3377–3385
26. Gach PC et al (2011) Isolation and manipulation of living adherent cells by micromolded magnetic rafts. Biomicrofluidics 5:032002
27. Detwiler DA et al (2012) Polystyrene-coated micropallets for culture and separation of primary muscle cells. Anal Bioanal Chem 402(3):1083–1091
28. Gach PC et al (2012) Microfabricated arrays for splitting and assay of clonal colonies. Anal Chem 84(24):10614–10620
29. Guo S et al (2012) Ultrasound-induced release of micropallets with cells. Appl Phys Lett 101(16):163703
30. Pai JH et al (2013) Efficient division and sampling of cell colonies using microcup arrays. Analyst 138(1):220–228
31. Wang Y et al (2013) Capture and 3D culture of colonic crypts and colonoids in a microarray platform. Lab Chip 13(23):4625–4634
32. Wang Y et al (2013) Array of biodegradable microrafts for isolation and implantation of living, adherent cells. RSC Adv 3(24):9264–9272
33. Dickinson AJ et al (2013) Automated capillary electrophoresis system for fast single-cell analysis. Anal Chem 85(9):4797–4804
34. Dobes NC et al (2013) Laser-based directed release of array elements for efficient collection into targeted microwells. Analyst 138(3):831–838
35. Gach PC et al (2014) Micropallet arrays for the capture, isolation and culture of circulating tumor cells from whole blood of mice engrafted with primary human pancreatic adenocarcinoma. Biosens Bioelectron 15(54):476–83
36. Mazutis L et al (2013) Single-cell analysis and sorting using droplet-based microfluidics. Nat Protocol 8(5):870–891
37. van den Brink FT et al (2011) Parallel single-cell analysis microfluidic platform. Electrophoresis 32:3094–3100
38. Arakawa T et al (2011) High-throughput single-cell manipulation system for a large number of target cells. Biomicrofluidics 5:014114
39. Kobel S et al (2010) Optimization of microfluidic single cell trapping for long-term on-chip culture. Lab Chip 10:857–863
40. Ainla A et al (2010) A microfluidic pipette for single-cell pharmacology. Anal Chem 82:4529–4536
41. Wheeler AR et al (2003) Microfluidic device for single-cell analysis. Anal Chem 75:3581–3586
42. Hsu HY et al (2010) Phototransistor-based optoelectronic tweezers for dynamic cell manipulation in cell culture media. Lab Chip 10:165–172

43. Hu W et al (2012) Hydrogel microrobots actuated by optically generated vapour bubbles. Lab Chip 12:3821–3826
44. Sakar MS et al (2010) Single cell manipulation using ferromagnetic composite microtransporters. Appl Phys Lett 96:043705
45. Kim SH et al (2011) Electroactive microwell arrays for highly efficient single-cell trapping and analysis. Small 7(22):3239–3247
46. Doh I et al (2012) Deformation measurement of individual cells in large populations using a single-cell microchamber array chip. Appl Phys Lett 100:173702
47. Scott MA et al (2012) Ultra-rapid laser protein micropatterning: screening for directed polarization of single neurons. Lab Chip 12(12):2265–2276
48. Cheng Q et al (2011) Surface chemical patterning for long-term single-cell culture. J Biomed Mater Res A 96(3):507–12
49. Leong TG et al (2008) Self-loading lithographically structured microcontainers: 3D patterned, mobile microwells. Lab Chip 8:1621–1624
50. Jamal M et al (2010) Directed growth of fibroblasts into three dimensional micropatterned geometries via self-assembling scaffolds. Biomaterials 31:1683–1690
51. Randall CL et al (2012) Self-folding materials and devices for biomedical applications. Trends Biotechnol 30(3):138–146
52. Feinberg AW et al (2007) Muscular thin films for building actuators and powering devices. Science 317:1366–1370
53. Azam A et al (2011) Self-folding micropatterned polymeric containers. Biomed Microdevices 13:51–58
54. Jamal M et al (2013) Bio-origami hydrogel scaffolds composed of photocrosslinked PEG bilayers. Adv Healthc Mater 2:1142–1150
55. Gimi B et al (2005) Self-assembled three dimensional Radio Frequency (RF) shielded containers for cell encapsulation. Biomed Microdevices 7(4):341–345
56. Tomaselli KJ et al (1987) Interactions of a neuronal cell line (PC12) with Laminin, Collagen IV, and Fibronectin: identification of integrin-related glycoproteins involved in attachment and process outgrowth. J Cell Biol 105:2347–2358
57. Ishihara K et al (2000) Photoinduced graft polymerization of 2-methacryloyloxyethyl phosphorylcholine on polyethylene membrane surface for obtaining blood cell adhesion resistance. Colloids Surf B 18:325–335
58. Zamanian B et al (2010) Interface-directed self-assembly of cell-laden microgels. Small 6(8):937–944
59. Gracias DH et al (2002) Fabrication of micrometer-scale, patterned polyhedra by self-assembly. Adv Mater 14(3):235–238
60. Tan W, Desai TA (2004) Layer-by-layer microfluidics for biomimetic three-dimensional structures. Biomaterials 25:1355–1364
61. Hutmacher DW et al (2010) Can tissue engineering concepts advance tumor biology research? Trends Biotechnol 28(3):125–133
62. Tsang VL, Bhatia SN (2004) Three-dimensional tissue fabrication. Adv Drug Deliv Rev 56:1635–1647
63. Kachouie NN et al (2010) Directed assembly of cell-laden hydrogels for engineering functional tissues. Organogenesis 6(4):234–244
64. Hoffman AS (2002) Hydrogels for biomedical applications. Adv Drug Deliv Rev 54:3–12
65. Iatridis JC (2009) Function follows form. Nat Mater 8:923–924
66. Whitesides GM, Grzybowski B (2002) Self-assembly at all scales. Science 295:2418–2421
67. Whitesides GM, Boncheva M (2002) Beyond molecules: self-assembly of mesoscopic and macroscopic components. Proc Natl Acad Sci U.S.A 99(8):4769–4774
68. Mastrangeli M et al (2009) Self-assembly from milli- to nanoscales: methods and applications. J Micromech Microeng 19:083001
69. Terfort A et al (1997) Three-dimensional self-assembly of millimetre-scale components. Nature 386:162–164

70. Onoe H et al (2007) Three-dimensional sequential self-assembly of microscale objects. Small 3(8):1383–1389
71. Grzybowski BA et al (2009) Self-assembly: from crystals to cells. Soft Matter 5:1110–1128
72. Gao Y et al (2012) Imaging enzyme-triggered self-assembly of small molecules inside live cells. Nat Comm 3:1033–1049
73. Mastrangeli M et al (2011) Modeling self-assembly across scales: the unifying perspective of smart minimal particles. Micromachines 2:82–115
74. Hosokawa K et al (1996) Two-dimensional micro-self-assembly using the surface tension of water. Sens Actuators A: Phys 57(2):117–125
75. Bishop KJM et al (2009) Nanoscale forces and their uses in self-assembly. Small 5(14):1600–1630
76. Chung SE et al (2008) Guided and fluidic self-assembly of microstructures using railed microfluidic channels. Nat Mater 7:581–587
77. Legrain A et al (2011) Elastocapillary folding of three dimensional micro-structures using water pumped through the wafer via a silicon nitride tube. In: Proceedings of the 22nd Micromechanics and Microsystems Technology Europe Workshop, pp 250–253, Toensberg, Norway, 19–22 June 2011
78. Syms RRA et al (2003) Surface tension-powered self-assembly of microstructures—the state-of-the-art. J Microelectromech Syst 12(4):387–416
79. Tavacoli JW et al (2013) The fabrication and directed self-assembly of micronsized superparamagnetic non-spherical particles. Soft Matter 9:9103–9110
80. Iwase E, Shimoyama I (2005) Multistep sequential batch assembly of three-dimensional ferromagnetic microstructures with elastic hinges. J Microelectromech Syst 14(6):1265–1271
81. Shechter E et al (2007) towards batch fabrication and assembly of 3D microstructures: a sequential assembly planner with new hard magnet configuration. In: Proceedings of the 2007 IEEE/RSJ international conference on intelligent robots and systems, pp 584–589, San Diego, USA, 29 October–2 November, 2007
82. Ghosh S, Puri IK (2013) Soft polymer magnetic nanocomposites: microstructure patterning by magnetophoretic transport and self-assembly. Soft Matter 9:2024–2029
83. Lee SW, Bashir R (2003) Dielectrophoresis and electrohydrodynamics-mediated fluidic assembly of silicon resistors. Appl Phys Lett 83(18):3833–3835
84. Zhou M et al (2009) Self-assembled peptide-based hydrogels as scaffolds for anchorage dependent cells. Biomaterials 30:2523–2530
85. Andersen ES et al (2009) Self-assembly of a nanoscale DNA box with a controllable lid. Nature 459:73–77
86. Qi H et al (2013) DNA-directed self-assembly of shape-controlled hydrogels. Nat Comm 4:2275–2284
87. Juul S et al (2013) Temperature-controlled encapsulation and release of an active enzyme in the cavity of a self-assembled DNA nanocage. ACS Nano 7(11):9724–9734
88. Randall CL et al (2007) 3D lithographically fabricated nanoliter containers for drug delivery. Adv Drug Deliv Rev 59:1547–1561
89. Shim TS et al (2012) Controlled origami folding of hydrogel bilayers with sustained reversibility for robust microcarriers. Angew Chem Int Ed 51:1420–1423
90. Xu F et al (2011) The assembly of cell-encapsulating microscale hydrogels using acoustic waves. Biomaterials 32:7847–7855
91. Du Y et al (2008) Directed assembly of cell-laden microgels for fabrication of 3D tissue constructs. Proc Natl Acid Sci U.S.A 105(28):9522–9527
92. Fernandez JG, Khademhosseini A (2010) Micro-masonry: construction of 3D structures by microscale self-assembly. Adv Mater 22:1–4
93. Du Y et al (2010) Surface-directed assembly of cell-laden microgels. Biotechnol Bioeng 105(3):655–662
94. Leong TG et al (2010) Three dimensional fabrication at small size scales. Small 6(7):792–806

95. Brittain ST et al (2001) Microorigami: fabrication of small, three-dimensional, metallic structures. J Phys Chem B 105:347–350
96. Azam A et al (2009) Compactness determines the success of cube and octahedron self-assembly. PLoS One 4(2):e4451
97. Ionov L (2011) Soft microorigami: self-folding polymer films. Soft Matter 7:6786–6791
98. Suzuki K et al (2007) Self-assembly of three dimensional micro mechanisms using thermal shrinkage of polyimide. Microsyst Technol 13:1047–1053
99. Jamal M et al (2011) Differentially photo-crosslinked polymers enable self-assembling microfluidics. Nat Comm 2:527–532
100. Vaccaro PO et al (2003) Valley-fold and mountain-fold in the micro-origami technique. Microelectr J 34:447–449
101. Smela E et al (1995) Controlled folding of micrometer-size structures. Science 268:1735–1738
102. Laflin KE et al (2012) Laser triggered sequential folding of microstructures. Appl Phys Lett 101:131901
103. Richter K et al (2010) The heat shock response: life on the verge of death. Mol Cell 40:253–266
104. Park J et al (2013) Design for a lithographically patterned bioartificial endocrine pancreas. Art Org 37(12):1059–1067
105. Vasiev I et al (2013) Self-folding nano- and micropatterned hydrogel tissue engineering scaffolds by single step photolithographic process. Microelectron Eng 108:76–81
106. Ananthakrishnan R, Ehrlicher A (2007) The forces behind cell movement. Int J Biol Sci 3(5):303–317
107. Aoki K et al (2003) Microassembly of semiconductor three-dimensional photonic crystals. Nat Mater 2:117–121
108. Zhang Y et al (2013) Microassembly of MEMS actuators and sensors via micro-masonry. Paper presented at the 26th international conference on Micro Electro Mechanical Systems, Taipei International Convention Center, Taipei, 20–24 January 2013
109. Arai F, Fukuda T (1997) A new pick up and release method by heating for micromanipulation. In: Proceedings of the 10th international conference on micro electro mechanical systems, Nagoya, Japan, 26–30 January 1997
110. Tsuchiya K et al (1999) Micro assembly and micro bonding in nano manufacturing world. In: Nelson BJ, Breguet JM (eds) Proceedings of SPIE conference on microrobotics and microassembly, vol 3834, pp 132–140
111. Cecil J et al (2005) A review of gripping and manipulation techniques for micro-assembly applications. Int J Prod Res 43(4):819–828
112. Savia M, Koivo HN (2009) Contact micromanipulation–survey of strategies. IEEE/ASME Trans Mechatron 14(4):504–514

Chapter 11
Photofablication Techniques for 3D Tissue Construct

Shinji Sugiura, Fumiki Yanagawa and Toshiyuki Kanamori

Abstract Living cells in human body exhibit their function in microstructured three dimensional (3D) environments composed of soluble factors, extracellular matrix (ECM), and neighboring cells. Fabricating biological 3D *in vitro* models of tissue and organ is a critical step for developing new strategies for cell based assay in drug discovery. To date, a number of research groups have developed micropatterning of hydrogels using photocurable and photodegradable hydrogels. This chapter reviews recent development on the techniques for microscale 3D fabrication of biomaterials by means of photolithography. Also, this chapter introduces an activated-ester-type photocleavable crosslinker, which we recently synthesized to generate photodegradable hydrogels using biocompatible materials such as polyethylene glycol and gelatin. This new type of crosslinker enabled convenient preparation of photodegradable hydrogel by two component mixing reaction. The hydrogels were degraded by micropatterned light irradiation, local light irradiation, and two-photon excitation. This simple and convenient approach to prepare and fabricate photodegradable hydrogels is creating new opportunity for novel cell manipulation and 3D tissue engineering techniques.

Keywords Tissue engineering · Light · Hydrogel · Biomaterials · Photodegradation

11.1 Introduction

In order to develop new cell based assay technique for drug discovery, researchers in academia and industry require *in vitro* models that could reproduce biological events *in vivo* models. Current two dimensional (2D) *in vitro* platforms are useful for investigating the molecular responses in tissue and organ based on physiological and pathological studies. Recent advances in molecular biology can identify the proteins, receptors and ligands responses 2D *in vitro* models. However, these

S. Sugiura (✉) · F. Yanagawa · T. Kanamori
Research Center for Stem Cell Engineering, National Institute of Advanced Industrial Science and Technology (AIST), Tsukuba, Ibaraki, Japan
e-mail: shinji.sugiura@aist.go.jp

T. Arai et al. (eds.), *Hyper Bio Assembler for 3D Cellular Systems*,
DOI 10.1007/978-4-431-55297-0_11

2D *in vitro* platforms do not precisely simulate the complicated cell-cell and cell-extracellular matrix (ECM) interactions, while cells *in vivo* communicate with each other in a microstructured three dimensional (3D) environment in response to soluble factors, ECM molecules and intercellular contact dependent signals. Therefore, cell culture on 2D substrates, which has been employed in conventional biological research, does not adequately recapitulate the 3D nature of native cellular microenvironment [1, 2]. In addition, spatially uniform and static materials lack the intricate spatial and temporal aspects of *in vivo* systems. Cells dynamically respond to the local microenvironment during diverse processes such as tissue morphogenesis, stem cell differentiation, cancer progression [3, 4]. Recapitulating such dynamic microenvironments 3D *in vitro* platforms would have high potential impact in cell biology by providing an excellent model for systematic differentiation of stem cells, and for understanding of tissue regeneration, ultimately leading to more rational tissue engineering strategies. Recently, the approaches for fabricating 3D *in vitro* models using biocompatible hydrogels based on top-down and bottom-up methods have been used in stem cell research, regenerative medicine and tissue engineering. In particular, photoresponsible hydrogels such as photocurable and photodegradable hydrogels have been employed to reproduce complicated structures for mimicking complicated 3D tissue and organ.

11.2 Photocurable Hydrogels

Photocrosslinkable polymers are popular materials for controlling not only hydrogel microstructure and stiffness but also cellular behavior on and within the hydrogels. A number of research groups have generated micropatterned hydrogels for tissue engineering applications with photocrosslinked polymers, including derivatives of poly(ethylene glycol) diacrylate (PEGDA) [5–7], polysaccharides [8–10], and proteins (e.g., gelatin) [11, 12]. In order to construct 3D Microarchitecture, photolithography and stereolithography have been applied to fabricate hydrogels using these photocurable materials [12, 13–17, 18]. Also, the approaches can provide the 3D *in vitro* platforms for controlling systematic differentiation of stem cells [13, 19]. Although Photocurable Hydrogels are useful materials to construct complex engineered tissues, they may not be readily applicable to generating complex 3D vascularized tissue. Eventually, either microengineered 3D porous or perfusable structures in these hydrogels to perform long-term cell culture are necessary for supplying oxygen, nutrients and other soluble factors for cell growth inside engineered 3D thick tissues [7, 12, 13, 15, 16, 17, 18, 20]. To construct perfusable 3D thick tissue, fabrication of porous or hollow structure in hydrogels is desirable. For this application, use of photodegradable hydrogels, which are degraded by light irradiation, can be another possible approach.

11.3 Photodegradable Hydrogels

Recently, photodegradable hydrogels have attracted significant attention due to their tunable mechanical, chemical properties and their use in the fabrication of 3D microstructures [21, 22] in biomaterials and tissue engineering research [23]. Such photodegradable hydrogels have garnered substantial attention from the biomaterials and tissue engineering research fields [24, 25]. The physical and chemical properties of photodegradable hydrogels can be both temporally and spatially controlled by irradiation with light (single- and two-photon) [26, 27], and this process is compatible with living cells [21, 28].

To date, despite reporting many approaches, current methods to prepare photodegradable hydrogels are restricted to reactions between synthetic molecules, such as radical reactions [21, 27, 29, 30–32], Michael-type conjugations [33], and orthogonal click reactions [22, 34]. In these reactions, one class of convenient hydrogel preparation methods is multicomponent mixing reaction, in which two or more molecular-scale components react with each other and gelation takes place spontaneously after mixing. Such systems are highly tunable as either component can be easily modified to alter the hydrogel performance or to introduce additional functionalities. In the previous study, Michael-type conjugations [34] and orthogonal click reactions [22, 33] have been applied to fabrication of photodegradable hydrogels, by multicomponent mixing reaction. However, these reactions require chemical modification of the component materials. In addition, most of the resulting photodegradable hydrogels were composed of derivatives of PEG or polysaccharides [22, 34], which are biologically inactive; cells can neither bind to nor degrade them. This lack of cell-responsive features possibly affects the viability and activities of cells, and greatly limits the cultural types of cells and applicable culture conditions. Although cell-binding sequences such as Arg-Gly-Asp and other peptides can be incorporated into photodegradable hydrogels by using a combination of synthetic homo- and heterofunctional crosslinkers [21, 22], development of simple systems to prepare biologically functional photodegradable hydrogels can facilitate widespread use of this technology.

11.4 Activated-Ester-Type Photocleavable Crosslinker

We recently reported an activated-ester-type photocleavable crosslinker, namely photocleavable *N*-hydroxysuccinimide (NHS) tetra-arm polyethylene glycol (NHS-PC-4armPEG), which facilitated the formation of photodegradable hydrogels by means of a one-step, two-component mixing reaction with a biocompatible polymer containing amino moieties (amino-terminated tetra-arm PEG (amino-4armPEG) or gelatin) [35]. The NHS-PC-4armPEG is composed of the following functional groups (Fig. 11.1): (i) poly(ethylene glycol) as a water-soluble main polymer chain; (ii) nitrobenzyl groups, which are cleavable in response to light irradiation; and

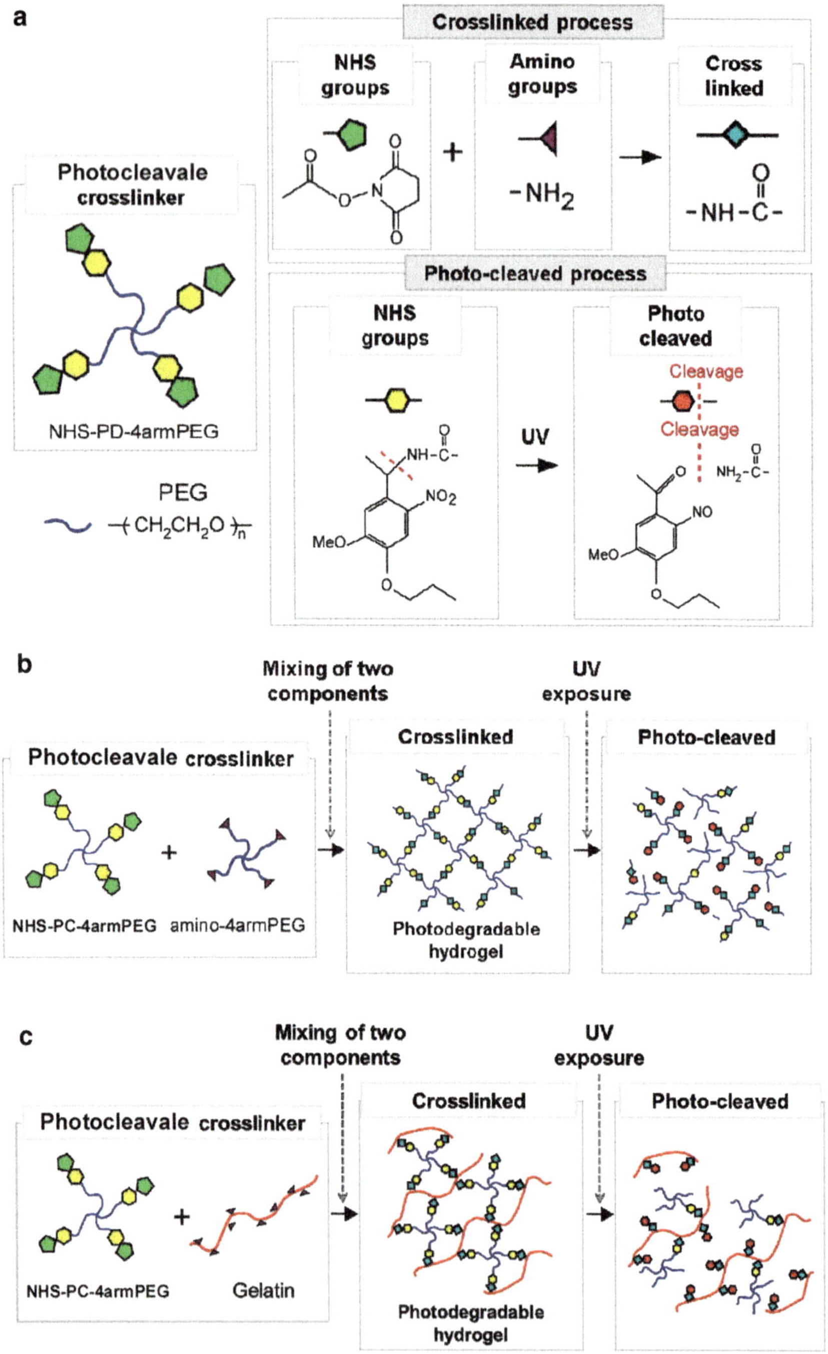

Fig. 11.1 Shematic diagram of, **a** chemical structure and cleavage process of photocleavable crosslinker. Shematic diagram of degradation process of photo-degradable hydrogels with, **b** amino-4arm PEG and, **c** gelatin

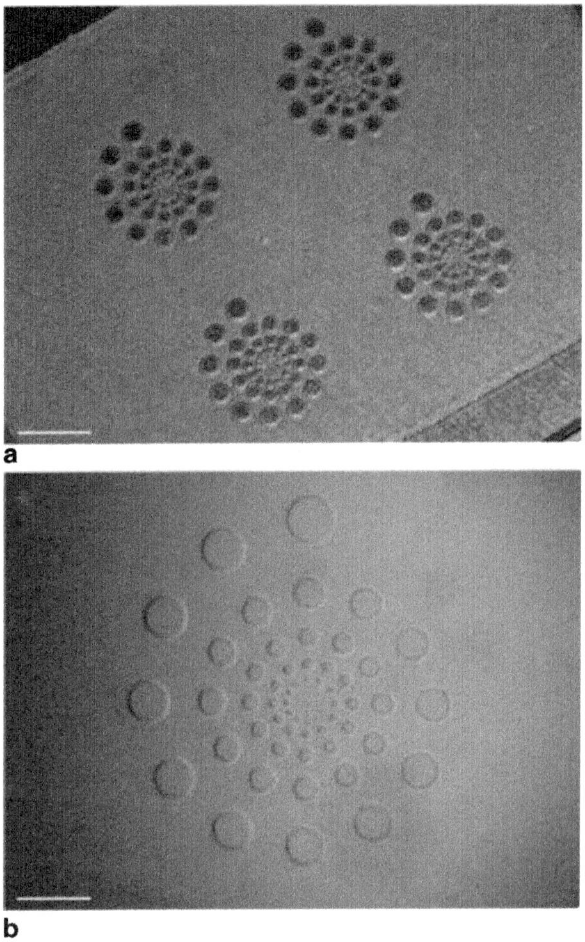

Fig. 11.2 Microscopic images of patterned photodegradable hydrogels crosslinked with amino-4arm PEG (a), and gelatin (b). Bars = 1 mm (a) and 500 μm (b)

(iii) NHS ether groups, which react with primary amine groups and form peptide bonds under physiological conditions.

To demonstrate hydrogel micropatterning by photolithography, prepolymer solution which included 5.0 mM amino-4arm polyethylene glycol (amino-4arm-PEG, Mw = 9,617) or 2.5 % (w/v) gelatin was mixed with 0.25–5.0 mM NHS-PC-4armPEG solution. After mixing of the two components, a drop of mixture was placed onto an amino coated glass slide covered by a cover slip and separated by a cover slip or pet film spacers to provide hydrogel units with controlled thicknesses. Subsequently, film masks were placed on a cover slip between light source and hydrogel, and then each sample were exposed to light. Light irradiation through film mask degraded exposed regions in hydrogels and created micropatterned structure (Fig. 11.2), whereas unexposed regions remained intact. Patterning on

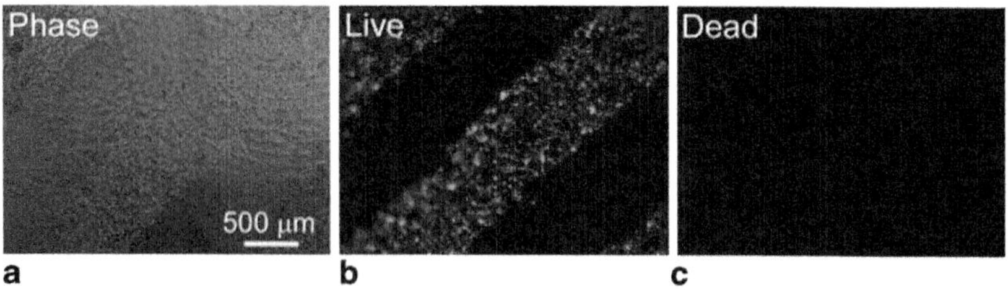

Fig. 11.3 Cell encapsulation (HepG2) in patterned photodegradable hydrogels prepared with gelatin and photocleavable crosslinker. Live/dead test was carried out after 2 days of culture.

photodegradable hydrogels with microscale precision (20–500 μm) by photolithography was achieved. To estimate morphology of hydrogels degradation, microparticles suspension was put onto patterned hydrogels and the 3D images were captured using confocal microscope. As expected, the depth of degraded regions increased proportionally to light exposure time. That is, morphology of degraded regions was a function of the amount of light exposure energy. We have also demonstrated cell micropatterning in the photodegradable hydrogel (Fig. 11.3). Human liver carcinoma (HepG2) cells has been successfully encapsulated in gelatin based photodegradable hydrogel and patterned by micropatterned light irradiation with maintaining their viability.

11.5 Conclusion

Our new Photocleavable crosslinker can be utilized to prepare photodegradable hydrogel using biomaterials by two-component mixing. Theoretically, NHS group can crosslink any molecules with primary amine. Therefore, many types of components including synthetic polymer, peptide, and cells can be crosslinked in the photodegradable hydrogels (Fig. 11.4). The prepared photodegradable hydrogel were precisely fabricated with desired morphology by micropatterned light irradiation through photomask, local light irradiation from maskless light irradiation system, and two-photon reaction. We recently applied our photodegradable hydrogel and maskless light irradiation system to single cell manipulation in the hydrogels [36]. These approaches could provide variety of tools to fabrication, manipulation, and evaluation of cells and 3D tissues in the future.

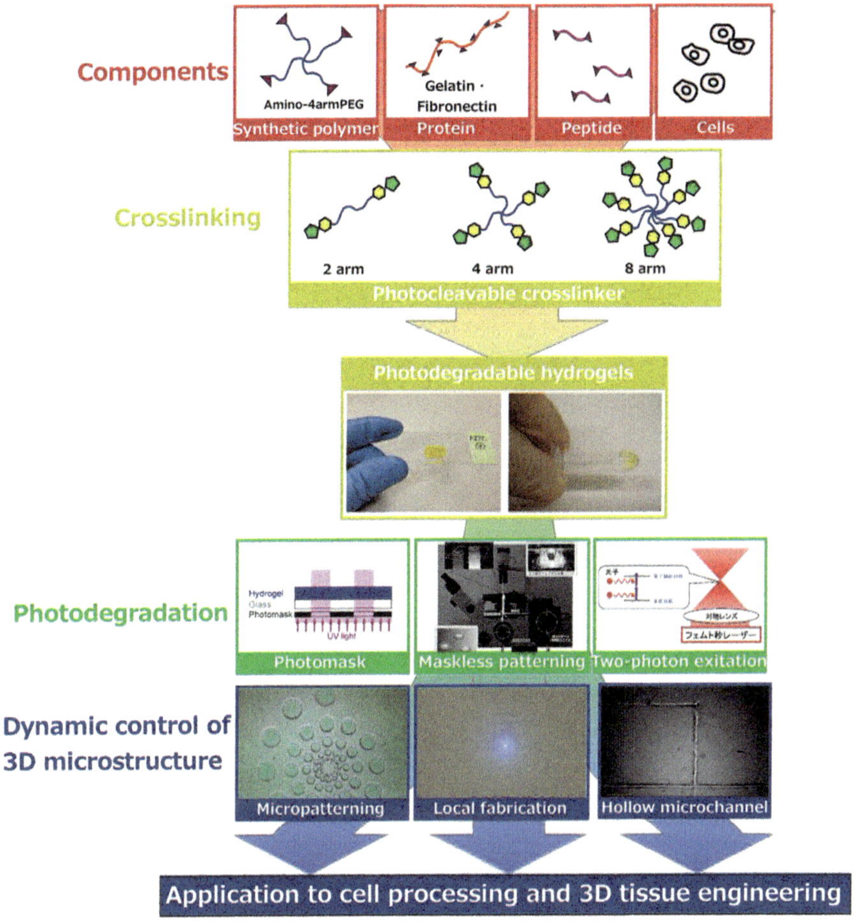

Fig. 11.4 Potential application of photodegradable hydrogels

References

1. Pampaloni F, Reynaud EG, Stelzer EHK (2007) The third dimension bridges the gap between cell culture and live tissue. Nat Rev Mol Cell Biol 8:839–845
2. Tibbitt MW, Anseth KS (2009) Hydrogels as extracellular matrix mimics for 3D cell culture. Biotechnol Bioeng 103:655–663
3. Daley WP, Peters SB, Larsen M (2008) Extracellular matrix dynamics in development and regenerative medicine. J Cell Sci 121:255–264
4. Lopez JI, Mouw JK, Weaver VM (2008) Biomechanical regulation of cell orientation and fate. Oncogene 27:6981–6993
5. Slaughter BV, Khurshid SS, Fisher OZ, Khademhosseini A, Peppas NA (2009) Hydrogels in regenerative medicine. Adv Mater 21:3307–3329

6. Du Y, Lo E, Ali S, Khademhosseini A (2008) Directed assembly of cell-laden microgels for fabrication of 3D tissue constructs. Proc Natl Acad Sci U S A 105:9522–9527
7. Yanagawa F, Kaji H, Jang Y-H, Bae H, Yanan D, Fukuda J, Qi H, Khademhosseini A (2011) Directed assembly of cell-laden microgels for building porous three-dimensional tissue constructs. J Biomed Mater Res A 97A:93–102
8. Marklein RA, Burdick JA (2010) Spatially controlled hydrogel mechanics to modulate stem cell interactions. Soft Matter 6:136–143
9. Camci-Unal G, Cuttica D, Annabi N, Demarchi D, Khademhosseini A (2013) Synthesis and characterization of hybrid hyaluronic acid-gelatin hydrogels. Biomacromolecules 14:1085–1092
10. Shin H, Olsen BD, Khademhosseini A (2012) The mechanical properties and cytotoxicity of cell-laden double-network hydrogels based on photocrosslinkable gelatin and gellan gum biomacromolecules. Biomaterials 33:3143–3152
11. Van den Bulcke AI Bogdanov B De Rooze N Schacht EH Cornelissen M Berghmans H (2000) Structural and rheological properties of methacrylamide modified gelatin hydrogels. Biomacromolecules 1:31–38
12. Nichol JW, Koshy ST, Bae H, Hwang CM, Yamanlar S, Khademhosseini A (2010) Cell-laden microengineered gelatin methacrylate hydrogels. Biomaterials 31:5536–5544
13. Qi H, Du Y, Wang L, Kaji H, Bae H, Khademhosseini A (2010) Patterned differentiation of individual embryoid bodies in spatially organized 3D hybrid microgels. Adv Mater 22:5276–5281
14. Aubin H, Nichol JW, Hutson CB, Bae H, Sieminski AL, Cropek DM, Akhyari P, Khademhosseini A (2010) Directed 3D cell alignment and elongation in microengineered hydrogels. Biomaterials 31:6941–6951
15. Zorlutuna P, Jeong JH, Kong H, Bashir R (2011) Stereolithography-based hydrogel microenvironments to examine cellular interactions. Adv Funct Mater 21:3642–3651
16. Chan V, Zorlutuna P, Jeong JH, Kong H, Bashir R (2010) Three-dimensional photopatterning of hydrogels using stereolithography for long-term cell encapsulation. Lab Chip 10:2062–2070
17. Hammoudi TM, Lu H, Temenoff JS (2010) Long-term spatially defined coculture within three-dimensional photopatterned hydrogels. Tissue Eng Part C 16:1621–1628
18. Khetan S, Burdick JA (2010) Patterning network structure to spatially control cellular remodeling and stem cell fate within 3-dimensional hydrogels. Biomaterials 31:8228–8234
19. Sugiura S, Cha J M, Yanagawa F, Zorlutuna P, Bae H, Khademhosseini A (2013) Dynamic three-dimensional micropatterned cell co-cultures within photocurable and chemically degradable hydrogels. J Tissue Eng Regen Med. doi: 10.1002/term.1843
20. Gauvin R, Chen Y-C, Lee JW, Soman P, Zorlutuna P, Nichol JW, Bae H, Chen S, Khademhosseini A (2012) Microfabrication of complex porous tissue engineering scaffolds using 3D projection stereolithography. Biomaterials 33:3824–3834
21. Kloxin AM, Kasko AM, Salinas CN, Anseth KS (2009) Photodegradable hydrogels for dynamic tuning of physical and chemical properties. Science 324:59–63
22. DeForest CA, Anseth KS (2011) Cytocompatible click-based hydrogels with dynamically tunable properties through orthogonal photoconjugation and photocleavage reactions. Nat Chem 3:925–931
23. Lutolf MP (2012) Materials science: cell environments programmed with light. Nature 482:477–478
24. Blow N (2009) Cell culture: building a better matrix. Nat Meth 6:619–U694
25. Jay SM, Saltzman WM (2009) Shining light on a new class of hydrogels. Nat Biotechnol 27:543–544
26. Kloxin AM, Tibbitt MW, Kasko AM, Fairbairn JA, Anseth KS (2010) Tunable hydrogels for external manipulation of cellular microenvironments through controlled photodegradation. Adv Mater 22:61–66
27. Wong DY, Griffin DR, Reed J, Kasko AM (2010) Photodegradable hydrogels to generate positive and negative features over multiple length scales. Macromolecules 43:2824–2831

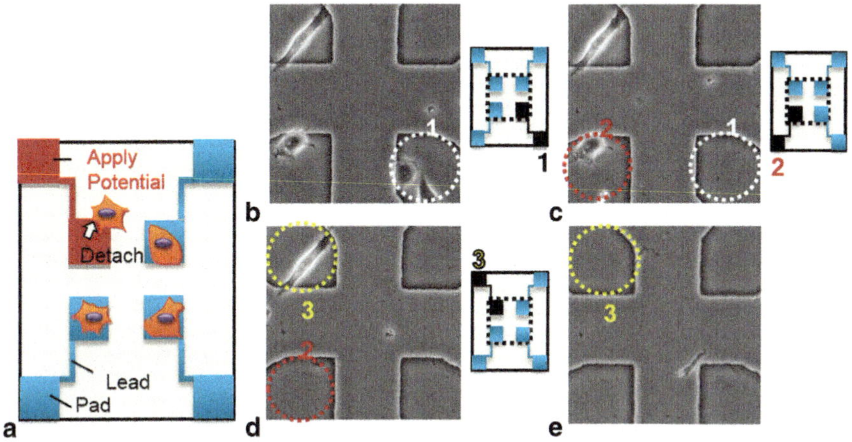

Fig. 12.1 Spatially resolved detachment of single cells using micropatterned electrodes. **a** Illustration of micropatterned ITO electrodes modified with alkanethiol SAMs. A cell on an electrode was detached by applying a negative potential. **b** The *circle* labeled 1 shows the initially activated electrode. **c** The cell on the electrode labeled 1 was detached. The *circle* labeled 2 indicates the subsequently activated electrode. **d** The cell on the electrode labeled 2 was detached. The *circle* labeled 3 indicates the activated electrode. **e** The cell on the electrode labeled 3 was detached

of cells along with the desorption of SAMs [13]. Selective detachment of cells using this technique was achieved by micropatterning electrode arrays on a glass surface (Fig. 12.1a) [15]. Instead of gold electrodes, transparent indium tin oxide (ITO) electrodes were used. Halation makes it difficult to observe cells on gold electrode arrays because of the large difference in light permeability between uncoated and gold-coated regions of glass. Cells on ITO micropatterns were individually detached with single-cell resolution (Fig. 12.1b). These techniques might be promising tools for collecting specific cells based on shape and location, a capability that is beyond typical cell sorting techniques, such as fluorescent-activated cell sorting and magnetic-activated cell sorting.

12.3 Formation and Collection of Spheroids

Multicellular aggregates such as spheroids have been investigated as cancer models for drug development and as cellular building blocks for tissue engineering applications. Compared with conventional monolayer culture, spheroid cultures are considered to provide microenvironments more closely related to *in vivo* conditions [16, 17]. For example, primary hepatocytes and articular chondrocytes maintained their organ-specific functions when they were cultured as spheroids but not when grown in monolayer cultures [18]. Various spheroid fabrication techniques that employ nonadherent surfaces [19], spinner flasks [20], hanging drop methods [21], micropatterning of cell-adhesive/non-adhesive surfaces [22, 23], and micromolding [24, 25] have been reported. Microwells consisting of thermoresponsive hydrogel

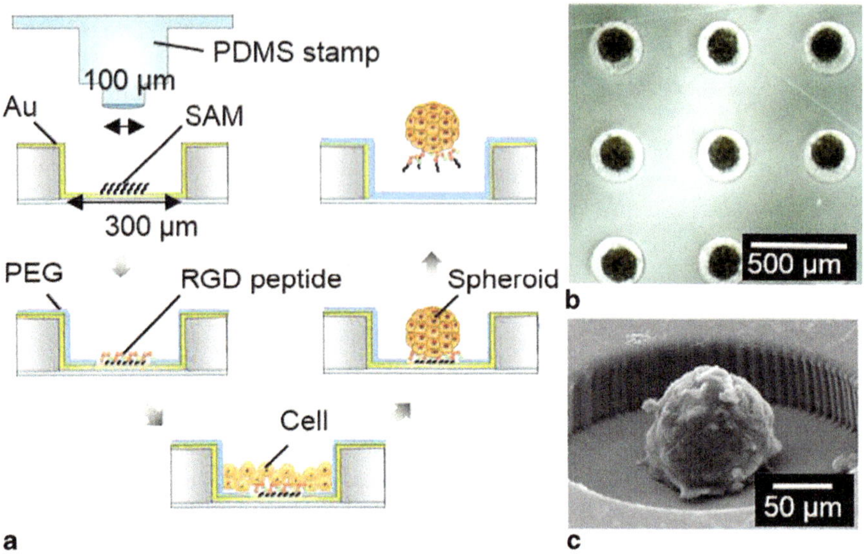

Fig. 12.2 Fabrication and collection of spheroids through an electrochemical reaction. **a** Schematic diagram. **b** Hepatocytes forming spheroids with a uniform diameter in each microwell. **c** Scanning electron microscope (SEM) image of a spheroid in a microwell

have been used not only to generate spheroids but also to subsequently release them via a temperature-induced change in the hydrogel [26]. Spheroids have been harvested also with an electrochemical approach [13]. The technique combined an electrochemical reaction with photolithography and micro-contact printing, as shown in Fig. 12.2a. Microwells were fabricated with photolithography and covered with a thin gold layer. Then, the central region of the microwells was modified with RGD peptide-terminated alkanethiol SAMs using microcontact printing. The entire region, except the alkanethiol spots, was modified with poly(ethylene glycol) to prevent cell attachment. After 3 days in culture, cells seeded in the microwells had spontaneously formed spheroids on the cell-adhesive region at the center of the microwells (Fig. 12.2b, c). The spheroids were subsequently collected by applying a negative potential. In the field of bioprinting, spheroids have been printed as an ink on a substrate in order to fabricate larger cellular tissues. These approaches to fabricating and collecting uniform spheroids might become useful tools for bioprinting and other cellular build-up approaches.

12.4 Detachment and Stacking of Cell Sheets

The most promising approach for cell sheet engineering is the use of thermoresponsive polymers [27]. Cell sheet engineering using this approach has become sophisticated in recent years and has been evaluated in diverse clinical trials, including

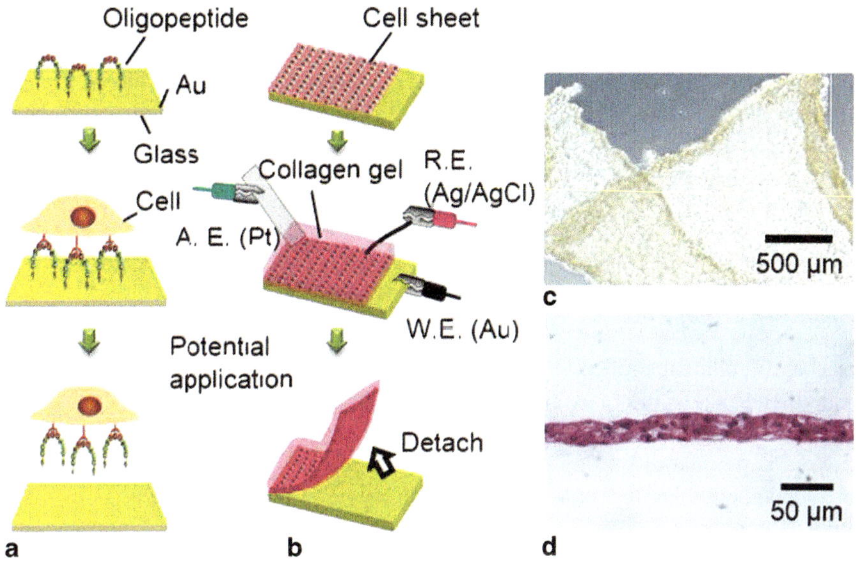

Fig. 12.3 Stacking of cell sheets by electrochemical stimuli. **a** Electrochemical detachment of cells along with the reductive desorption of the oligopeptide. **b** Detachment of cell sheets. **c** Electrochemically harvested mono-layer fibroblast sheet. **d** Three-layer hepatocyte sheet fabricated by stacking harvested cell sheets one by one

diseases of the cornea, myocardium, and esophagus. The major advantage of cell sheet engineering is that intact cell sheets can be obtained in which cell-cell junctions and deposited extracellular matrices are maintained, in contrast to their destruction during typical cell collection using trypsin. Detached cell sheets can be readily stacked with other cell sheets in succession to form a multilayered cell sheet [13]. In addition to thermoresponsive polymers, other stimulus-responsive surfaces have been proposed for engineering stacked cell sheets [13, 28–30]. Polyelectrolyte films have been used as sacrificial layers. Polyelectrolytes spontaneously adsorb onto oppositely charged substrates through electrostatic interactions and form thin polymer films. Change in the local pH at the substrate/film interface, which was caused by water electrolysis, led to dissolution of the polyelectrolyte film and detachment of cell sheets [31].

Using an electroresponsive system (Fig. 12.3a, b), similarly to the thermoresponsive system, cell sheets with the dimensions of 10 mm × 10 mm were collected (Fig. 12.3c) and stacked to fabricate multi-layered cell sheets (Fig. 12.3d). Compared to the thermoresponsive system, an advantage of the electroresponsive system is that cell sheet detachment can be accomplished more quickly. Typically, cell sheet harvesting took 30–70 min [32] with the thermoresponsive system, but took only 10 min in the electroresponsive system [13]. To improve the supply of oxygen and nutrients to dense cell sheets, polydimethylsiloxane (PDMS) membranes [24, 33] or cell culture inserts have been used [34]. PDMS possesses excellent biocompatibility and gas permeability properties [35]. Hepatoblastoma cells adhered, grew, and formed cell sheets with a thickness of > 50 μm on a PDMS substrate [33], primarily because of the improved oxygen supply. Further increase in the thickness of a cell sheet requires that microvessels supply oxygen through perfusion. Cocultivation of endothelial cells in cardiac cell sheets has shown that endothelial cells form vascular networks in cell sheets, thereby making it possible to fabricate thicker cardiac cell sheets [36].

12.5 Engineering Vascularized Tissues

Successful clinical applications have been limited to thin or avascular tissues, such as skin, cartilage, and bladder [37, 38], and fabricating viable 3-dimensional constructs remains a major challenge in tissue engineering. One of the major limiting factors is the inability to deliver nutrients and oxygen via vascular networks that are fundamental in most important organs in the body such as the liver, kidney, and lung. The shortage of oxygen supply poses the risk of hypoxia and necrotic cell death in the core of the tissue constructs [39].

A number of studies in recent decades have reported the fabrication of vascularized tissues *in vitro*, and endothelial cells have played a pivotal role in the formation of functional vascular networks [40]. Most previous approaches, however, relied on self-organization of endothelial cells in hydrogels and cell sheets [41, 42], which were typically too small to perfuse culture medium throughout tissue constructs in order to supply oxygen and nutrients. Thus, even when luminal structures with endothelial cells were fabricated, the supply of oxygen and nutrients was mostly limited to molecular diffusion [43]. Fabricating perfusable vascular networks might be vital for the fabrication of thick and cell-dense tissue constructs *in vitro* [44–46].

More recently, several approaches for the *in vitro* fabrication of perfusable endothelialized vascular structures have been reported [47–50]. In these approaches, endothelial cells were seeded into previously formed microchannels in hydrogels. Sufficient oxygen and nutrients for cells encapsulated in the hydrogels were provided by perfusion of culture medium through the endothelialized channels. One drawback of such approaches is that the flow of culture medium was stopped for a while in order to allow endothelial cells to attach to the inner surface of the channel, thereby making it difficult to avoid hypoxic injury to cells embedded at a high cell density in hydrogels. Prompt initiation of culture medium flow will be required to satisfy their oxygen demand.

One promising feature of electrochemical approaches is that cells can be rapidly detached from cylindrical rods to fabricate endothelial cell-lined vascular-like structures (Fig. 12.4a) [51]. By applying a negative potential to the gold rods for 5 min, endothelial cell monolayers were transferred onto the internal surface of microchannels in collagen gel. After extracting the rods from the gel, culture medium was perfused immediately. In the subsequent perfusion culture, it was seen that the transferred endothelial cells sprouted and formed luminal structures in the gel

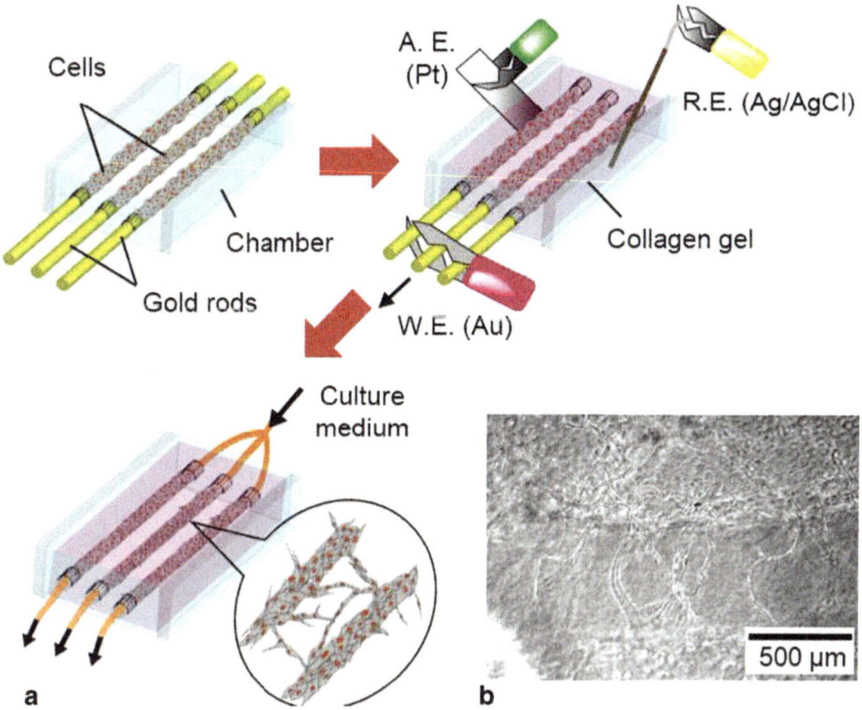

Fig. 12.4 Fabrication of vascular like structures. **a** The procedures. The oligopeptide was chemically adsorbed onto gold rods via gold–thiolate bonds. Endothelial cells attached to the surface were detached during the reductive desorption of the oligopeptide, and then the rods were extracted from the chamber. The capillaries were connected to a microsyringe pump in order to perfuse culture medium. **b** Migration of transferred endothelial cells in collagen gel. The luminal structures of endothelial cells extended to the neighboring vascular-like structures and bridged them

(Fig. 12.4b). This approach has potential for engineering vascularized tissues that are capable of supplying sufficient oxygen and nutrients to entire tissue constructs.

12.6 Conclusion

This chapter has summarized approaches for modulating cell adhesion on a culture surface using external stimuli, such as thermal, electrical, photo, and magnetic fields. Electrochemical approaches for cell detachment are described in detail. These approaches have been used for several interesting purposes, including the detachment of specific cells at single-cell resolution under a microscope, the collection of spheroids, the fabrication of stacked multi-layered cell sheets, and engineering perfusable vascular structures in tissue constructs. These fundamental approaches will contribute to the further advancement of the fields of tissue engineering and regenerative medicine.

References

1. Lahann J, Mitragotri S, Tran TN, Kaido H, Sundaram J, Choi IS, Hoffer S, Somorjai GA, Langer R (2003) A reversibly switching surface. Science 299(5605):371–374
2. Wildt B, Wirtz D, Searson PC (2009) Programmed subcellular release for studying the dynamics of cell detachment. Nat Methods 6(3):211–213. doi:10.1038/nmeth.1299
3. Egami M, Haraguchi Y, Shimizu T, Yamato M, Okano T (2014) Latest status of the clinical and industrial applications of cell sheet engineering and regenerative medicine. Arch Pharmacal Res 37(1):96–106. doi:10.1007/s12272-013-0299-8
4. Fukuda J, Khademhosseini A, Yeh J, Eng G, Cheng J, Farokhzad OC, Langer R (2006) Micropatterned cell co-cultures using layer-by-layer deposition of extracellular matrix components. Biomaterials 27(8):1479–1486. doi:10.1016/j.biomaterials.2005.09.015
5. Cole MA, Voelcker NH, Thissen H, Griesser HJ (2009) Stimuli-responsive interfaces and systems for the control of protein-surface and cell-surface interactions. Biomaterials 30(9):1827–1850. doi:10.1016/j.biomaterials.2008.12.026
6. Yang J, Yamato M, Shimizu T, Sekine H, Ohashi K, Kanzaki M, Ohki T, Nishida K, Okano T (2007) Reconstruction of functional tissues with cell sheet engineering. Biomaterials 28(34):5033–5043. doi:10.1016/j.biomaterials.2007.07.052
7. Elloumi-Hannachi I, Yamato M, Okano T (2010) Cell sheet engineering: a unique nanotechnology for scaffold-free tissue reconstruction with clinical applications in regenerative medicine. J Int Med 267(1):54–70. doi:10.1111/j.1365-2796.2009.02185.x
8. Nishida K, Yamato M, Hayashida Y, Watanabe K, Yamamoto K, Adachi E, Nagai S, Kikuchi A, Maeda N, Watanabe H, Okano T, Tano Y (2004) Corneal reconstruction with tissue-engineered cell sheets composed of autologous oral mucosal epithelium. N Engl J Med 351(12):1187–1196
9. Akiyama H, Ito A, Kawabe Y, Kamihira M (2010) Genetically engineered angiogenic cell sheets using magnetic force-based gene delivery and tissue fabrication techniques. Biomaterials 31(6):1251–1259. doi:10.1016/j.biomaterials.2009.11.017
10. Tamura M, Yanagawa F, Sugiura S, Takagi T, Sumaru K, Matsui H, Kanamori T (2014) Optical cell separation from three-dimensional environment in photodegradable hydrogels for pure culture techniques. Scientific reports 4:4793. doi:10.1038/srep04793
11. Yeo W-S, Mrksich M (2006) Electroactive self-assembled monolayers that permit orthogonal control over the adhesion of cells to patterned substrates. Langmuir 22(25):10816–10820
12. Kim M, Lee JY, Shah SS, Tae G, Revzin A (2009) On-cue detachment of hydrogels and cells from optically transparent electrodes. Chemical communications (39):5865–5867. doi:10.1039/b909169f
13. Inaba R, Khademhosseini A, Suzuki H, Fukuda J (2009) Electrochemical desorption of self-assembled monolayers for engineering cellular tissues. Biomaterials 30(21):3573–3579. doi:10.1016/j.biomaterials.2009.03.045
14. Yeo W-S, Mrksich M (2006) Electroactive self-assembled monolayers that permit orthogonal control over the adhesion of cells to patterned substrates. Langmuir (the ACS J Surf colloids) 22(25):10816–10820. doi:10.1021/la061212y
15. Fukuda J, Kameoka Y, Suzuki H (2011) Spatio-temporal detachment of single cells using microarrayed transparent electrodes. Biomaterials 32 (28):6663–6669. doi:10.1016/j.biomaterials.2011.05.068
16. Fukuda J, Nakazawa K (2005) Orderly arrangement of hepatocyte spheroids on a microfabricated chip. Tissue Eng 11(7–8):1254–1262
17. Griffith LG, Swartz MA (2006) Capturing complex 3D tissue physiology in vitro. Nat Rev Mol Cell Biol 7(3):211–224. doi:10.1038/nrm1858
18. Bierwolf J, Lutgehetmann M, Feng K, Erbes J, Deichmann S, Toronyi E, Stieglitz C, Nashan B, Ma PX, Pollok JM (2011) Primary Rat Hepatocyte Culture on 3D Nanofibrous Polymer Scaffolds for Toxicology and Pharmaceutical Research. Biotechnol Bioeng 108(1):141–150. doi:10.1002/bit.22924

19. Kunz-Schughart LA, Kreutz M, Knuechel R (1998) Multicellular spheroids: a three-dimensional in vitro culture system to study tumour biology. Int J Exp Pathol 79(1):1–23. doi:10.1046/j.1365-2613.1998.00051.x
20. Yamada K, Kamihira M, Hamamoto R, Iijima S (1998) Efficient induction of hepatocyte spheroids in a suspension culture using a water-soluble synthetic polymer as an artificial matrix. J Biochem 123(6):1017–1023
21. Kelm JM, Timmins NE, Brown CJ, Fussenegger M, Nielsen LK (2003) Method for generation of homogeneous multicellular tumor spheroids applicable to a wide variety of cell types. Biotechnol Bioeng 83(2):173–180. doi:10.1002/bit.10655
22. Fukuda J, Sakai Y, Nakazawa K (2006) Novel hepatocyte culture system developed using microfabrication and collagen/polyethylene glycol microcontact printing. Biomaterials 27(7):1061–1070
23. Wang W, Itaka K, Ohba S, Nishiyama N, Chung U-i, Yamasaki Y, Kataoka K (2009) 3D spheroid culture system on micropatterned substrates for improved differentiation efficiency of multipotent mesenchymal stem cells. Biomaterials 30(14):2705–2715. doi:10.1016/j.biomaterials.2009.01.030
24. Anada T, Fukuda J, Sai Y, Suzuki O (2012) An oxygen-permeable spheroid culture system for the prevention of central hypoxia and necrosis of spheroids. Biomaterials 33(33):8430–8441. doi:10.1016/j.biomaterials.2012.08.040
25. Markovitz-Bishitz Y, Tauber Y, Afrimzon E, Zurgil N, Sobolev M, Shafran Y, Deutsch A, Howitz S, Deutsch M (2010) A polymer microstructure array for the formation, culturing, and high throughput drug screening of breast cancer spheroids. Biomaterials 31(32):8436–8444. doi:10.1016/j.biomaterials.2010.07.050
26. Tekin H, Anaya M, Brigham MD, Nauman C, Langer R, Khademhosseini A (2010) Stimuli-responsive microwells for formation and retrieval of cell aggregates. Lab on a chip 10(18):2411–2418. doi:10.1039/c004732e
27. Ohashi K, Yokoyama T, Yamato M, Kuge H, Kanehiro H, Tsutsumi M, Amanuma T, Iwata H, Yang J, Okano T, Nakajima Y (2007) Engineering functional two- and three-dimensional liver systems in vivo using hepatic tissue sheets. Nat Med 13(7):880–885. doi:10.1038/nm1576
28. Kakegawa T, Mochizuki N, Sadr N, Suzuki H, Fukuda J (2013) Cell-Adhesive and Cell-Repulsive Zwitterionic Oligopeptides for Micropatterning and Rapid Electrochemical Detachment of Cells. Tissue Eng Part A 19(1–2):290–298. doi:10.1089/ten.tea.2011.0739
29. Guillaume-Gentil O, Semenov OV, Zisch AH, Zimmermann R, Voeroes J, Ehrbar M (2011) pH-controlled recovery of placenta-derived mesenchymal stem cell sheets. Biomaterials 32(19):4376–4384. doi:10.1016/j.biomaterials.2011.02.058
30. Ito A, Ino K, Kobayashi T, Honda H (2005) The effect of RGD peptide-conjugated magnetite cationic liposomes on cell growth and cell sheet harvesting. Biomaterials 26(31):6185–6193
31. Guillaume-Gentil O, Akiyama Y, Schuler M, Tang C, Textor M, Yamato M, Okano T, Voeroes J (2008) Polyelectrolyte coatings with a potential for electronic control and cell sheet engineering. Adv Mater 20(3):560–565 doi:10.1002/adma.200700758
32. Hyeong Kwon O Kikuchi A, Yamato M, Okano T (2003) Accelerated cell sheet recovery by co-grafting of PEG with PIPAAm onto porous cell culture membranes. Biomaterials 24(7):1223–1232
33. Evenou F, Hamon M, Fujii T, Takeuchi S, Sakai Y (2011) Gas-permeable membranes and Co-Culture with Fibroblasts Enable High-density Hepatocyte Culture as Multilayered Liver Tissues. Biotechnol Prog 27(4):1146–1153. doi:10.1002/btpr.626
34. Murakami D, Yamato M, Nishida K, Ohki T, Takagi R, Yang J, Namiki H, Okano T (2006) The effect of micropores in the surface of temperature-responsive culture inserts on the fabrication of transplantable canine oral mucosal epithelial cell sheets. Biomaterials 27(32):5518–5523
35. Evenou F, Fujii T, Sakai Y (2010) Spontaneous formation of highly functional three-dimensional multilayer from human hepatoma Hep G2 cells cultured on an oxygen-permeable polydimethylsiloxane membrane. Tissue Eng Part C Methods 16(2):311–318. doi:10.1089/ten.tec.2009.0042

36. Sekine H, Shimizu T, Sakaguchi K, Dobashi I, Wada M, Yamato M, Kobayashi E, Umezu M, Okano T (2013) In vitro fabrication of functional three-dimensional tissues with perfusable blood vessels. Nat commun 4:1399
37. Wisser D, Steffes J (2003) Skin replacement with a collagen based dermal substitute, autologous keratinocytes and fibroblasts in burn trauma. Burns 29(4):375–380. doi:10.1016/s0305-4179(03)00013-5
38. Atala A (2009) Engineering organs. Curr Opin Biotechnol 20(5):575–592. doi:http://dx.doi.org/10.1016/j.copbio.2009.10.003
39. Curcio E, Macchiarini P, De Bartolo L (2010) Oxygen mass transfer in a human tissue-engineered trachea. Biomaterials 31(19):5131–5136. doi:10.1016/j.biomaterials.2010.03.013
40. Levenberg S, Rouwkema J, Macdonald M, Garfein ES, Kohane DS, Darland DC, Marini R, van Blitterswijk CA, Mulligan RC, D'Amore PA, Langer R (2005) Engineering vascularized skeletal muscle tissue. Nat Biotechnol 23(7):879–884
41. Koike N, Fukumura D, Gralla O, Au P, Schechner JS, Jain RK (2004) Creation of long-lasting blood vessels. Nature 428(6979):138–139. doi:10.1038/428138a
42. Zisch AH, Lutolf MP, Ehrbar M, Raeber GP, Rizzi SC, Davies N, Schmokel H, Bezuidenhout D, Djonov V, Zilla P, Hubbell JA (2003) Cell-demanded release of VEGF from synthetic, biointeractive cell-ingrowth matrices for vascularized tissue growth. FASEB J 17(13):2260–2262. doi:10.1096/fj.02-1041fje
43. Ko HCH, Milthorpe BK, McFarland CD (2007) Engineering thick tissues—The vascularisation problem. Eur Cells Mat 14:1–18
44. Novosel EC, Kleinhans C, Kluger PJ (2011) Vascularization is the key challenge in tissue engineering. Adv Drug Deliv Rev 63(4–5):300–311. doi:10.1016/j.addr.2011.03.004
45. Lovett M, Lee K, Edwards A, Kaplan DL (2009) Vascularization strategies for tissue engineering. Tissue Eng Part B Rev 15(3):353–370. doi:10.1089/ten.teb.2009.0085
46. Khademhosseini A, Langer R (2007) Microengineered hydrogels for tissue engineering. Biomaterials 28(34):5087–5092. doi:10.1016/j.biomaterials.2007.07.021
47. Miller JS, Stevens KR, Yang MT, Baker BM, Nguyen D-HT, Cohen DM, Toro E, Chen AA, Galie PA, Yu X, Chaturvedi R, Bhatia SN, Chen CS (2012) Rapid casting of patterned vascular networks for perfusable engineered three-dimensional tissues. Nat Mater 11(9):768–774. doi:10.1038/nmat3357
48. Zheng Y, Chen J, Craven M, Choi NW, Totorica S, Diaz-Santana A, Kermani P, Hempstead B, Fischbach-Teschl C, Lopez JA, Stroock AD (2012) In vitro microvessels for the study of angiogenesis and thrombosis. Proc Nat Acad Sci USA 109(24):9342–9347. doi:10.1073/pnas.1201240109
49. Bellan LM, Singh SP, Henderson PW, Porri TJ, Craighead HG, Spector JA (2009) Fabrication of an artificial 3-dimensional vascular network using sacrificial sugar structures. Soft Matter 5(7):1354–1357. doi:10.1039/b819905a
50. Takei T, Sakai S, Ono T, Ijima H, Kawakami K (2006) Fabrication of endothelialized tube in collagen gel as starting point for self-developing capillary-like network to construct three-dimensional organs in vitro. Biotechnol Bioeng 95(1):1–7. doi:10.1002/bit.20903
51. Seto Y, Inaba R, Okuyama T, Sassa F, Suzuki H, Fukuda J (2010) Engineering of capillary-like structures in tissue constructs by electrochemical detachment of cells. Biomaterials 31(8):2209–2215. doi:10.1016/j.biomaterials.2009.11.104

Chapter 13
Quantitative Evaluation of Cell-Hydrogel Adhesion by Advanced Optical Techniques

Hiroshi Y. Yoshikawa

Abstract This chapter describes new methodologies to quantitatively evaluate cell-hydrogel adhesion using advanced optical techniques. The techniques enable one to gain two quantitative measures of cell-hydrogel adhesion, contact area and adhesion strength, which contribute to the rational design of biomaterials in tissue engineering.

Keywords Cell adhesion · Hydrogel · Interferometric microscopy · Laser-induced pressure wave

13.1 Introduction

Hydrogel is an indispensable material as structural support for cells in nature (i.e., ECM) as well as in tissue engineering. In addition, recent studies revealed that stiffness of hydrogels can influence many cell functions such as motility [1] and differentiation [2]. Such mechano-response was found even at a tissue level. For instance, synchronized beating in culture cardiomyocyte tissue on hydrogel substrate can significantly be modulated with hydrogel elasticity [3]. Local stiffening of liver due to fibrosis results in significant changes in functions within whole organs [4]. These results indicate that hydrogel with adequate stiffness is required to regenerate functional tissues. However, the choice of such hydrogel is still very dependent on trial-and-error although various hydrogels with tunable elasticity have been proposed to mimic natural microenvironments. One of the reasons for the difficulty in the rational design of hydrogel is lack of quantitative measures for cell-hydrogel affinity (adhesion). In fact, although molecular mechanism of cell-hydrogel adhesion was systematically studied, quantitative insights into cell-hydrogel adhesion at a cellular level remain still unknown.

H. Y. Yoshikawa (✉)
Department of Chemistry, Saitama University,
Shimo-okubo 255, Sakura-ku, Saitama 338-8570 Saitama, Japan
e-mail: hiroshi@mail.saitama-u.ac.jp

T. Arai et al. (eds.), *Hyper Bio Assembler for 3D Cellular Systems*,
DOI 10.1007/978-4-431-55297-0_13

Recently, I have been developing new methodologies to quantitatively evaluate cell-hydrogel adhesion under the collaboration with various scientists [5–8]. Here, two advanced optical techniques were utilized to obtain quantitative measures of cell-hydrogel adhesion, contact area and adhesion strength. In this account, I explain a comprehensive overview of the two optical techniques [7, 8] and their potentials for providing quantitative measures for rational tissue engineering.

13.2 Quantitative Evaluation of Cell-Hydrogel Contacts by Advanced Interferometric Microscopy

In general, due to small gap between cell membrane and hydrogel at adhesion sites (< 100 nm), visualization of physical contact between cells and hydrogel by optical microscopy is challenging. To date, total internal reflection fluorescence (TIRF) microscopy has been widely used to visualize cell adhesion molecules (e.g., integrin) near substrate. However, height resolution of TIRF (100–200 nm) [9] is not sufficient to identify the physical contact between cells and substrate. In addition, the existence of adhesion molecules does not necessarily guarantee physical contact between cells and hydrogel. It should be also noted that physical contact mediated by other generic interactions such as electrostatic interaction cannot be visualized by the fluorescence imaging. Alternatively, the leading technique for visualizing cell-substrate contacts is RICM, which detects interference of linearly polarized light reflected at cell-liquid (i.e., cell membrane-liquid) and substrate-liquid interfaces [10–14]. If glass substrate is used, RICM is a powerful label-free technique to measure the distance between cell membrane and substrates with a resolution of ~ 2 nm, [14] which is much finer than surface-sensitive fluorescence techniques such as TIRF microscopy. However, in case of hydrogel, contrast of RICM images becomes very poor because intensity and coherence of light are lowered by the following two reasons. Firstly, the refractive index of hydrogels is very close to that of water, which results in low reflectivity at hydrogel-liquid interfaces. Secondly, gels for cell mechano-response study should not be thin monolayer but much thicker (>several μm). As the results, RICM had not been applied to soft, thick hydrogels, which is widely used for cell mechano-response study and tissue engineering.

Very recently, I have developed an interferometric optical microscopy for the high contrast visualization of cell-hydrogel contact [8]. As schematically illustrated in Fig. 13.1a, conventional RICM setup utilizes the antiflex method; cross polarizers combined with an objective lens equipped with a quarter-wave plate (Antiflex EC Plan-Neofluar, 63 ×, Numerical aperture = 1.25, Oil Ph3, Zeiss, Gottingen, Germany). A mercury lump with a monochromatic filter (typically $\lambda = 546$ nm) has been widely used as a light source. To enhance the contrast of interference images obtained from cell-hydrogel contact, the RICM setup was modified by three points written below (Fig. 13.1b). (1) A confocal unit (FV300, Olympus, Tokyo, Japan) was attached to an inverted microscope (IX70, Olympus, Tokyo, Japan). A confocal aperture in the unit can significantly reduce stray light originating from outside of cell-hydrogel contact zones, e.g., reflection and scattering from glass, gels, cell

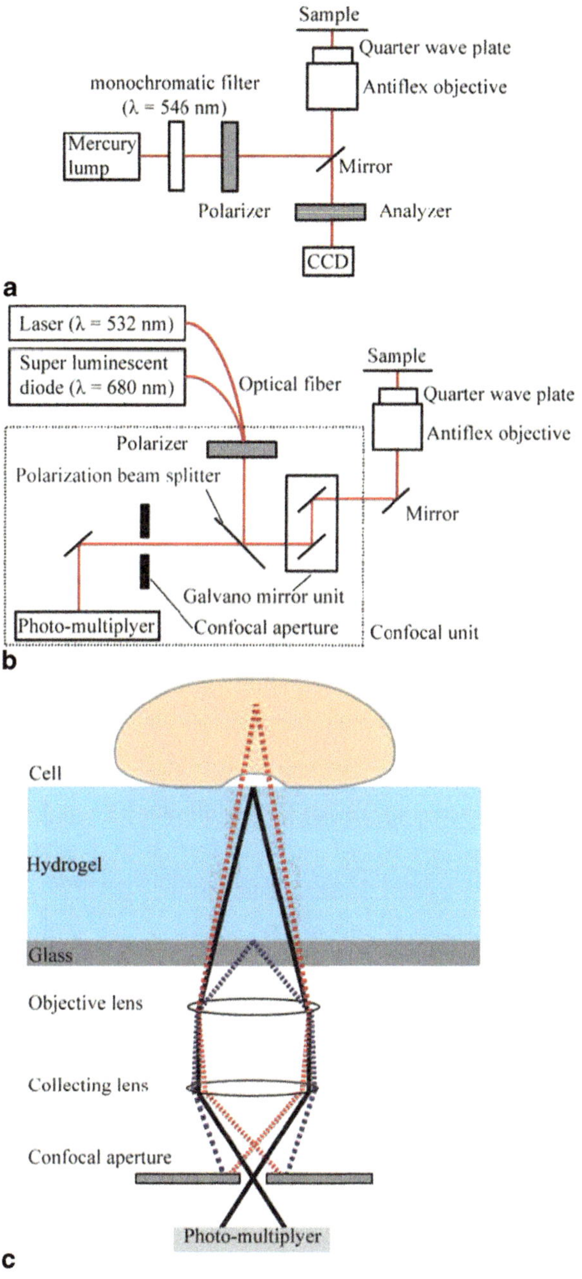

Fig. 13.1 Schematic illustration of experimental setup of **a** conventional RICM, **b** advanced RICM. **c** A confocal aperture reduces untargeted stray light originating from glass (*blue dashed line*), cell cytoplasm, and organelle (*red dashed line*). Reprinted with permission from ref. 8. Copyright (2014) American Chemical Society

cytoplasm, and organelle (Fig. 13.1c). (2) High throughput optics mentioned in the following were adopted to increase intensity hence to further improve the signal to noise ratio. Firstly, a polarization beam splitter (PBS, custom-ordered, Olympus Tokyo, Japan) that reflects s-polarized light and transmits p-polarized light over 95 % at the light wavelength was used. The use of the PBS instead of a half mirror and an analyzer can drastically increase light intensity of both illumination and detection in the antiflex setup, while a half mirror splits light at the both sides. Secondly, a polarizer made of aligned silver nanoparticles (colorPol VISIR CWO2, CODIXX AG, Barleben, Germany) was used, because it can enhance the signals by an order of magnitude than the polarizer based on polymer films [15]. (3) As a light source, a diode-pumped solid state laser (λ = 532 nm, 300 mW, SAPPHIRE 532–300-CW-CDPH, Coherent Inc., Santa Clara, USA) or a super luminescent diode (SLD) (λ = 680 nm, 5 mW, coherent length ~10 μm, ASLD68–050-B-FA, Amonics, Hong-Kong, China) was used. The use of such an intense monochromatic light source instead of halogen lamps should provide clearer interference patterns. In addition, since coherent length of the solid-state laser and SLD is generally shorter than a gas laser such as a He-Ne laser, [16] the use of light sources with such "moderate" coherence is expected to reduce untargeted interference signals except for cell-hydrogel contact zones [16, 17]. In fact, Sazaki et al. demonstrated the high-contrast visualization of ice and protein crystal surfaces at atomic resolution by using a SLD as a light source for scanning confocal microscopy combined with differential interference contrast microscopy [18].

Figure 13.2a, 13.1b, 13.2c show RICM images of a polystyrene latex bead with 100 μm in diameter (Φ) on polyacrylamide (PAAm) gel (E = 13.2 ± 0.5 kPa, thickness = 6 ~ 7 μm) taken by a conventional RICM setup and by the advanced RICM system, respectively. The advanced RICM provided much clearer interference patterns than those by a conventional RICM system. It should be also noted that both, laser and SLD, show clear Newton's ring-like fringes without untargeted interference such as speckle, which is often observed using a He-Ne laser as a light source [16, 17]. From the intensity profile (Fig. 13.2e), the height profile of a bead was reconstructed (Fig. 13.2d) by the following Eq. (13.1) [13, 14] of

$$\frac{2I-(I_{\max}+I_{\min})}{-(I_{\max}-I_{\min})}=\frac{\sin y}{y}\cos\left\{\frac{4\pi n}{\lambda}\left[h(1-\sin^2(\frac{\alpha}{2})\right]\right\}. \tag{13.1}$$

Here, I is measured intensity. I_{max} and I_{min} are maximum and minimum intensity. λ is wavelength of light, h separation between substrate and sample, and n refractive index of medium (~1.333). y represents $2\pi n \sin^2(\alpha/2)/\lambda$, where α is a half angle of the cone illumination (55°). The height profile of a bead reconstructed from interference signals is almost equivalent to height profile of a theoretically assumed sphere (Φ = 100 μm) up to 2 μm above the surface. The results clearly demonstrate that the advanced RICM system can provide with clearer interferometric patterns and quantitatively reconstructed height profile of objects on hydrogels.

In the next step, cells adherent on hydrogels were visualized using a conventional RICM setup and the advanced RICM system. Figure 13.3a shows a bright field image of a mouse metastatic melanoma cell (B16–F10) on a PAAm gel (E = 13.2 ± 0.5 kPa) functionalized with fibronectin. Cells were cultured for 3 h in Leibovitz's L-15 medium without serum and then fixed with a 3.7 w/v% DPBS-formaldehyde solution. With a conventional RICM, the cell came out brighter than

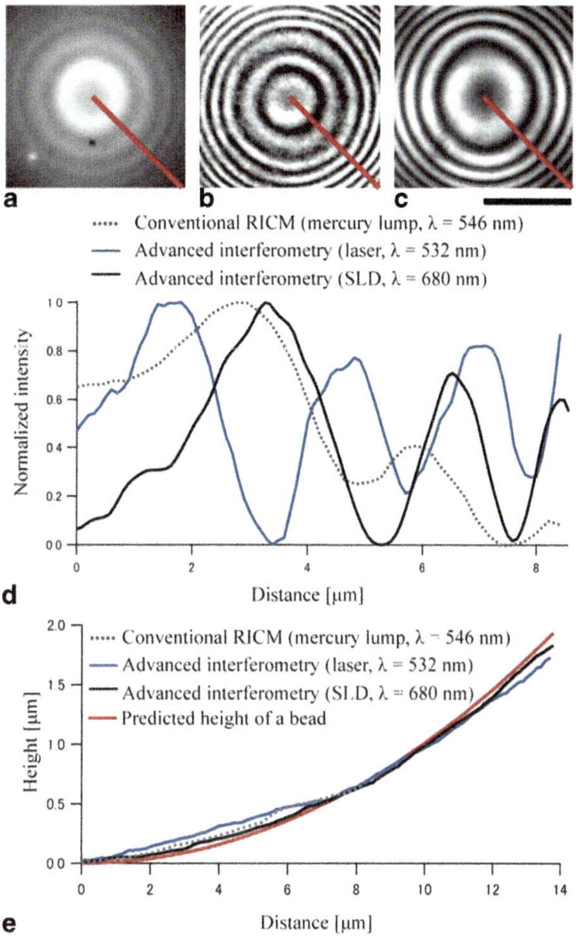

Fig. 13.2 Interference images of polystyrene latex bead 100 μm in diameter on PAAm gels ($E = 13.2 \pm 0.5$ kPa) taken by **a** conventional RICM and advanced RICM with a light source of **b** laser or **c** SLD. Scale bars: 10 μm. Size of confocal aperture is 60 μm. **d** Intensity profiles along the *red lines* in **a**–**c** and **e** reconstructed and predicted height profiles of a bead. Reprinted with permission from ref. 8. Copyright (2014) American Chemical Society

its surrounding region, indicating that scattered light from outside of a focal plane (e.g., cell cytoplasm) obscured interference signals (Fig. 13.3b). On the other hand, the advanced RICM yielded a black cell body in sharp contrast against the bright surrounding region (Fig. 13.3e, 13.3d, 13.3e). To assess influence of stray light on the image contrast of our interferometry, RICM images of the cell were taken with different confocal aperture diameters. The image contrast became higher by reducing the aperture diameter from 300 to 60 μm, corresponding to from 2.10 to 0.42 Airy unit, which represents the theoretically derived Airy disk diameter. The advanced RICM could visualize local structures of cell filopodia that are in tight contact with the hydrogel surface. These results clearly indicate that the advanced RICM can clearly visualize adhesion zone of cells adherent to 6 ~ 7 μm thick hydrogels, which is not possible by conventional RICM.

In the final step, the advanced RICM was applied to evaluate the impact of substrate elasticity on cell adhesion. B16-F10 cells were cultured for 3 h on PAAm gels

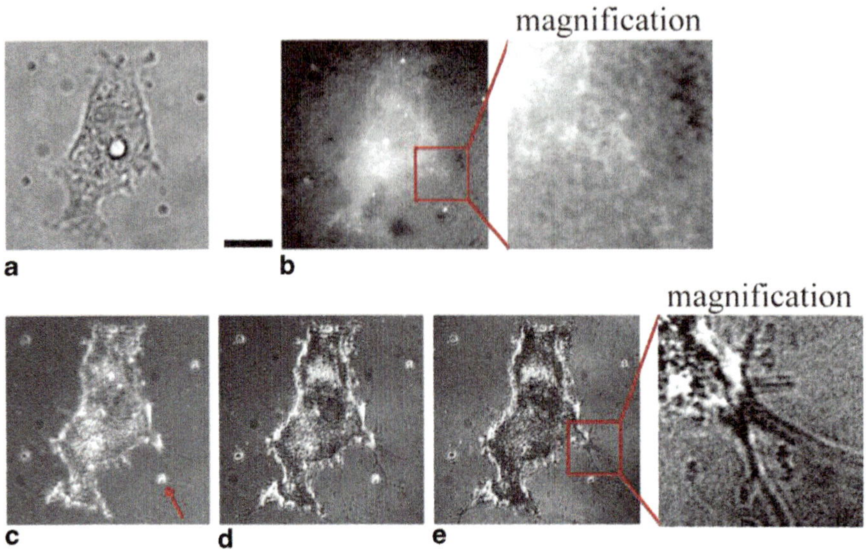

Fig. 13.3 A B16-F10 cell on a PA gel ($E = 13.2 \pm 0.5$ kPa) measured by **a** bright field microscopy, **b** conventional RICM, and **c**–**e** advanced RICM. Diameter of confocal apertures is **c** 300 μm, **d** 150 μm, and **e** 60 μm. SLD ($\lambda = 680$ nm) was used as a light source for the advanced interferometric microscopy. Images at the *right* of **b** and **e** are magnification of cell periphery. *Scale bars*: 10 μm. The *bright dots* like the one indicated by a *red arrow* in **c** are polystyrene beads embedded in PAAm gels. Reprinted with permission from ref. 8. Copyright (2014) American Chemical Society

with $E = 9.8$, 13.2, and 62.5 kPa and then fixed. Bright field microscopy images (Fig. 13.4a) implies that B16-F10 exhibited a more pronounced spreading on stiffer gels, and the zones of cell-hydrogel tight contacts are identified as black patches in RICM images (Fig. 13.4b). Such black area can be typically seen for B16-F10 cells on hard glass substrate coated with fibronectin, which binds specifically to its receptor such as integrin. Figure 13.4c shows zone for cell-hydrogel tight contacts estimated by image analysis [8]. It clearly shows patch-like adhesion structures, which were typically observed in RICM images of cells adhered to flat glass substrate [11, 14]. The estimated tight contact area is monotonically increased from 74 to 578 μm^2 according to the increase in the stiffness of hydrogel from 9.8 ± 0.4 kPa to 62.5 ± 4.7 kPa, suggesting mechano-response of B16-F10 cells.

In summary, the advanced RICM system drastically enhanced the contrast in interferometric patterns from cell-hydrogel contact zones. Physical contacts between cells and hydrogels could be successfully determined. I foresee that the advanced RICM will provide quantitative insights into mechanical interactions between cells and hydrogels, which contribute to the rational design of biomaterials in tissue engineering.

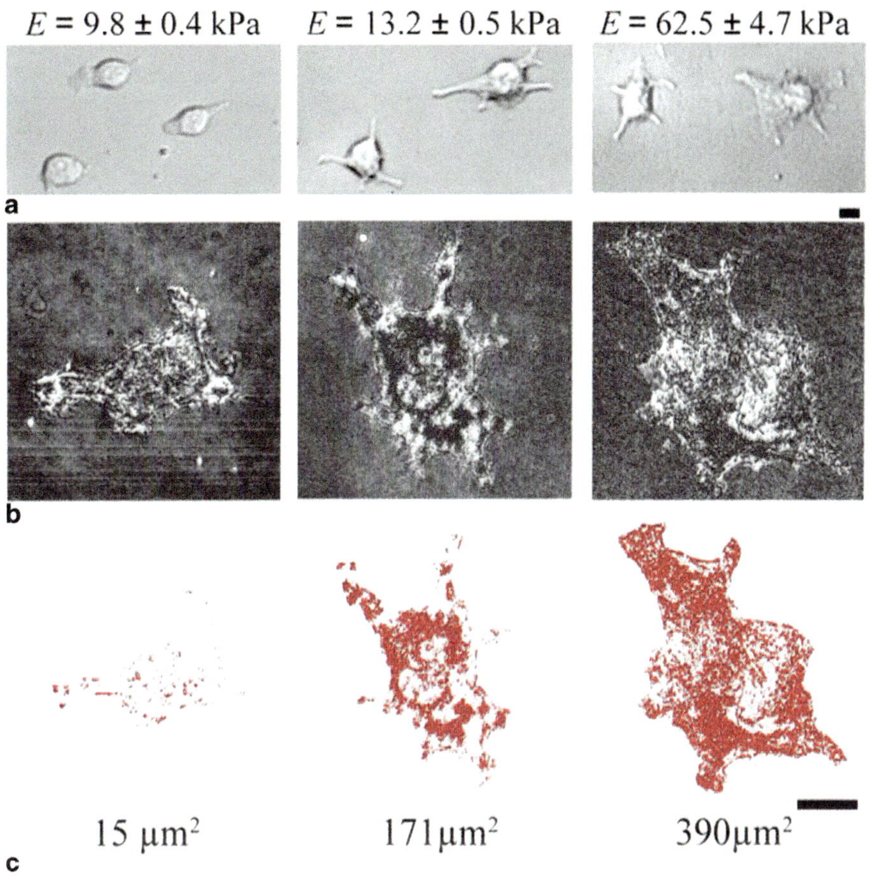

Fig. 13.4 B16-F10 cells on PAAm gels with three different stiffness ($E = 9.8 \pm 0.4$ kPa, $E = 13.2 \pm 0.5$ kPa, $E = 62.5 \pm 4.7$ kPa) were measured by **a** bright field microscopy and **b** advanced RICM. Laser ($\lambda = 532$ nm) was utilized as a light source. **c** Tight adhesion area. Diameter of a confocal aperture was 60 μm. Scale bars: 10 μm. Reprinted with permission from ref. 8. Copyright (2014) American Chemical Society

13.3 Quantitative Evaluation of Adhesion Strength Between Cells and Hydrogel by Using Laser-Induced Pressure Wave

Gelatin is one of the most widely used biomaterials for tissue engineering. Recently, Kidoaki et al. reported styrenized gelatin gel, whose elastic modulus can be tuned by photo-crosslinking conditions [19]. In a recent account, they reported that elastic pattern of gelatin can be used for inducing mechanotaxis of cells [19, 20]. In addition, synchronized beating in culture cardiomyocyte tissue on photocurable gelatin can significantly be influenced with gelatin elasticity [3]. These clearly indicate

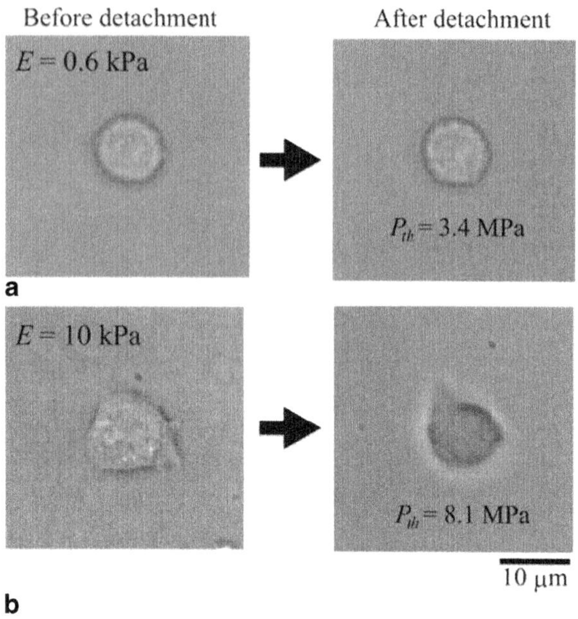

Fig. 13.5 Bright field images of an individual cell on gelatin gel at E = **a** 0.6 kPa and **b** 10 kPa before (*left*) and after (*right*) its detachment when subjected to a shock wave above a certain minimum pressure. Reprinted with permission from ref. 7. Copyright (2013) American Chemical Society

that elasticity of gelatin substrate modulates cell-gelatin interaction, which is a key for regulating cell and tissue functions. However, the quantitative insights into the "strength" of interactions between cells and photocurable gelatin still remained unknown.

Recently, adhesion strength of myoblast (C2C12) cells and photocurable gelatin was quantitatively measured by the self-developed cell detachment assay, [7] where the critical pressure for the cell detachment can be determined by pressure waves induced by picosecond laser pulses [5, 6]. The critical pressure required for the detachment of adherent cells P_{th} can be used as a "mechanical" measure to quantitatively assess cell-substrate interaction (adhesion). Figure 13.5 represents bright field images of myoblast (C2C12) cells on gelatin gel before and after the laser-induced cell detachment. Here, the critical detachment pressures of P_{th} = 3.4 MPa or 8.1 MPa can be determined for gels with E = 0.6 and 10 kPa, respectively (Fig. 13.5). Figure 13.6a shows the average pressure as a function of gel elasticity, E, at t = 3 h. The average of P_{th} increased from 4.1 to 34 MPa according to the increase in the stiffness, 0.2 to 140 kPa. The monotonic increase in the adhesion strength of C2C12 cells is consistent with the past studies by micropipetts, [21, 22] but the transition from weak adhesion (P_{th} < 5 MPa) to strong adhesion (P_{th} > 20 MPa) takes place in a narrower E range (1 kPa ≤ E ≤ 20 kPa). To gain more quantitative insights into the dependence of cell adhesion on gelatin mechanics, the plots were fitted with the empirical Hill equation.

$$P_{th} = bE^m / [(E_{1/2-adhesion})^m + E^m) \tag{13.2}$$

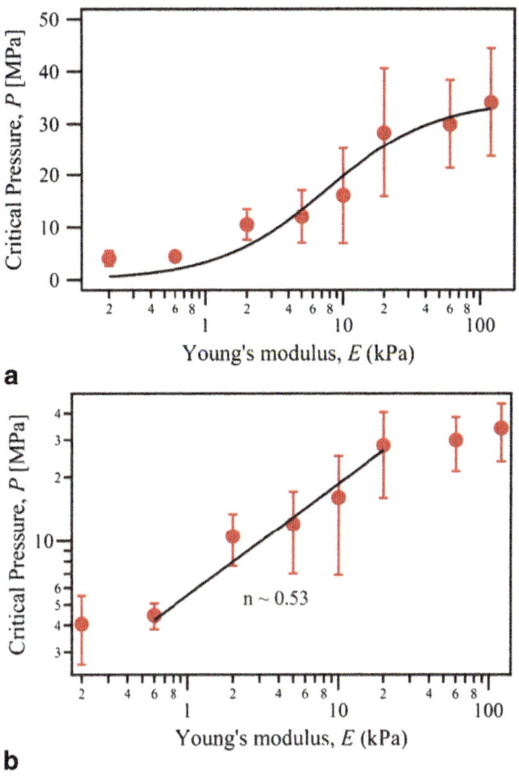

Fig. 13.6 Evaluation of adhesion strength of C2C12 cells on gelatin gels. **a** The average of critical pressure as a function of gel elasticity at $t \sim 3$ h. The *black solid line* is the fitting with the Hill equation (Eq. 13.2). The *error bar* represents standard deviation. **b** Log-log plots converted from the plot of (A). The *black solid line* represents the fitting with the power law function (Eq. 13.3). Reprinted with permission from ref. 7. Copyright (2013) American Chemical Society

As indicated by the solid line, the fit with the cooperativity coefficient taken from image analysis of projected area ($m = 1.1$) [7] can well represent the experimental results. The half-saturation level $E_{1/2\text{- adhesion,}}$ was ~7.6 kPa,

Other studies already reported that various cellular processes can phenomenologically be analyzed with power laws, such as the dependence of viscosity of cytoplasm on shear rate, [23] the dependence of projected area of smooth muscle cells on the substrate elasticity, [24] and the dependence of transit time of leukocytes through narrow pores [25]. Figure 13.6b represents the log-log plot of P_{th} vs. E for the regime in which the transition was observed; 0.6 kPa$\leq E \leq$20 kPa. The solid line corresponds to the power law:

$$P_{th} = cE^n \tag{13.3}$$

The obtained power law exponent, $n \sim 0.53$, may be attributed to the establishment of firm adhesion by the formation of stress fibers bound to focal adhesions via talin [26, 27]. Here, contraction force originated from focal adhesions activates the Src kinase, which should lead to an increase in the adhesion strength [28]. In fact, the range of elastic modulus (0.6 kPa$\leq E \leq$20 kPa) is in good agreement with the regime where the stress fiber formation was formed [7]. The observed power law dependences suggest the transition from weakly adhered to strongly adhered regime can be attributed to the stress fiber formation. The traction force exerted by cells at $E \sim 1$ kPa may reach the critical threshold for the activation of mechanoreceptors. On the other hand, the plateau regime with $E \geq 20$ kPa suggests that the activation of mechanosensors of C2C12 cells is saturated.

In summary, the laser-induced cell detachment assay revealed that the adhesion strength exhibits a transition from weak adhesion to strong adhesion within the distinct elasticity range ($E \sim 1$–20 kPa). Actually, elastic modulus of environments of myoblast cells in nature is ~ 10 kPa. Engler et al. reported that an optimal elastic modulus of $E \sim 12$ kPa maximizes myosin striations in muscle [22]. Interestingly, the optimal elastic modulus is very close to that at the middle of adhesion strength transition ($E_{1/2\text{- }adhesion} \sim 7.6$ kPa). This clearly indicates that adhesion strength can be a quantitative measure to assess optimal cell-substrate interaction for muscle tissue formation. I believe that the approach with the combination of hydrogel and the pressure wave assays is powerful to reveal the optimal interaction between cells and hydrogel for tissue engineering.

13.4 Conclusion

The advanced optical techniques mentioned in this chapter can provide quantitative measures for cell-hydrogel adhesion. The advanced RICM can visualize cell-hydrogel contact on the basis on distance information reconstructed from interferrometric signals. The laser-induced cell detachment assay can provide with adhesion strength of cell-hydrogel, which is a direct, mechanical measure for the interaction between cells and hydrogel. The systematic application of these techniques will contribute to the understanding of mechanical interactions between cells and hydrogel, which subsequently leads to the rational design of biomaterials in tissue engineering.

Acknowledgments I thank Prof. Motomu Tanaka for the tight collaboration for developing the advanced optical techniques. I also thank Prof. Satoru Kidoaki for the support of photocurable gelatin and Prof. Gen Sazaki for the support of the advanced RICM.

References

1. Pelham RJ, Wang YL (1997) Cell locomotion and focal adhesions are regulated by substrate flexibility. Proc Natl Acad Sci USA 94(25):13661–13665
2. Engler AJ, Sen S, Sweeney HL, Discher DE (2006) Matrix elasticity directs stem cell lineage specification. Cell 126(4):677–689. doi:10.1016/j.cell.2006.06.044
3. Horning M, Kidoaki S, Kawano T, Yoshikawa K (2012) Rigidity matching between cells and the extracellular matrix leads to the stabilization of cardiac conduction. Biophys J 102(3):379–387. doi:10.1016/j.bpj.2011.12.018

4. Wells RG (2008) The role of matrix stiffness in regulating cell behavior. Hepatology 47(4):1394–1400
5. Yoshikawa HY, Rossetti FF, Kaufmann S, Kaindl T, Madsen J, Engel U, Lewis AL, Armes SP, Tanaka M (2011) Quantitative evaluation of mechanosensing of cells on dynamically tunable hydrogels. J Am Chem Soc 133(5):1367–1374. doi:10.1021/Ja1060615
6. Yoshikawa HY, Cui J, Kratz K, Matsuzaki T, Nakabayashi S, Marx A, Engel U, Lendlein A, Tanaka M (2012) Quantitative evaluation of adhesion of osteosarcoma cells to hydrophobic polymer substrate with tunable elasticity. J Phys Chem B 116(28):8024–8030. doi:10.1021/jp212385p
7. Yoshikawa HY, Kawano T, Matsuda T, Kidoaki S, Tanaka M (2013) Morphology and adhesion strength of myoblast cells on photocurable gelatin under native and non-native micromechanical environments. J Phys Chem B 117(15):4081–4088
8. Matsuzaki T, Sazaki G, Suganuma M, Watanabe T, Yamazaki T, Tanaka M, Nakabayashi S, Yoshikawa HY (2014) High contrast visualization of cell-hydrogel contact by advanced interferometric optical microscopy. J Phys Chem Lett 5(1):253–257
9. Funatsu T, Harada Y, Tokunaga M, Saito K, Yanagida T (1995) Imaging of single fluorescent molecules and individual ATP turnovers by single myosin molecules in aqueous-solution. Nature 374(6522):555–559
10. Curtis AS (1964) The mechanism of adhesion of cells to glass. a study by interference reflection microscopy. J Cell Biol 20:199–215
11. Izzard CS, Lochner LR (1976) Cell-to-substrate contacts in living fibroblasts: an interference reflexion study with an evaluation of the technique. J Cell Sci 21(1):129–159
12. Gingell D, Todd I (1979) Interference reflection microscopy. A quantitative theory for image interpretation and its application to cell-substratum separation measurement. Biophys J 26(3):507–526. doi:10.1016/S0006-3495(79)85268-6
13. Rädler J, Sackmann E (1993) Imaging optical thickness and separation distances of phospholipid vesicles at solid surfaces. J Phys II France 3:724–748
14. Limozin L, Sengupta K (2009) Quantitative reflection interference contrast microscopy (RICM) in soft matter and cell adhesion. Chemphyschem 10(16):2752–2768
15. Wen R, Lahiri A, Azhagurajan M, Kobayashi S, Itaya K (2010) A new in situ optical microscope with single atomic layer resolution for observation of electrochemical dissolution of AU(111). J Am Chem Soc 132(39):13657–13659
16. Suzuki Y, Sazaki G, Matsumoto M, Nagasawa M, Nakajima K, Tamura K (2009) First direct observation of elementary steps on the surfaces of glucose Isomerase crystals under high pressure. Cryst Growth Des 9(10):4289–4295
17. Sazaki G, Matsui T, Tsukamoto K, Usami N, Ujihara T, Fujiwara K, Nakajima K (2004) In situ observation of elementary growth steps on the surface of protein crystals by laser confocal microscopy. J Cryst Growth 262(1–4):536–542
18. Sazaki G, Zepeda S, Nakatsubo S, Yokoyama E, Furukawa Y (2010) Elementary steps at the surface of ice crystals visualized by advanced optical microscopy. Proc Natl Acad Sci USA 107(46):19702–19707
19. Kidoaki S, Matsuda T (2008) Microelastic gradient gelatinous gels to induce cellular mechanotaxis. J Biotechnol 133(2):225–230
20. Kawano T, Kidoaki S (2011) Elasticity boundary conditions required for cell mechanotaxis on microelastically-patterned gels. Biomaterials 32(11):2725–2733
21. Griffin MA, Engler AJ, Barber TA, Healy KE, Sweeney HL, Discher DE (2004) Patterning, prestress, and peeling dynamics of myocytes. Biophys J 86(2):1209–1222. doi:10.1016/S0006-3495(04)74195-8
22. Engler AJ, Griffin MA, Sen S, Bonnetnann CG, Sweeney HL, Discher DE (2004) Myotubes differentiate optimally on substrates with tissue-like stiffness: pathological implications for soft or stiff microenvironments. J Cell Biol 166(6):877–887
23. Tsai MA, Waugh RE, Keng PC (1996) Cell cycle-dependence of HL-60 cell deformability. Biophys J 70(4):2023–2029

24. Engler A, Bacakova L, Newman C, Hategan A, Griffin M, Discher D (2004) Substrate compliance versus ligand density in cell on gel responses. Biophys J 86(1):388a
25. Nossal R (1998) Cell transit analysis of ligand-induced stiffening of polymorphonuclear leukocytes. Biophys J 75(3):1541–1552
26. Burridge K, Connell L (1983) A new protein of adhesion plaques and ruffling membranes. J Cell Biol 97(2):359–367
27. Kaufmann S, Piekenbrock T, Goldmann WH, Barmann M, Isenberg G (1991) Talin binds to actin and promotes filament nucleation. FEBS Lett 284(2):187–191
28. Vogel V, Sheetz M (2006) Local force and geometry sensing regulate cell functions. Nat Rev Mol Cell Biol 7(4):265–275. doi:10.1038/nrm1890

Chapter 14
Cell Scooper: A Device for the Rapid Transfer of Living Cell Sheet

Kenjiro Tadakuma, Nobuyuki Tanaka, Yuji Haraguchi, Mitsuru Higashimori, Makoto Kaneko, Tatsuya Shimizu, Masayuki Yamato and Teruo Okano

Abstract In this study, we developed a device that could easily, rapidly, and completely transfer cell sheets from one material to another or transplant cell sheets onto the dorsal subcutaneous tissues of rats without leaving residual cells. Because the manipulation is as simple as pipetting, technical expertise is not required to transfer cell sheets very rapidly (the transfer time was 3.7 ± 1.6 s) using the device compared with that of a conventional method using a pipette (430 ± 180 s). After transfer by the device, C2C12 skeletal myoblast sheets showed active cell metabolism, cell viability, and very high production of vascular endothelial growth factor and stromal-derived factor-1α, indicating transfer without cell damage. Cardiac cell sheets after transfer showed spontaneous and synchronous beating, indicating intact cell-cell junctions and ion channel proteins on the cell opsurface. In addition, the device allowed us to transfer C2C12 cell sheets onto soft, rugged and curved surfaces such as human hands. Furthermore, cardiac cell sheets adhered rapidly and tightly onto the dorsal subcutaneous tissues of rats. This transfer/transplantation device may be a powerful tool in cell sheet-based tissue engineering and regenerative medicine.

Keywords Mechanism · Device · Scooping · Transfer

14.1 Introduction

Cell-based therapy and regenerative medicine have been progressing rapidly and a number of clinical trials have already been performed [1–3]. However, injection of dissociated cells shows poor survival of transplanted cells and, thus, such a transplantation method might impede the expected therapeutic effects. To overcome

M. Yamato (✉) · N. Tanaka · Y. Haraguchi · T. Shimizu · T. Okano
Graduate School of Medicine, Tokyo Women's Medical University Tokyo, Japan
e-mail: yamato.masayuki@twmu.ac.jp

T. Okano
e-mail: tokano@abmes.twmu.ac.jp

K. Tadakuma · M. Higashimori · M. Kaneko
Graduate School of Engineering, Osaka University, Tokyo, Japan

T. Arai et al. (eds.), *Hyper Bio Assembler for 3D Cellular Systems*,
DOI 10.1007/978-4-431-55297-0_14

these problems, tissue engineering has been developed as the next generation of cell therapy, and clinical trials have already been performed [4–8]. Our laboratory has developed a scaffold-free tissue engineering methodology, which is called "cell sheet engineering", using a temperature-responsive culture surface, and cell sheet-based tissue engineering has already been successfully applied for regeneration of various damaged tissues [9–15]. Cell sheet transplantation shows significantly more effective tissue regeneration and therapeutic effects than those observed by injection of dissociated cells [16–18]. In addition, clinical trials using autologous cell sheets have already been performed to replace several tissues including cornea epithelial, esophageal and myocardial tissues [19–22].

On the other hand, generally, single-layer cell sheets were quite fragile and easily crumpled when picking up the cell sheets from culture medium with forceps etc. Therefore, we have been trying to develop manipulators/methods, which can manipulate the cell sheet easily and simply. The cell sheet is transferred from a temperature-responsive culture surface to another surface or in vivo tissues by several techniques/methods using pipettes, support membranes, plunger-like devices, and other [14, 15, 17, 19–35]. However, a unifying transfer/transplantation method of cell sheets has not yet been established and the degree of success by these transfer methods depends largely on the skill and experience of investigators/technicians. Therefore, development of a system for easy transfer/transplantation of cell sheets, in which technical expertise is not essential, is required for advancing cell sheet-based tissue engineering and regenerative medicine, and to ensure research results are more reproducible. In this study, we developed a device that easily, rapidly, and completely transferred/transplanted cell sheets without cell damages in vitro and in vivo [44].

14.2 Materials and Methods

All animal experiments were performed according to the Guideline of Tokyo Women's Medical University on Animal Use, The Principles of Laboratory Animal Care formulated by the National Society for Medical Research, and the Guide for the Care and Use of Laboratory Animals prepared by the Institute of Laboratory Animal Resources and published by the National Institutes of Health (NIH Publication No. 86-23, revised 1985).

14.2.1 Cell Sheet Transfer/Transplantation Device

A cell sheet transfer/transplantation device was developed in this study. The device was mainly composed of two parts; a scooping part and a handling part (Fig. 14.1a). The scooping part was further composed of an inner plate made of aluminum and

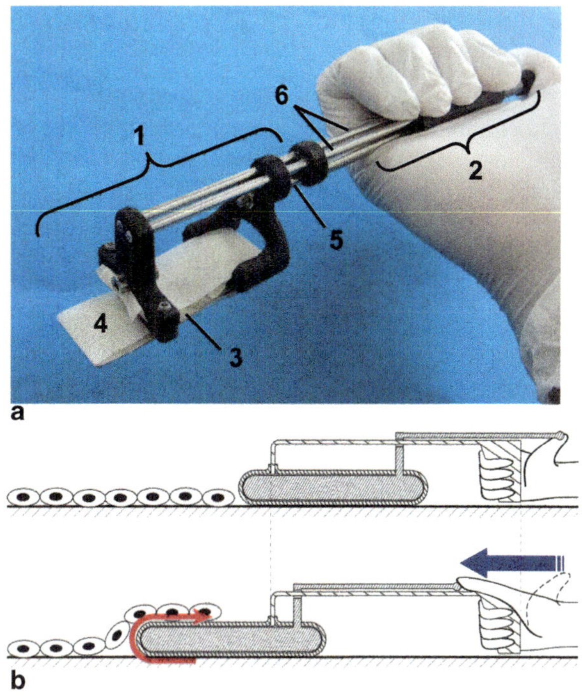

Fig. 14.1 Cell sheet transfer/transplantation device. An *upper* photograph **a** shows the device, which has several parts; a scooping part (*1*), a handling part (*2*), an inner plate (*3*), an outer movable belt (*4*), a pushing rod (*5*), and stainless rods (*6*). The mechanism of cell sheet scooping by the device is schematically illustrated in **b**. The size of scooping part and cell sheet in (*b*) was largely exaggerated for easy understanding

an outer polytetrafluoroethylene-glass cross (AS ONE, Osaka, Japan) that covered the inner plate. The inner plate was connected to a movable pushing rod in the handling part. The outer cross was also fixed to the handling part via two stainless rods. When the pushing rod was pushed by hand, the inner plate and the outer cross were extended by pushing the rod in the direction of the tip of the device.

The mechanism of cell sheet scooping by the device was as follows. (1) After the device was sterilized with rubbing alcohol, both the inner plate and outer cross were retracted into the device. (2) The tip of the device was extended and tilted toward the near edge of a cell sheet. (3) The pushing rod was pushed by hand, and then the inner plate was slid in the direction of the tip, and the outer cross was moved out with the movement of the inner plate and rolled up at the tip simultaneously. After contacting the cell sheet, the device could scoop the cell sheet by the movement of the outer cross. (4) After moving the device with the cell sheet, the tip of the device was placed onto another surface and then pulled into the device to release the cell sheet onto the surface. The mechanism of cell sheet scooping by the device is schematically illustrated in Fig. 14.1b. The outer cross, which contacts cell sheets and dish surfaces directly, of the device is coated polytetrafluoroethylene, which is a nonadherent, low friction, low wearing, and FDA approval material. Therefore, it is expected that the manipulation of the devise does not affect the cell sheet and the surface when the cell sheet was scooped and released.

14.2.2 Preparation of C2C12 Cell Sheets and Cardiac Cell Sheets

C2C12 mouse skeletal myoblasts (Dainippon Sumitomo Pharma, Osaka, Japan) were cultured in Dulbecco's modified Eagle's medium (DMEM; Sigma-Aldrich, St. Louis, MO, USA) supplemented with 10 % fetal bovine serum (FBS; Japan Bio Serum, Nagoya, Japan) and 1 % penicillin/streptomycin (Invitrogen, Carlsbad, CA, USA). C2C12 cell sheets were fabricated as described previously [35]. Briefly, 6.0 × 105 C2C12 cells were plated onto a 35-mm temperature-responsive culture dish (Upcell; CellSeed, Tokyo, Japan) and cultured at 37 °C. After 3 days, the culture dish was transferred to a CO2 incubator set at 20 °C for recovering a C2C12 cell sheet. A C2C12 cell sheet was photographed by a digital camera (GR Digital; Ricoh, Tokyo, Japan). For cardiac cell sheets, cardiac cells were isolated from the ventricles of 1-day-old Sprague-Dawley (SD) rats (CLEA, Tokyo, Japan), and prepared as described previously [11, 15, 17, 24, 25]. 2.4 × 106 rat cardiac cells were plated onto a 35-mm temperature-responsive culture dish and cultured at 37 °C. After 4 days of cultivation, a cardiac cell sheet was recovered by reducing the culture temperature (20 °C). The recovered cell sheets were used for next transfer/transplantation experiments.

14.2.3 Transfer of C2C12 Cell Sheets and Cardiac Cell Sheets by the Device

For confirming the mechanism of device visually, after a C2C12 cell sheet was stained with 0.001 % neutral red solution, which was prepared by the dilution of 0.1 % neutral red solution (Tokyo Chemical Industry, Toknyo, Japan) in the culture medium, for 15 min, the stained cell sheets were scooped and released by using the device. A C2C12 or cardiac cell sheet on a dish was transferred to another culture dish using the device or a conventional method using a pipette [15]. After transfer of a C2C12 cell sheet, the cell sheet was incubated at 37 °C for adherence to the culture dish. After the incubation, fresh culture medium was added to the cells, followed by incubation at 37 °C for 22 h. Then, the culture medium was collected and used for cell metabolic and damage analyses, and enzyme-linked immunosorbent assays (ELISAs). A C2C12 cell sheet was also transferred onto a human hand covered with a glove (JMS, Tokyo, Japan). The manipulations were recorded by a digital video camera (Handycam HDR-CX500V; Sony, Tokyo, Japan). Transfer times were measured by a stop-watch (Casio, Tokyo, Japan).

After the transfer, cardiac cell sheets were observed under a phase-contrast microscope (ET300; Nikon, Tokyo, Japan), and images were recorded by a digital video camera (DCR-TRV900; Sony) with CCD camera equipment (HV-D28S; Nikon).

14.2.4 Measurement of Glucose Consumption, Lactate Production, and Released Lactate Dehydrogenase Activity

The metabolic activities of transferred C2C12 cell sheets were monitored by measuring glucose consumption and lactate production in the culture medium. The release of lactate dehydrogenase (LDH) from cultured cells is used as a common index of cell injury and death. For measuring the values, culture medium samples were collected after cultivation of C2C12 cell sheets for 22 h. The concentrations of glucose and lactate, and LDH activities were determined by hexokinase UV method, lactic oxidase method, and LDH assay kit (Sicaliquid LDH J) (Kanto Chemical, Tokyo, Japan), respectively, as described previously [36, 37].

14.2.5 ELISAs

A cytokine, vascular endothelial growth factor (VEGF), and a chemokine, stromal-derived factor-1α (SDF-1α), secreted from transferred C2C12 cell sheets for 22 h into the culture supernatant were quantitated by commercially available ELISA kits (R&D Systems, Minneapolis, MN, USA).

14.2.6 Transplantation of a Cardiac Cell Sheet by the Device

Cardiac cell sheets were transplanted onto the dorsal subcutaneous tissue of SD rats as described previously [34]. The rats were anesthetized by inhalation of isoflurane (up to 3.5 %). The dorsal skins were cut and opened, and then cardiac cell sheets were transplanted onto the dorsal subcutaneous tissues using the device. The tissue constructs were covered with silicone rubbers membrane (0.5 mm thick), and the skin incisions were closed. After 2 days, the transplanted portions were reopened and recorded by a digital camera (GR Digital) and a digital video camera (Handycam HDR-CX500V).

14.2.7 Histological Analysis

Cell sheets on dishes were fixed with 4 % paraformaldehyde. Specimens were embedded in paraffin, sectioned, and stained with hematoxylin and eosin. Prepared specimens were examined under a microscope (ELIPSE TE2000-U; Nikon).

14.2.8 Data Analysis

Data were expressed as the mean ± SD. The unpaired Student's t test was performed to compare two groups. A value of $p < 0.05$ was considered statistically significant.

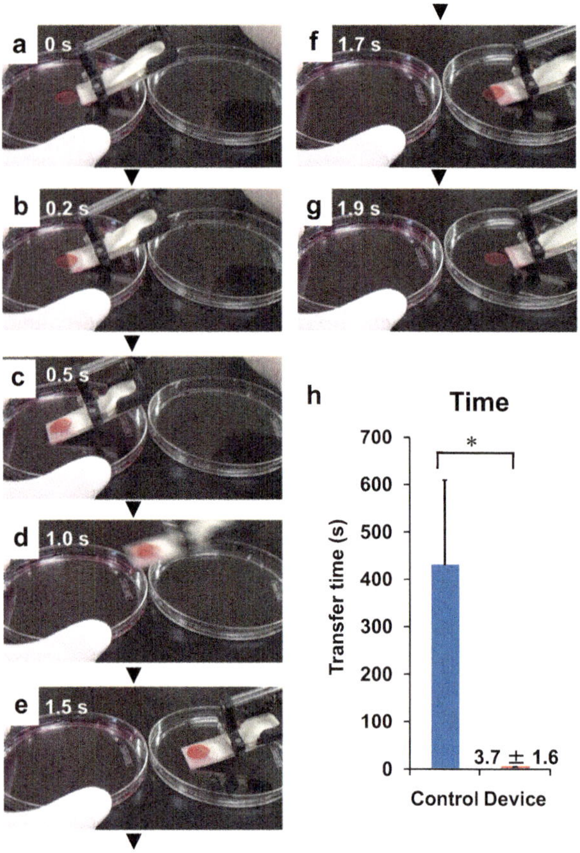

Fig. 14.2 Transfer a C2C12 cell sheet stained with neutral *red* from a dish to another culture dish using the transfer device. Scooping (**a**–**c**), transfer (**d**), and cell sheet-release (e–g). Numbers in the photographs indicate elapsed times. Comparison of transfer time by using the device (device) or a conventional method using a pipette (control) **h**. Each data point represents the mean ± SD ($n = 3$), *$p < 0.05$

14.3 The Stiffness-Based Deformability

14.3.1 C2C12 Cell Sheet Transfer from Dishes to Other Surfaces Using the Device

We attempted to transfer C2C12 cell sheets from dishes to other surfaces using the device. As shown in Fig. 14.2a–g, a C2C12 cell sheet on a culture dish could be easily transferred to another culture dish without leaving residual cells by a simple manipulation, similar to a pipette manipulation. The device even allowed us to transfer a single-layer cell sheet without any breakages (Fig. 14.3a). In more than one hundred trials, no unsuccessful transfer occurred. Histological analysis showed that a C2C12 cell sheet could be transferred as a contiguous cell sheet with tight cell-cell junctions (Fig. 14.3b). In addition, even an inexperienced researcher successfully transferred C2C12 cell sheets very rapidly using the device (the transfer time was 3.7 ± 1.6 s, $n = 3$), whereas the inexperienced researcher required a longer time to

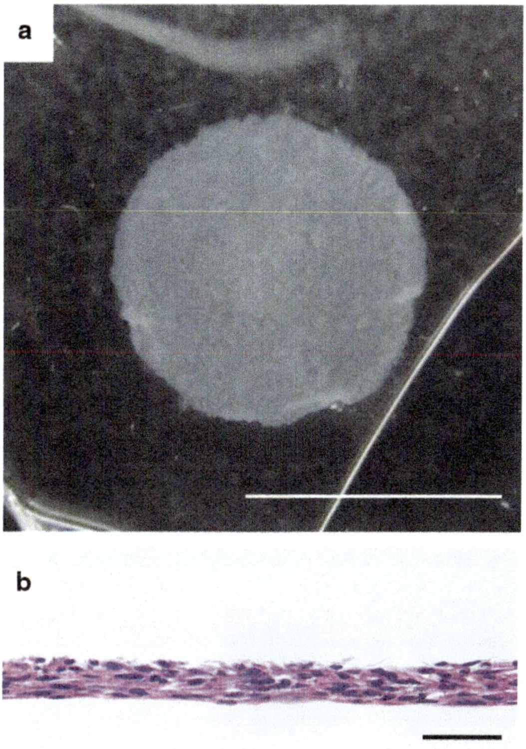

Fig. 14.3 Morphological and histological observations of a C2C12 cell sheet transferred using the device. An *upper* photograph shows a C2C12 cell sheet just after transfer **a** and a *lower* photograph shows *cross-sectional* observation of the cell sheet after 1 day of cultivation on a culture dish **b**. *White* and *black bars* represent 10 and 50 μm, respectively

transfer a cell sheet with a conventional method using a pipette (430 ± 180 s, $n=3$) (Fig. 14.2h) [15]. Of course, the manipulation time of the cell sheet transfer was shortened by the skillful and experience of researchers/technicians. On the other hand, as shown in this study, the development of simple cell sheet manipulation devices as well as the skill and experience of the researchers/technicians is important to advance tissue engineering and regenerative medicine, and ensure research results are more reproducible.

Next, we examined whether the cell sheets were transferred by the device without cell damage. Cell sheets transferred by the device and those by the conventional method showed similar cell metabolisms, namely glucose consumption (Fig. 14.4a) and lactate production (Fig. 14.4b), which are indexes of the bioactivity of cells, and the release of LDH (Fig. 14.4c), a common index of cell injury and death, indicating that the cell sheets could be transferred using the device without any cell damage, similar to that using the conventional method. Furthermore, C2C12 cell sheets transferred using the device produced large amounts of VEGF and SDF-1α, similar to that of cell sheets transferred by the conventional method (Fig. 14.5). Autologous skeletal myoblast sheets are already used in various damaged myocardial animal models, resulting in inhibition of detrimental myocardial remodeling and improvement of myocardial functions [38]. It is generally believed

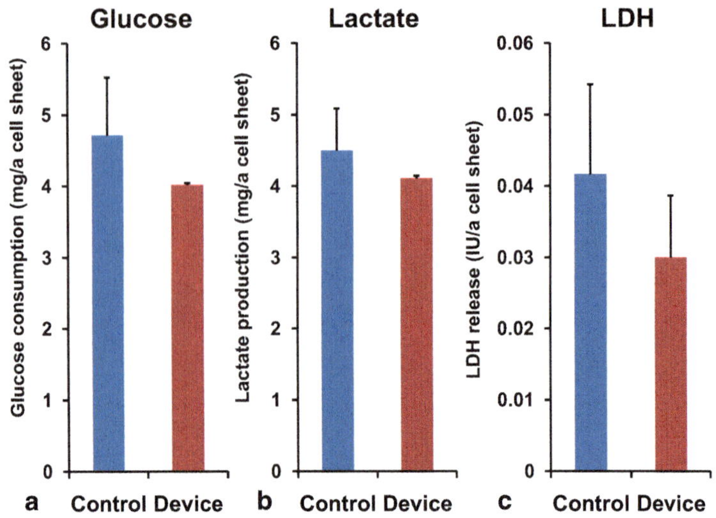

Fig. 14.4 Comparison of glucose consumption (**a**), lactate production (**b**), and lactate dehydrogenase (LDH) release (**c**) by C2C12 cell sheets transferred by using the device (device) or the conventional method (control). Total glucose consumption was calculated by subtracting the glucose concentration in the medium after cultivation for 22 h from that before incubation. The values of lactate production and LDH release were calculated by subtracting the backgrounds of lactate concentration and LDH activity in the fresh medium from those after cultivation for 22 h, respectively. Each data point represents the mean ± SD ($n = 3$)

Fig. 14.5 Comparisons of VEGF (**a**) and SDF-1a (**b**) secretions from C2C12 cell sheets transferred using the device (device) and the conventional method (control). The concentrations of VEGF and SDF-1a in culture supernatants (conditioned for 22 h) were determined by ELISAs. The values of those secretions were calculated by subtracting the backgrounds of VEGF/SDF-1a concentration in the fresh medium from those after cultivation for 22 h. Values represent the mean ± SD ($n = 6$)

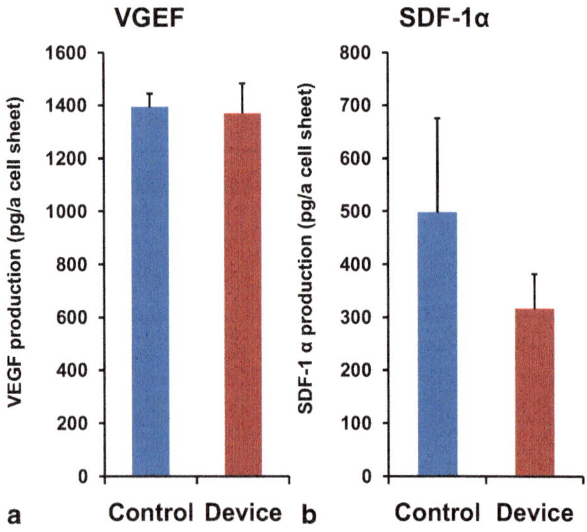

that the paracrine effects of bioactive factors, including VEGF and SDF-1α secreted from transplanted skeletal myoblast sheets, contribute toward the improvement of damaged myocardium after transplantation [16, 38]. VEGF induces strong angiogenesis and SDF-1α recruits several kinds of stem/progenitor cells, such as hematopoietic stem cells and endothelial progenitor cells, which express CXC chemokine receptor 4, the receptor of SDF-1α, and clinical trials for the repair of damaged myocardium using these bioactive factors have already been performed [39–40]. Our results also confirmed that the device allowed the transfer of bioactive cell sheets without cell damage. Moreover, a C2C12 cell sheet was compeletly transferred onto a human hand covered with a glove using the device, indicating that the device can transfer cell sheets onto soft, rugged and curved surfaces, as well as hardy and flat materials.

14.3.2 Cardiac Cell Sheet Transfer from Dishes to Other Surfaces Using the Device

Similar to C2C12 cell sheets, cardiac cell sheets on a dish could also be easily transferred to another culture dish using the device. Transferred cardiac cell sheet exhibited spontaneous and synchronous beating. In native cardiac tissue, electrical coupling between cardiomyocytes occurs via cell-cell contacts called gap junctions that mediate the exchange of small molecules and ions between neighboring cells. The electrical activities of cardiomyocytes occur via ion flows that mediate several ion channels on cardiomyocyte membranes, including sodium, calcium and potassium channels, which are critical for spontaneous and synchronous beating [41]. The spontaneous and synchronous beating of transferred cardiac cell sheet suggested that cardiomyocytes within the cell sheet had intact gap junctions and ion channels on the cell surface. Namely intact cell sheets without the damage of cell surface proteins and cell-cell junctions could be scooped and transferred by the devise.

Finally, transplantation of a cardiac cell sheet onto the dorsal subcutaneous tissue of rats was performed. As shown in Fig. 14.6, the cardiac cell sheet could also be easily transplanted onto the tissue. At 10 min after transplantation, the transplanted cardiac cell sheet could not be washed out by the dropping of a saline solution, indicating that the cell sheet adhered tightly and rapidly onto the tissue. The rapid and tight adhesion of transferred cell sheets onto the tissue suggested that the devise allowed us to be able to transplant cell sheets containing intact ECM matrix. Therefore, it is thought that intact cell sheets could be scooped and transplanted without the damage of ECMs by the devise. At 2 days after transplantation, the cardiac cell sheet became red, suggesting blood flow within the cell sheet (Fig. 14.7).

At present, clinical trials using autologous skeletal myoblast sheets are in progress [21]. In addition, pulsatile cardiac cell sheets have been fabricated with human induced pluripotent stem cells for the purpose of clinical usage [37, 42], and cardiac cell sheets have already been investigated in a porcine model [43]. The newly developed device allowed us to transfer skeletal myoblast sheets and cardiac cell

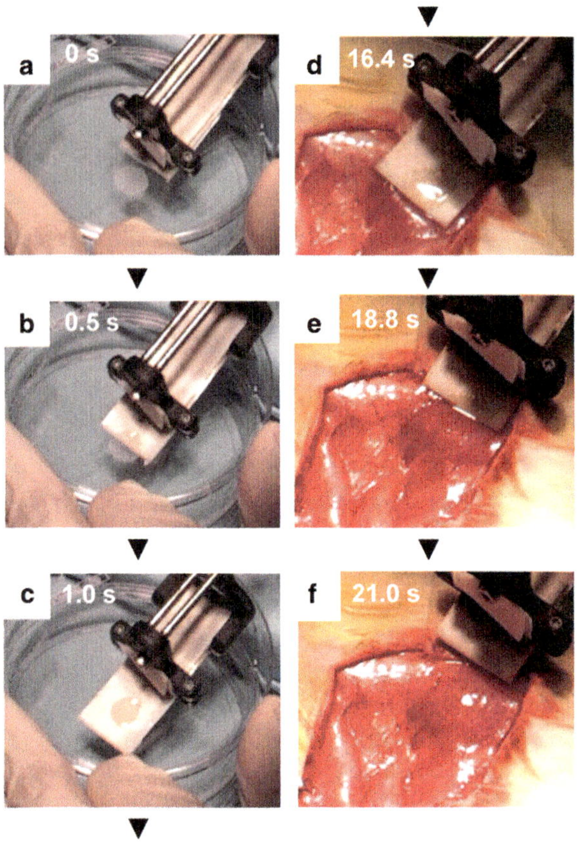

Fig. 14.6 Transplantation of a cardiac cell sheet onto the subcutaneous tissue of a rat using the device. Numbers in the photographs indicate elapsed times

sheets without cell damage. The skeletal myoblast sheets produced large amounts of paracrine factors nsevaluate the potential for clinical use, a large animal (porcine) experiment is ongoing.

In conclusion, we developed a device that can easily and rapidly transfer cell sheets between surfaces without leaving residual cells and undergoing cell damage. In addition, the device can be used for easy transplantation of cell sheets onto animal tissues. Because the device manipulation is simple, similar to a pipette manipulation, the device is not dependent on the skill and experience of the manipulators. Therefore, we are confident that this device can be used a powerful tool in the fields of cell sheet-based tissue engineering and regenerative medicine.

Acknowledgement Upon the development of the cell-sheet transfer/transplantation device in this study, the authors referred the basic concept of SWTL, a food-handling devise, developed by Furukawakikou Co., LTd. (Niigata, Japan). The authors are grateful to the members of Furukawakikou. This work was supported by grants from a new area of Hyper Bio Assembler for 3D Cellular Systems (BioAssembler) Project, a Grant-in-Aid for Japan Society for the Promotion of Science (JSPS) Fellows (23•7758), the Global Center of Excellence Program, Multidisciplinary Education and Technology and Research Center for Regenerative Medicine (MERCREM), Innovation

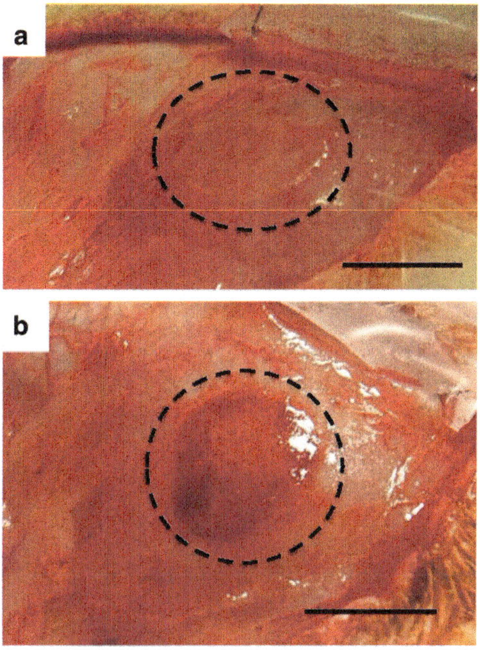

Fig. 14.7 Transplanted cardiac cell sheet onto the subcutaneous tissue of a rat using the device. Photographs (**a**) and (**b**) show a cardiac cell sheet (*dashed circles*) transplanted onto the subcutaneous tissue of a rat after 1 min and 2 d, respectively. *Black bars* represent 10 mm

Center for Fusion of Advanced Technologies in the Special Coordination Funds for Promoting Science, and the High-Tech Research Center Program from the Ministry of Education, Culture, Sports, Science, and Technology (MEXT), Japan, and JSPS through the "Funding Program for World-Leading Innovative R&D on Science and Technology (FIRST Program)," initiated by the Council for Science and Technology Policy (CSTP).

References

1. Menasché P (2008) Skeletal myoblasts and cardiac repair. J Mol Cell Cardiol 45:545–553
2. Atala A, Lanza R, Thomson JA, Nerem R (2011) Principles of regenerative medicine, 2nd edn. Academic Press, San Diego
3. Bolli R, Chugh AR, D'Amario D, Loughran JH, Stoddard MF, Ikram S, Beache GM, Wagner SG, Leri A, Hosoda T, Sanada F, Elmore JB, Goichberg P, Cappetta D, Solankhi NK, Fahsah I, Rokosh DG, Slaughter MS, Kajstura J, Anversa P (2011) Cardiac stem cells in patients with ischaemic cardiomyopathy (SCIPIO): initial results of a randomised phase 1 trial. Lancet 378:1847–1857
4. Zimmermann WH, Melnychenko I, Wasmeier G, Didié M, Naito H, Nixdorff U, Hess A, Budinsky L, Brune K, Michaelis B, Dhein S, Schwoerer A, Ehmke H, Eschenhagen T (2006) Engineered heart tissue grafts improve systolic and diastolic function in infarcted rat hearts. Nat Med 12:452–458
5. Chachques JC, Trainini JC, Lago N, Masoli OH, Barisani JL, Cortes-Morichetti M, Schussler O, Carpentier A (2007) Myocardial assistance by grafting a new bioartificial upgraded myocardium (MAGNUM clinical trial): one year follow-up. Cell Transplant 16:927–934
6. Shinoka T, Breuer C (2008) Tissue-engineered blood vessels in pediatric cardiac surgery. Yale J Biol Med 81:161–166

7. Lee K, Chan CK, Patil N, Goodman SB (2009) Cell therapy for bone regeneration-bench to bedside. J Biomed Mater Res Part B Appl Biomater 89:252–263
8. Iwasa J, Engebretsen L, Shima Y, Ochi M (2009) Clinical application of scaffolds for cartilage tissue engineering. Knee Surg Sports Traumatol Arthrosc 17:561–577
9. Yamada N, Okano T, Sakai H, Karikusa F, Sawasaki Y, Sakurai Y (1990) Thermo-responsive polymeric surface: control of attachment and detachment of cultured cells. Makromol Chem Rapid Commun 11:571–576
10. Okano T, Yamada H, Sakai H, Sakurai Y (1993) A novel recovery system for cultured cells using plasma-treated polystyrene dishes grafted with poly (N-isopropylacrylamide). J Biomed Mater Res 27:1243–1251
11. Shimizu T, Yamato M, Isoi Y, Akutsu T, Setomaru T, Abe K, Kikuchi A, Umezu M, Okano T (2002) Fabrication of pulsatile cardiac tissue grafts using a novel 3-dimensional cell sheet manipulation technique and temperature-responsive cell culture surfaces. Circ Res 90:e40–e48
12. Matsuda N, Shimizu T, Yamato M, Okano T (2007) Tissue engineering based on cell sheet technology. Adv Mater 19:3089–3099
13. Yang J, Yamato M, Shimizu T, Sekine H, Ohashi K, Kanzaki M, Ohki T, Nishida K, Okano T (2007) Reconstruction of functional tissues with cell sheet engineering. Biomaterials 28:5033–5043
14. Obokata H, Yamato M, Tsuneda S, Okano T (2011) Reproducible subcutaneous transplantation of cell sheets into recipient mice. Nat Protoc 6:1053–1059
15. Haraguchi Y, Shimizu T, Sasagawa T, Sekine H, Sakaguchi K, Kikuchi T, Sekine W, Sekiya S, Yamato M, Umezu M, Okano T (2012) Fabrication of functional three-dimensional tissues by stacking cell sheets in vitro. Nat Protoc 7:850–858
16. Memon IA, Sawa Y, Fukushima N, Matsumiya G, Miyagawa S, Taketani S, Sakakida SK, Kondoh H, Aleshin AN, Shimizu T, Okano T, Matsuda H (2005) Repair of impaired myocardium by means of implantation of engineered autologous myoblast sheets. J Thorac Cardiovasc Surg 130:1333–1341
17. Sekine H, Shimizu T, Dobashi I, Matsuura K, Hagiwara N, Takahashi M, Kobayashi E, Yamato M, Okano T (2011) Cardiac cell sheet transplantation improves damaged heart function via uperior cell survival in comparison with dissociated cell injection. Tissue Eng Part A 17:2973–2980
18. Wei F, Qu C, Song T, Ding G, Fan Z, Liu D, Liu Y, Zhang C, Shi S, Wang S (2011) Vitamin C treatment promotes mesenchymal stem cell sheet formation and tissue regeneration by elevating telomerase activity. J Cell Physiol 227:3216–3224
19. Nishida K, Yamato M, Hayashida Y, Watanabe K, Yamamoto K, Adachi E, Nagai S, Kikuchi A, Maeda N, Watanabe H, Okano T, Tano Y (2004) Corneal reconstruction with tissue-engineered cell sheets composed of autologous oral mucosal epithelium. N Engl J Med 351:1187–1196
20. Ohki T, Yamato M, Ota M, Takagi R, Murakami D, Kondo M, Sasaki R, Namiki H, Okano T, Yamamoto M (2012) Prevention of esophageal stricture after endoscopic submucosal dissection using tissue-engineered cell sheets. Gastroenterology 143:582–588
21. Sawa Y, Miyagawa S, Sakaguchi T, Fujita T, Matsuyama A, Saito A, Shimizu T, Okano T (2012) Tissue engineered myoblast sheets improved cardiac function sufficiently to discontinue LVAS in a patient with DCM: report of a case. Surg Today 42:181–184
22. Burillon C, Huot L, Justin V, Nataf S, Chapuis F, Decullier E, Damour O (2012) Transplantation for the treatment of corneal limbal epithelial stem cell deficiency. Invest Ophthalmol Vis Sci 53:1325–1331
23. Nishida K, Yamato M, Hayashida Y, Watanabe K, Maeda N, Watanabe H, Yamamoto K, Nagai S, Kikuchi A, Tano Y, Okano T (2004) Functional bioengineered corneal epithelial sheet grafts from corneal stem cells expanded ex vivo on a temperature-responsive cell culture surface. Transplantation 77:379–385
24. Sekine H, Shimizu T, Kosaka S, Kobayashi E, Okano T (2006) Cardiomyocyte bridging between hearts and bioengineered myocardial tissues with mesenchymal transition of mesothelial cells. J Heart Lung Transplant 25:324–332

25. Haraguchi Y, Shimizu T, Yamato M, Kikuchi A, Okano T (2006) Electrical coupling of cardiomyocyte sheets occurs rapidly via functional gap junction formation. Biomaterials 27:4765–4774
26. Ohashi K, Yokoyama T, Yamato M, Kuge H, Kanehiro H, Tsutsumi M, Amanuma T, Iwata H, Yang J, Okano T, Nakajima Y (2007) Engineering functional two- and three-dimensional liver systems in vivo using hepatic tissue sheets. Nat Med 13:880–885
27. Tsuda Y, Shimizu T, Yamato M, Kikuchi A, Sasagawa T, Sekiya S, Kobayashi J, Chen G, Okano T (2007) Cellular control of tissue architectures using a three-dimensional tissue fabrication technique. Biomaterials 28:4939–4946
28. Kanzaki M, Yamato M, Yang J, Sekine H, Takagi R, Isaka T, Okano T, Onuki T (2008) Functional closure of visceral pleural defects by autologous tissue engineered cell sheets. Eur J Cardiothorac Surg 34:864–869
29. Kobayashi H, Shimizu T, Yamato M, Tono K, Masuda H, Asahara T, Kasanuki H, Okano T (2008) Fibroblast sheets co-cultured with endothelial progenitor cells improve cardiac function of infarcted hearts. J Artif Organs 11:141–147
30. Iwata T, Yamato M, Tsuchioka H, Takagi R, Mukobata S, Washio K, Okano T, Ishikawa I (2009) Periodontal regeneration with multi-layered periodontal ligament-derived cell sheets in a canine model. Biomaterials 30:2716–2723
31. Maeda M, Yamato M, Kanzaki M, Iseki H, Okano T (2009) Thoracoscopic cell sheet transplantation with a novel device. J Tissue Eng Regen Med 3:255–259
32. Yaji N, Yamato M, Yang J, Okano T, Hori S (2009) Transplantation of tissue-engineered retinal pigment epithelial cell sheets in a rabbit model. Biomaterials 30:797–803
33. Asakawa N, Shimizu T, Tsuda Y, Sekiya S, Sasagawa T, Yamato M, Fukai F, Okano T (2010) Pre-vascularization of in vitro three-dimensional tissues created by cell sheet engineering. Biomaterials 31:3903–3909
34. Sasagawa T, Shimizu T, Sekiya S, Haraguchi Y, Yamato M, Sawa Y, Okano T (2010) Design of prevascularized three-dimensional cell-dense tissues using a cell sheet stacking manipulation technology. Biomaterials 31:1646–1654
35. Haraguchi Y, Sekine W, Shimizu T, Yamato M, Miyoshi S, Umezawa A, Okano T (2010) Development of a new assay system for evaluating the permeability of various substances through three-dimensional tissue. Tissue Eng Part C Methods 16:685–692
36. Sekine W, Haraguchi Y, Shimizu T, Umezawa A, Okano T (2011) Thickness limitation and cell viability of multi-layered cell sheets and overcoming the diffusion limit by a porous-membrane culture insert. J Biochip Tissue Chip S2:001
37. Haraguchi Y, Matsuura K, Shimizu T, Yamato M, Okano T (2013) Simple suspension culture system of human iPS cells maintaining their pluripotency for cardiac cell sheet engineering. J Tissue Eng Regen Med. doi:10.1002/term.1761
38. Haraguchi Y, Shimizu T, Yamato M, Okano T (2012) Cell therapy and tissue engineering for cardiovascular disease. stem cells. Trans Med 1:136–141
39. Penn MS, Pastore J, Miller T, Aras R (2011) SDF-1 in myocardial repair. Gene Ther 19:583–587
40. Lavu M, Gundewar S, Lefer DJ (2011) Gene therapy for ischemic heart disease. J Mol Cell Cardiol 50:742–750
41. Boron WF, Boulpaep EL (2003) Medical physiology. In: Radisic M, Michael VM (eds) The cardiovascular system (Chapter IV). Elsevier Science, Philadelphia, pp 421–590
42. Matsuura K, Wada M, Shimizu T, Haraguchi Y, Sato F, Sugiyama K, Konishi K, Shiba Y, Ichikawa H, Tachibana A, Ikeda U, Yamato M, Hagiwara N, Okano T (2012) Creation of human cardiac cell sheets using pluripotent stem cells. Biochem Biophys Res Commun 425:321–327
43. Kawamura M, Miyagawa S, Miki K, Saito A, Fukushima S, Higuchi T, Kawamura T, Kuratani T, Daimon T, Shimizu T, Okano T, Sawa Y (2012) Feasibility, safety, and therapeutic efficacy of human induced pluripotent stem cell-derived cardiomyocyte sheets in a porcine ischemic cardiomyopathy model. Circulation 126:S29–S37

Part III
Sociocytology

Chapter 15
Hydrogel-Based Microenvironment for Modulating Gland Tissue Morphogenesis

Takuya Matsumoto

Abstract Biological tissue fabrication in vitro is garnering attention these days for its potential usage as a new therapeutic tool and for understanding the mechanisms underlying tissue development. At this stage, Researchers have already developed several methods to fabricate three-dimensional (3D) cell constructs with desired size and morphology. Now, we seek to determine how we can transform this simple 3D cell construct into a functional biological tissue in vitro. Applying external stimuli to the cell construct may be one way to achieve this goal. In this context, we modified biocompatible hydrogel materials to reproduce the biomimetic physical and chemical environment. Hydrogels with different mechanical stiffness were prepared to physically mimic environments, and RGD-modified hydrogels were prepared to chemically mimic environments. To understand the effect of these synthesized environments on tissue generation and growth, the organotypic culture of mouse salivary gland tissue was studied using the environments. The potential use of the biomimetic environments synthesized by biocompatible hydrogel material for modulating tissue growth is discussed in this chapter.

Keywords Biomimetic environment · Hydrogel · Cell manipulation · Tissue manipulation · Submandibular gland

15.1 Introduction

In vitro biological tissue fabrication using the bottom-up approach has been garnering attention these days. A number of methods to fabricate three-dimensional (3D) cellular constructs have been introduced for this purpose. For example, a conventional method called the droplet method has been utilized for fabricating cell spheroids [1]. A polydimethylsiloxane (PDMS) micro-chamber system fabricated using a micro-electro mechanical system (MEMS) has been applied for

T. Matsumoto (✉)
Department of Biomaterials, Graduate School of Medicine, Dentistry and Pharmaceutical Sciences, Okayama University, 2-5-1 Shikata-Cho, Kita-ku, Okayama 700-8558, Japan
e-mail: tmatsu@md.okayama-u.ac.jp

T. Arai et al. (eds.), *Hyper Bio Assembler for 3D Cellular Systems*,
DOI 10.1007/978-4-431-55297-0_15

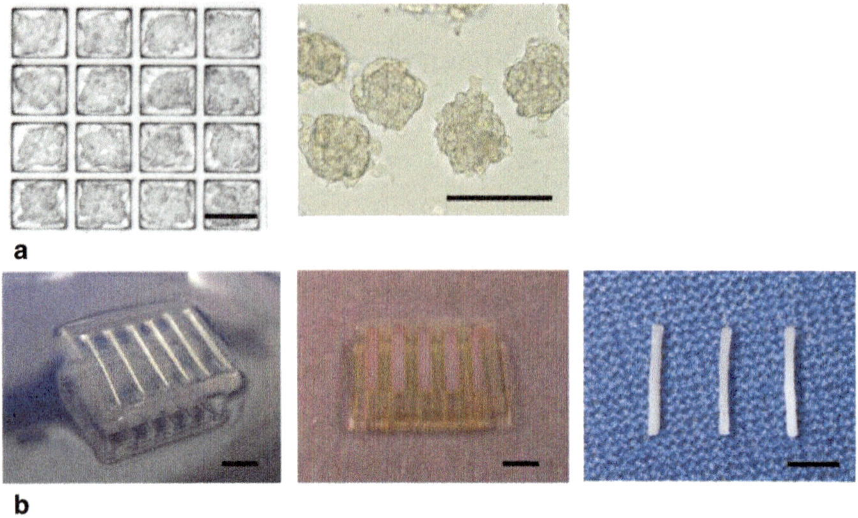

Fig. 15.1 **a** Three-dimensional (3D) cell spheroids fabricated by using micro-chamber system (Bar = 100 μm). **b** Three-D cell stick fabricated by using thermo-responsive hydrogel system (Bar = 5 mm)

the mass production of small cell spheroids [2]. Cell sheets are made by using a thermo-responsive polymer surface and piled up for the 3D cell constructs [3]. A thermo-responsive hydrogel, not a surface, has also been used for fabricating 3D cell constructs with desired size and morphology (Fig. 15.1) [4]. These 3D cell constructs would be useful as source materials for in vitro fabrication of biological tissues, but the problem is how to guide the 3D cell constructs into the biological tissue. Here, the fabricated biological tissue should not be simple aggregates of differentiated cells but rather organized cell constructs reproducing total functions with tissue-specific mechanical properties, size, and morphology. Developing precisely designed protocols for this remains a big challenge. One approach might be employing a biomimetic environment synthesized by the materials themselves. In this chapter, we focus on the fabrication of a hydrogel-based biomimetic environment and the modulation of grand tissue morphogenesis.

15.2 Tuning Microenvironment for Reproducing Biomimetic Stimuli

Advances in biomaterials science gave us the idea of materials having cellular and tissue biocompatibility [5]. Moreover, with the advances in tissue engineering studies, researchers have started paying more attention to 3D applications than to 2D [6]. Consequently, the fabrication of a biomimetic environment using biomaterials

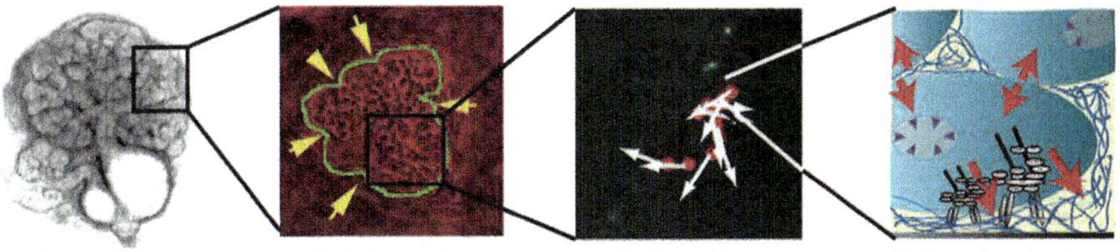

Fig. 15.2 Different manners of mechanical force are induced by molecular interactions, cell adhesion, cell motility, cell growth and differentiation, and tissue deformation during the process of tissue development

has been of interest these past few decades. To clarify briefly, a biomimetic environment is a chemically and physically reproduced condition that mimics the natural biological condition. For example, soluble factors such as growth factors and cytokines are chemical stimuli in one's body relating to tremendous biological activities of cells and tissues. In addition, the physical stimuli as well as the chemical stimuli in cells and tissues are also crucial. For example, the mechanical stiffness of cell/tissue surroundings differs in each type of tissue and alters according to age and gender. Also, cells are constantly migrating and proliferating during tissue development and growth [7]. Cells have a variety of mutual adhesion systems depending upon cell type and are thus subjected to a variety of mechanical forces both internally and externally, which induces a force-specific signal communication called mechanotransduction (Fig. 15.2) [8]. In previous studies, researchers focused on the mechanical stiffness and reproduced this mechanical condition by using hydrogel [9]. Results of such research have indicated that cellular functions are governed by the surrounding mechanical conditions. For example, when mesenchymal stem cells (MSCs) were cultured on soft substrates that resembled the stiffness of actual brain tissue, genetic profiling suggested that these cells underwent neuronal differentiation. On stiff substrates that mimicked actual bone stiffness, MSCs underwent osteogenesis [10]. These results strongly suggest that such mechanical conditions might be effective for in vitro tissue manipulation.

15.3 Biomimetic Mechanical Environment Constructed by Hydrogel and Its Application for Modulation of Gland Tissue Growth

Understanding the normal tissue developmental process is crucial to achieve in vitro biological tissue fabrication. The development of mouse submandibular gland (SMG) starts with a protrusion of the oral epithelium into the underlying mesenchyme at embryonic day 11 (E11) to form a solid epithelial bud. Branch and duct

Fig. 15.3 SMG branching morphogenesis initiates with cleft formation

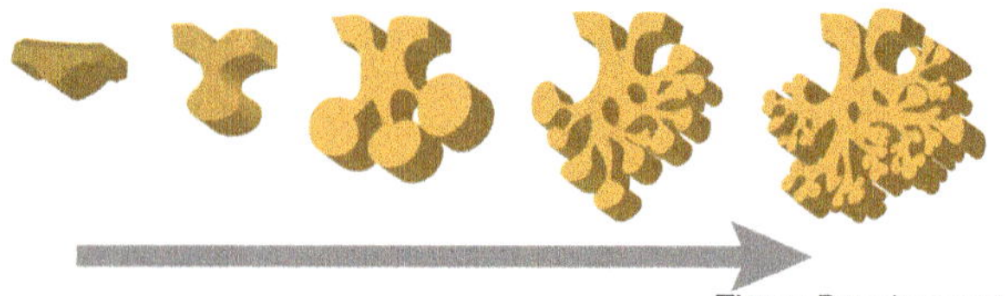

formation take place after this protrusion (Fig. 15.3) [11]. Branching processes greatly benefit glandular tissues in terms of creating a larger cellular area to meet the metabolic requirements within a limited space. This morphogenesis is regulated by the cross talk between epithelium and mesenchyme cells. To date, a number of studies have clarified the crucial roles of soluble factors for the branching morphogenesis of salivary glands [12–15]. For example, the ligands of the epidermal growth factor (EGF) family and the fibroblast growth factor (FGF) family play regulating roles in buddings and duct elongation in the developmental processes of SMG [16, 17]. Mitogen-activated protein kinase (MAPK), phospholipase Cγ1 (PLCγ1), and phosphatidyl-inositol-3-kinase (PI3K) caused by chemical stimuli including soluble factors are the main transducers in these processes. Extracellular matrix (ECM) components (e.g., collagen, laminin, and fibronectin) and their digestive enzymes (e.g., matrix metalloprotainase) are also essential factors for this growth.

Alginates are naturally derived polysaccharides that gel in the presence of divalent cations [18]. Alginates are widely used for cell encapsulation in vitro [19] and in vivo [20], especially for tissue engineering applications [21, 22], because of their biocompatibilities and gentle gelling properties. Due to their extreme hydrophilic properties, alginate materials have less affinity with proteins and mammalian cells [23]. Materials with less cellular affinity are crucial to investigate how mechanical stiffness alone affects tissue morphogenesis. Therefore, alginate hydrogels with different mechanical stiffness were utilized as tissue culture substrates to investigate the physical cue on the branching morphogenesis of SMG.

SMG tissue isolated from ICR mouse (E12.5) was cultured on alginate hydrogels with different mechanical stiffness (Young's modulus; 4–180 kPa). Results indicated that the softer gel (4 kPa) enhances the bud expansion and cleft formation of SMG whereas the stiffer gel (184 kPa) attenuates them (Fig. 15.4, 15.5). The mesenchymal tissue of SMG directly attaches to the gel surface under this tissue culture condition. Therefore, the surrounding mechanical stiffness should mainly affect the mesenchymal tissue. To confirm this, mesenchymal tissue was mechanically removed from SMG tissue and the remaining epithelial tissue was cultured under the same conditions. The cultured epithelial tissue did not grow under any mechanical conditions, indicating that the mesenchymal tissue senses the surrounding mechanics and guides the tissue growth under this condition. Consequently, by using western blotting and immunofluorescent staining, key roles of FGF7 and FGF10 altering their expressions by the surrounding mechanical stiffness, followed by the regulating SMG morphogenesis, was suggested. This was confirmed by the well-grown SMG when these growth factors were added to SMG cultured on the

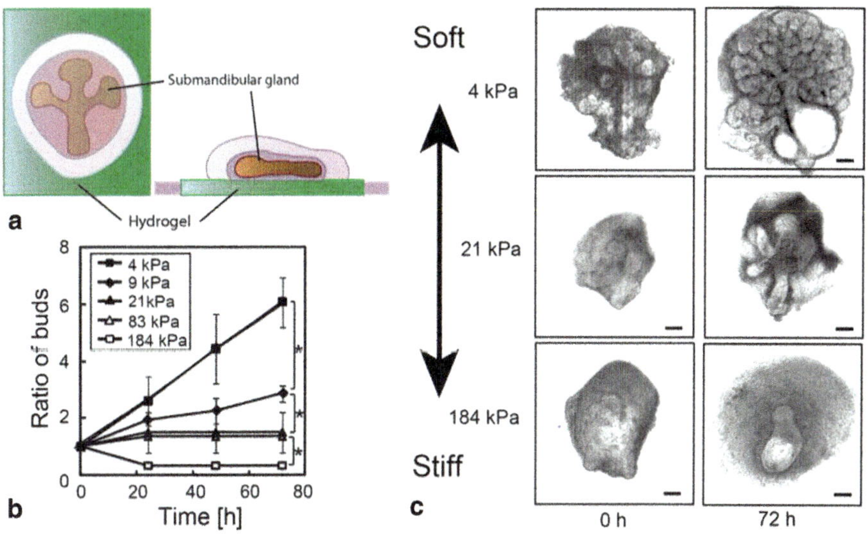

Fig. 15.4 Effect of surrounding mechanical stiffness on SMG morphogenesis. **a** Organotypic culture system used in this study (Bar = 100 μm). **b**, **c** Soft gel promotes SMG growth, while stiff gel attenuates it

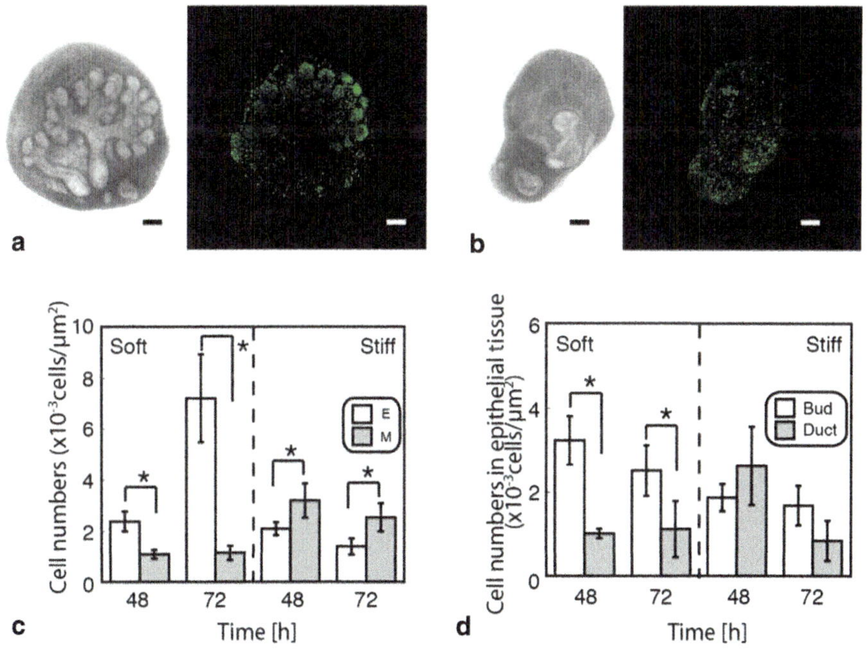

Fig. 15.5 Effect of surrounding mechanical stiffness on cell proliferation in SMG. BrdU positive cells are raveled with green (**a** SMG on soft gel, **b** SMG on stiff gel, Bar = 100 μm)

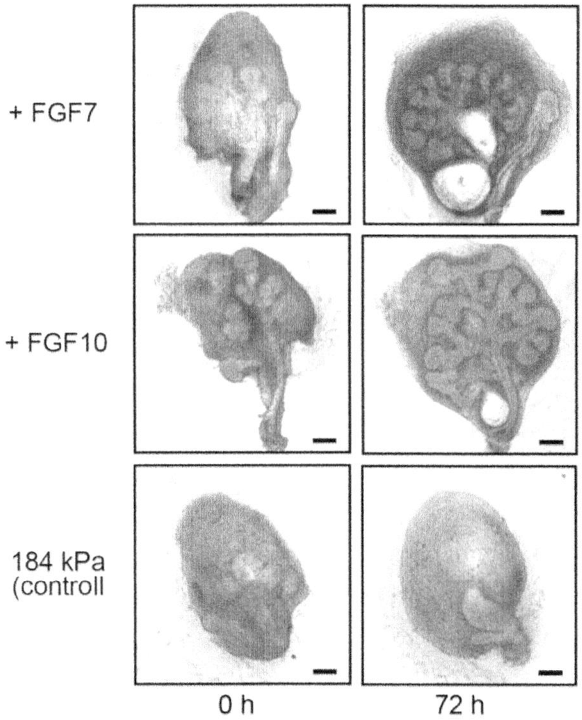

Fig. 15.6 SMG branching morphogenesis cultured on stiffer gel with FGF7 or FGF10 (184 kPa, Bar = 100 μm)

stiffer gel, which normally attenuates its growth (Fig. 15.6). Additional analysis using computer simulation suggested that cell deformation depending on the surrounding mechanical stiffness functioned as a trigger for modulating the growth factor expressions.

15.4 Biomimetic Cell-Adhesive Environment Constructed by Chemically Modified Hydrogel and Its Application for Modulation of Gland Tissue Growth

Fibronectin, a major extracellular matrix protein secreted by mesenchymal cells, is a key matrix protein because it can be found in many biological elements including tissue development [24], wound healing [25], and even cancer development [26]. Fibronectin binds with other organic molecules including collagen, fibrin, and syndecan and binds with cells via RGD (Arginine-Glycine-Aspartic Acid) motif, the ligand for integrins [27]. Previous reports have indicated that fibronectin is specifically precipitated at the cleft region in SMG [28]. Here, we considered that this fibronectin-related motif might become one of the regulating triggers for SMG morphogenesis. We therefore prepared an RGD-modified alginate hydrogel mimicking a cell's adhesive environment in order to modulate SMG branching. Initially, the RGD-modified alginate gel "sheet" was utilized as a tissue culture substrate. Strikingly, SMG cultured on the RGD-modified stiffer gel showed a larger cell adhesion area on the substrate and a better branching than that on the stiffer gel without RGD modification. Moreover, the SMG bud numbers increased with the increase of RGD ratio introduced in the alginate. Both FGF7 and FGF10 were also up-regulated under this condition. Since higher-introduced RGD modified alginate enhanced mesenchymal cell adhesion with higher cell deformation, the cell deformation was again considered as a key factor for modulating these factor expressions [29].

Since the effect of RGD was clearly observed, we next considered a different application of using this molecule on SMG morphogenesis. RGD-modified alginate hydrogel beads (ϕ=20–100 um) were prepared for this purpose. The beads were directly placed on the SMG cultured on the soft gel. Strikingly, the bead-placed region guided a cleft formation in SMG tissue. The ratio of guided branching in the cultured SMG was five times higher compared to the usage of non-RGD modified beads.

15.5 Conclusions

One of the most important strategies for in vitro tissue fabrication is developing methods to guide cell constructs into a biologically functional tissue with tissue specific size and morphology. Recently, several studies on in vitro tissue synthesis have been reported. For example, Sasai et al. fabricated optic cup and adeno-hypophysis by using embryonic stem cells encapsulated in type I collagen gel [30, 31]. Christiano et al. fabricated hair follicles [32], and Tsuji et al. generated tooth and salivary gland tissue by using organotypic culture composed of isolated cells with the collagen gel [33, 34]. In our own previous work, we reproduced endochondral ossification by using 3D cell constructs of mesenchymal stem cells [35]. However, at this point, these works indicate the distribution of functionally differentiated cells located in areas identical to native tissue. The fabrication of real biological tissue consisting of cells and matrix composites having tissue-specific size and morphology in vitro remains a big challenge. In this chapter, we focused on the synthesized microenvironment to modulate tissue growth and morphology. Newly developed synthesized environment would be promising to modulate cell and tissue development and growth.

References

1. Fischbach C, Chen R, Matsumoto T, Schmelzle T, Brugge JS, Polverini PJ, Mooney DJ (2007) Engineering tumors with 3D scaffolds. Nat Methods 4:855–60
2. Iwasaki A, Matsumoto T et al (2009) Adv Eng Mater 11:801

3. Sasagawa T, Shimizu T, Sekiya S, Haraguchi Y, Yamato M, Sawa Y, Okano T (2010) Design of prevascularized three-dimensional cell-dense tissues using a cell sheet stacking manipulation technology. Biomaterials 31:1646–1654
4. Sasaki J, Matsumoto T et al (2010) Fabrication of three-dimensional cell constructs using temperature-responsive hydrogel. Tissue Eng A 16:2497–2504
5. Lee KY, Kong HJ et al (2008) Quantifying interactions between cell receptors and adhesion ligand-modified polymers in solution. Macromol Biosci 8:140–145
6. Cima LG, Vacanti JP, Vacanti C, Ingber D, Mooney D, Langer R (1991) Tissue engineering by cell transplantation using degradable polymer substrates. J Biomech Eng 113:143–151
7. Campàs O, Mammoto T, Hasso S, Sperling RA, O'Connell D, Bischof AG, Maas R, Weitz DA, Mahadevan L, Ingber DE (2013) Quantifying cell-generated mechanical forces within living embryonic tissues. Nat Methods 11(2):183–189 (in press)
8. Ingber DE (2006) Cellular mechanotransduction: putting all the pieces together again. FASEB J 20(7):811–827
9. Kong HJ, Matsumoto T et al (2005) Non-viral gene delivery regulated by stiffness of cell adhesion substrates. Nat Mater 4(6):460–464
10. Engler AJ, Sen S et al (2006) Matrix elasticity directs stem cell lineage specification. Cell 126:677–689
11. Tucker AS (2007) Salivary gland development. Semin Cell Dev Biol 18:237–244
12. Nakanishi Y, Sugiura F, Kishi J, Hayakawa T (1986) Collagenase inhibitor stimulates cleft formation during early morphogenesis of mouse salivary gland. Dev Biol 113:201–206
13. Yang TL, Young TH (2008) The enhancement of submandibular gland branch formation on chitosan membranes. Biomaterials 29:2501–2508
14. Kadoya Y, Kadoya K, Durbeej M, Holmvall K, Sorokin L, Ekblom P (1995) Antibodies against domain E3 of laminin-1 and integrin alpha 6 subunit perturb branching epithelial morphogenesis of submandibular gland, but by different modes. J Cell Biol 129:521–534
15. Sakai T, Larsen M, Yamada KM (2003) Fibronectin requirement in branching morphogenesis. Nature 423:876–881
16. Steinberg Z, Myers C, Heim VM, Lathrop CA, Rebustini IT, Stewart JS et al (2005) FGFR2b signaling regulates ex vivo submandibular gland epithelial cell proliferation and branching morphogenesis. Development 132:1223–1234
17. Morita K, Nogawa H (1999) EGF-dependent lobule formation and FGF7-dependent stalk elongation in branching morphogenesis of mouse salivary epithelium in vitro. Dev Dyn 215:148–154
18. Smidsrød O, Skjåk-Bræk G (1990) Alginate as immobilization matrix for cells. Trends Biotechnol 8:71–78
19. Fremond B, Malandain C, Guyomard C, Chesne C, Guillouzo A, Campion JP (1993) Correction of bilirubin conjugation in the Gunn rat using hepatocytes immobilized in alginate gel beads as an extracorporeal bioartificial liver. Cell Transplant 2:453–460
20. Chang SC, Rowley JA, Tobias G, Genes NG, Roy AK, Mooney DJ et al (2001) Injection molding of chondrocyte/alginate constructs in the shape of facial implants. J Biomed Mater Res 55:503–511
21. Rowley JA, Madlambayan G, Mooney DJ (1999) Alginate hydrogels as synthetic extracellular matrix materials. Biomaterials 20:45–53
22. Atala A, Kim W, Paige KT, Vacanti CA, Retik AB (1994) Endoscopic treatment of vesicoureteral reflux with a chondrocyte-alginate suspension. J Urol 152:641–643
23. Smetana K Jr (1993) Cell biology of hydrogels. Biomaterials 14:1046–1050
24. George EL, Georges-Labouesse EN, Patel-King RS, Rayburn H, Hynes RO (1993) Defects in mesoderm, neural tube and vascular development in mouse embryos lacking fibronectin. Development. 119:1079–1091
25. Kohan M, Muro AF, White ES, Berkman N (2010) EDA-containing cellular fibronectin induces fibroblast differentiation through binding to alpha4beta7 integrin receptor and MAPK/Erk 1/2-dependent signaling. FASEB J 24:4503–4512

26. Knowles LM, Gurski LA, Engel C, Gnarra JR, Maranchie JK, Pilch J (2013) Integrin αvβ3 and fibronectin upregulate Slug in cancer cells to promote clot invasion and metastasis. Cancer Res 73:6175–6184
27. D'Souza SE, Ginsberg MH, Burke TA, Lam SC, Plow EF (1988) Localization of an Arg-Gly-Asp recognition site within an integrin adhesion receptor. Science 242:91–93
28. Sakai T, Larsen M, Yamada KM (2003) Fibronectin requirement in branching morphogenesis. Nature 423:876–881
29. Miyajima H, Matsumoto T, Sakai T, Yamaguchi S, An SH, Abe M, Wakisaka S, Lee KY, Egusa H, Imazato S (2011) Hydrogel-based biomimetic environment for in vitro modulation of branching morphogenesis. Biomaterials 32:6754–6763
30. Eiraku M, Takata N, Ishibashi H, Kawada M, Sakakura E, Okuda S, Sekiguchi K, Adachi T, Sasai Y (2011) Self-organizing optic-cup morphogenesis in three-dimensional culture. Nature 472:51–56
31. Suga H, Kadoshima T, Minaguchi M, Ohgushi M, Soen M, Nakano T, Takata N, Wataya T, Muguruma K, Miyoshi H, Yonemura S, Oiso Y, Sasai Y (2011) Self-formation of functional adenohypophysis in three-dimensional culture. Nature 480:57–62
32. Higgins CA, Chen JC, Cerise JE, Jahoda CA, Christiano AM (2013) Microenvironmental reprogramming by three-dimensional culture enables dermal papilla cells to induce de novo human hair-follicle growth. Proc Natl Acad Sci U S A 110:19679–19688
33. Nakao K, Morita R, Saji Y, Ishida K, Tomita Y, Ogawa M, Saitoh M, Tomooka Y, Tsuji T (2007) The development of a bioengineered organ germ method. Nat Methods 4:227–230
34. Ogawa M, Oshima M, Imamura A, Sekine Y, Ishida K, Yamashita K, Nakajima K, Hirayama M, Tachikawa T, Tsuji T (2013) Functional salivary gland regeneration by transplantation of a bioengineered organ germ. Nat Commun 4:2498
35. Sasaki J, Matsumoto T, Egusa H, Matsusaki M, Nishiguchi A, Nakano T, Akashi M, Imazato S, Yatani H (2012) In vitro reproduction of endochondral ossification using a 3D mesenchymal stem cell construct. Integr Biol (Camb) 4:1207–1214

Chapter 16
Bone Related Cell-Stimulating Scaffold Materials and a 3D Cellular Construct for Hard Tissue Regeneration

Osamu Suzuki and Takahisa Anada

Abstract The present article summarizes recently developed calcium phosphate-based scaffold materials that are capable of stimulating various cells involved in hard tissue formation as well as a method using newly developed cell culture devices to generate three dimensional (3D) cellular constructs. Synthetic octacalcium phosphate stimulates cellular functions of several cell types, including osteoblasts, chondrocytes, and pulp cells. The culture device promotes the easy assembly of chondrocytes in a 3D cellular construct. The scaffold materials, cellular constructs, and composite presented here are good candidates as implantable materials for the regeneration of hard tissue in bone defects depending on the severity of the tissue loss.

Keywords Octacalcium phosphate · Scaffold · Osteoblast · Spheroid · Polydimethylsiloxane · Cell culture device · Chondrocyte

16.1 Introduction

Bone tissue consists of cells related to bone formation and associated matrix materials, including collagen fibers mineralized with biomineral hydroxyapatite (HA) crystals and various growth factors [1]. Osteoblasts, osteoclasts, chondrocytes, endothelial cells, and bone marrow cells are involved in the remodeling processes that occur in bone tissues [1]. Recently, increased attention has been placed on developing three-dimensional (3D) scaffolds inoculated with exogenous cells or cell constructs without scaffold materials under well-controlled conditions that are suitable for increasing cell activity [2]. In this article, we summarize our methodology for constructing scaffolds particularly with synthetic biominerals, including scaffolds

O. Suzuki (✉) · T. Anada
Division of Craniofacial Function Engineering, Tohoku University Graduate School of Dentistry, 4-1 Seiryo-machi, Aoba-ku, Sendai 980-8575, Japan
e-mail: suzuki-o@m.tohoku.ac.jp

T. Arai et al. (eds.), *Hyper Bio Assembler for 3D Cellular Systems*,
DOI 10.1007/978-4-431-55297-0_16

that can stimulate exogenous cellular activity as well as 3D cellular constructs, both of which could be implantable. These approaches may be effective therapies for bone regeneration of large defects caused by various diseases, such as trauma or tumors, due to the limited autologous bone available in these patients [3]. The development of an implantable artificial scaffold and cellular construct that has biological performance comparable to autologous bone is the current goal in the field of bone tissue engineering [4, 5]. We show here that octacalcium phosphate (OCP) is a suitable synthetic biomineral analogue [6, 7] and that a polydimethylsiloxane (PDMS)-based cell culture device is a useful tool for not only developing the appropriate scaffold for growing the cell construct, but also developing a construct of the bone forming-related cells, chondrocytes [8].

16.2 Performance of Synthetic Biomaterials and the Device Used for Cellular Construct Development

We have developed biomaterials and cell culture devices to analyze cell-materials and cell-cell interactions for the regeneration of hard tissues [8–10]. Our previous studies showed that the interaction between calcium phosphate materials, in particular OCP, and various cells involved in hard tissue formation stimulates cellular functions such as proliferation and differentiation [11–15]. Based on the evidence, we hypothesized that the combination of calcium phosphate materials and 3D cellular constructs improves their cellular activities and leads to effective regeneration of hard tissues [4, 9, 16]. In this review we describe the development of composite materials as scaffolds for bone tissue regeneration and the fabrication of devices for the construction of 3D cellular aggregates [8, 9, 17].

We previously elucidated the effects of OCP on cell proliferation and differentiation in various cell types. Table 16.1 shows the summary of the cell-materials interaction investigated to date:

- OCP promotes an early stage of osteoblastic differentiation of mouse bone marrow derived stromal ST-2 cells [11].
- OCP promotes osteoclastic differentiation of mouse bone marrow cells by increasing receptor activator of NF-kB ligand (RANKL) expression by osteoblasts [14].
- OCP promotes odontoblastic differentiation of rat pulp cells [15].
- OCP suppresses chondrogenic differentiation of mouse chondrogenic ATDC5 cells [12].
- OCP stimulates mesenchymal cells (MSCs) to enhance new bone formation in rat critical-sized defects if implanted with OCP and collagen scaffold [4].

Table 16.1 Effect of OCP crystals on cellular activities of calcified tissue derived cells.

Cells examined	Cellular response	mRNA expression	References
Mouse bone marrow stromal ST-2 cells	Osteoblastic differentiation	Up-regulation of ALP[a], osterix and Col I	[11, 13]
Mouse bone marrow cells/osteoblastic cells	Osteoclast formation (in the absence of vitamin D_3)	Up-regulation of RANKL[b] in osteoblasts	[14]
Rat dental pulp cells	Odontoblastic differentiation	Up-regulation of DSPP[c], DMP[d], ALP	[15]
Chondrogenic ATDC5 cells	Inhibition of chondrogenic differentiation	Down-regulation of Sox6[e]	[12]

[a]Alkaline phosphatase; [b]Receptor activator of nuclear factor-kappa B ligand; [c]Dentin sialophosphoprotein; [d]Dentin matrix protein; [e]SRY (sex determining region Y)-box 6

16.3 Development of the Scaffold Materials for Tissue Regeneration and Inoculated Cells

16.3.1 Effect of Co-Precipitation of OCP With Gelatin Molecules on Their Composite Scaffolds [9]

The details for material preparation have been reported elsewhere [9]. Osteoblastic cells should interact with the co-precipitated OCP crystals as much as possible in order to maximize the cellular activity. To do so, OCP crystals were designed to be homogeneously dispersed in a gelatin (Gel) matrix. OCP-dispersed gelatin composite (OCP/Gel) was obtained through the co-precipitation of OCP from calcium and phosphate solutions under a supersaturated condition with respect to HA and OCP and at constant pH and temperature. OCP/Gel precipitates were molded, lyophilized, and treated dehydrothermally for the cross-linking step. The resultant materials consisted of highly porous (over 90 % in porosity) 3D structure and contained 40 % of OCP crystals by weight. The cellular adhesive property was examined using osteoblastic ST-2 cells inoculated onto OCP/Gel composites. Mouse bone marrow stromal ST-2 cells were seeded onto the OCP/Gel disks and the number of the viable cells attached was measured at 1 day post-inoculation. More than 60–70 % of the seeded cells were attached to the disk of OCP/Gel composites (Fig. 16.1), suggesting that host osteoblastic cells can attach, proliferate, and differentiate on OCP/Gel composite if implanted in a bone defect [9]. In fact, when the disks of OCP/Gel composite were implanted in rat critical-sized 9 mm diameter defects, new bone formation was observed together with accelerated biodegradation of the OCP/Gel composite by osteoclastic resorption [9]. Furthermore, the OCP/Gel composite was able to repair tibia bone defects of 5 mm in diameter, even in the early stages of implantation, accompanied by the relatively rapid material degradation and active new bone formation (Fig. 16.2). These results suggest that OCP and Gel can be successfully used as exogenous cellular scaffold materials.

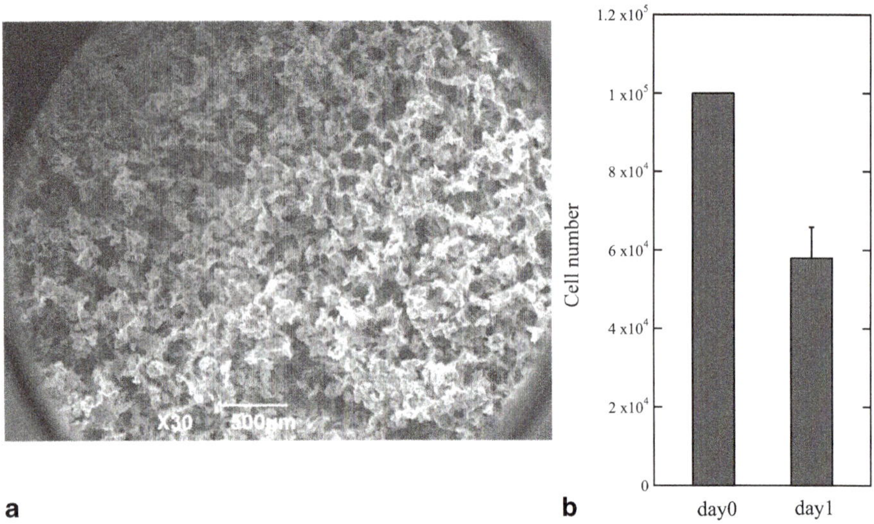

Fig. 16.1 **a** Scanning electron micrograph of OCP40/Gel composite. **b** Cell attachment to OCP40/Gel composite after 1 day of incubation. Reproduced from [9] with some modification with permission from Elsevier

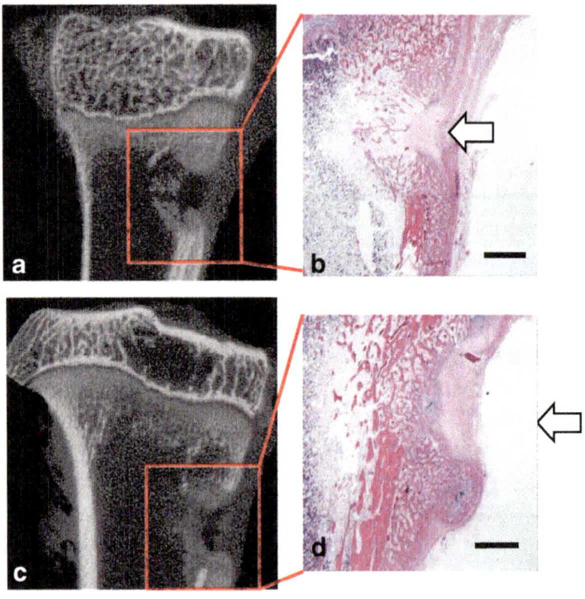

Fig. 16.2 Coronal view of micro CT and tissue section from rabbit tibia that was stained with hematoxylin and eosin. Control group (**a**), (**b**) and OCP40/Gel implantation group (**c**) and (**d**) at 2 weeks post-operation. Arrows: the site of the bone defect and OCP/Gel implantation. Bars: 3 mm. Reproduced from [17] with some modification with permission from Japanese Association of Inorganic Phosphorous Chemistry

Fig. 16.3 Scanning electron micrograph of nano-submicro OCP crystals. Bar = 1 μm

16.3.2 The Effect Of Micro-Nano OCP Crystal Inclusion in Gel Matrix [10]

A previous study reported that the microstructure of OCP crystals may be one factor that controls their bone regenerative properties when implanted in critical-sized mouse calvaria [18]. It has been shown that crystals that are several micrometers in length induce significantly greater enhancement of bone regeneration than OCP crystals greater than 10 μm in length [19]. Based on the experimental evidence, we investigated the effect of OCP crystals mechanically ground into micro-nano sized crystals dispersed in Gel matrix on bone regeneration if implanted in rat critical-sized calvaria defects [10]. The micro-nano OCP/Gel composite sufficiently provided a scaffold to support host osteoblastic cells as the material provided osteoconductivity in the defect [10]. Figure 16.3 shows the scanning electron micrograph of the OCP crystals ground mechanically. The micro-nano crystals were included together with larger OCP crystals. In fact, when the size of the particles was analyzed by dynamic light scattering in ethanol, it was found that the size was approximately 520 nm although there was a second distribution of approximately 24 μm [10].

16.3.3 Effect of OCP Crystals on MSC Differentiation by the OCP/Col/MSC Composite [4]

The effect of OCP crystals on differentiation of rat bone marrow-derived mesenchymal stem cells (MSCs) in collagen matrix (Col) was investigated by implanting OCP/Col/MSC in rat critical-sized calvaria defects [4]. Alkaline phosphatase (ALP)-expressing MSCs were obtained by incubating in osteogenic medium containing basic FGF, and then the cells were inoculated on OCP/Col and implanted in rat critical-sized calvaria defects. Eight weeks post-implantation, bone regeneration increased in the following order: Col, OCP/Col, and OCP/Col/MSCs. These results suggest that OCP crystals can enhance the differentiation of MSCs within an OCP/Col matrix in vivo [4].

16.3.4 Possible Scaffold Materials For Osteoblastic Cells

In order to develop osteoblastic cells for bone tissue regeneration, scaffold materials such as OCP crystals may effectively support cellular differentiation. The advancement of mineralization by this material under physiological conditions may control osteoblastic differentiation. Therefore, future studies will focus on designing scaffold materials for optimal distribution of cells and to determine which type of cells should be included for repair of the targeted hard tissues.

16.4 Spheroid Cell Culture Devices And Application With Chondrocytes

16.4.1 3D Cellular Constructs

Several studies have described the preparation of 3D cellular constructs [20]. Among them, much attention has focused on fabricating the cellular aggregates to have a spheroid shape that can act as building blocks for a large number of the cells [2]. It is thought that as the aggregate size increases, the spheroids begin to mimic the functions of living tissues compared to monolayer cell cultures regarding the morphology, cell-cell interactions, cell-matrix interactions, and signaling. This effect may be due to the 3D contact and close association of the cells with each other. Indeed, hepatocytes cultured in 3D exhibit long-term liver-specific functions [21–23]. The formation of osteoblast spheroids increases the expression of several proteins involved in bone formation [24]. It is well known that endochondral bone ossification initiates nodule formation with the aggregation of mesenchymal cells. The self-assembled mesenchymal cell aggregates cause dense cell-cell interactions and subsequent differentiation into chondrocytes [25]. A previous study described an approach to simulate mesenchymal condensation and cellular interactions, and mesenchymal stem cells (MSCs) or chondrocyte s have been examined for forming the pellet or spheroid for chondrogenesis. Moreover, spheroids have been verified as biological models in investigations of microtumors and metastases [26]. Moreover, spheroids have been used as drug screening assays as an alternative to animal models [27].

Several techniques for the culture of spheroids have been reported, including cultivation on nonadherent surfaces [28], cultivation in spinner flasks [29], the hanging drop method [30], micromolding technique [31], and a spheroid array on a micropatterned surface [32, 33]. However, conventional techniques are difficult to produce large amount of spheroids, control the spheroid size, and immobilize spheroids on the culture surface, which results in the requirement for invasive collection of cells.

16.4.2 Spheroid Culture Device Utilizing Thin Membrane Deformation Under Decompression (Prototype Device) [34]

We have previously developed a spheroid culture device to overcome the limitations of conventional methods through the utilization of thin membrane deformation by decreasing the pressure [34]. This culture device can be constructed with thousands of milli- to micro-sized hemispherical cavities by deforming a thin polydimethylsiloxane (PDMS) membrane under decompression. The resulting device can rapidly produce many spheroids at the same time within the hemispherical multicavities and can control the diameter of the spheroids within a narrow distribution. This approach also allows for the collection of spheroids from the device using a noninvasive approach. The device is constructed from PDMS, which is an acrylic resin plate, and a glass plate (Fig. 16.4). The PDMS chamber and membrane are assembled as a cell culture chamber. Figure 16.5 shows the mechanism of action of

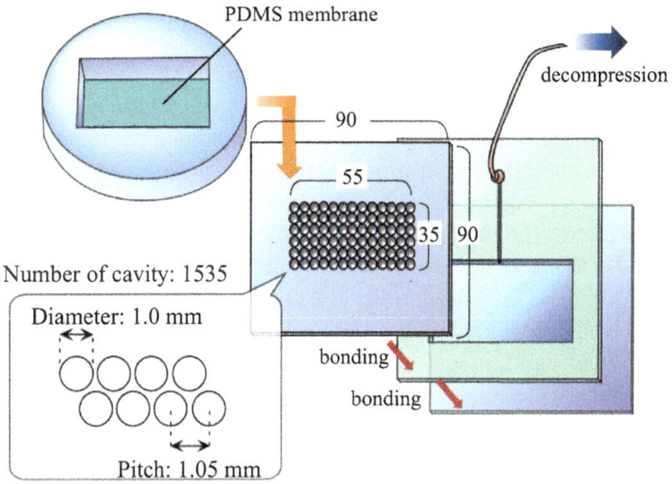

Fig. 16.4 Schematic illustration of components of the spheroid culture device. Redrawn from [34] with permission from Elsevier

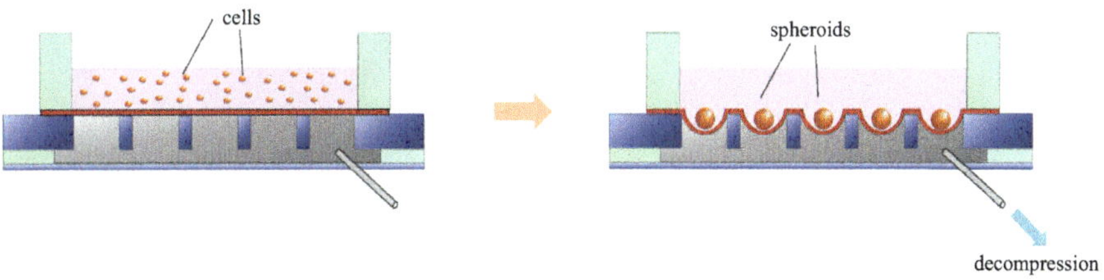

Fig. 16.5 Schematic illustration of the mechanism of action of the device. Modified from [34] with permission from Elsevier

the device. The multicavities in the acrylic resin plate are covered by the PDMS membrane of the culture chamber. These multicavities are then connected with a silicon tube to reduce pressure. Thus, the PDMS membrane can be deformed as a hemispherical cavity. A Pluronic F-127 solution is added to the culture chamber to prevent cell attachment and promote spheroid formation.The cell suspension is then added into the cell culture chamber, and a thin PDMS membrane is deformed by pressure reduction. The cells on the PDMS membrane self-assemble, which results in spheroid formation in the hemispherical cavity.

To test this device, human osteosarcoma MG63 cells and the human hepatoma cell line HepG2 were cultured in the cavities. We found that the cells spontaneously dropped into the cavities and a single spheroid formed in each. At the end of the culture, the cells were easily collected from the culture device by pressure release. Importantly, the recovery of cells from the device without enzymatic treatment provides a great advantage in applications for regenerative medicine.

16.4.3 Spheroid Formation Utilizing a PDMS-Based Culture Chip [8]

Our described spheroid culture device utilizes thin PDMS membrane deformation through negative pressure, but the decompression requires use of an external vacuum pump throughout the culture time in order to maintain the membrane deformation. In addition, the fabrication process is relatively time consuming. Therefore, we improved this procedure by using the prototype device as a master mold with a simple and rapid lithography technique [34]. Briefly, a PDMS prepolymer was poured into the PDMS negative mold and cured. The PDMS replica was peeled off from the mold and then used in the cell culture. To prevent cell adhesion to the PDMS surface and accelerate the formation of spheroids, the surface of the PDMS chip was covalently modified with polyethyleneglycol diacrylate (PEGDA) [35]. The culture chips were inlayed into a 6 cm diameter polystyrene dish to prevent contamination of the cultured cells (Fig. 16.6). The chips were designed to be com-

Fig. 16.6 Photograph of PDMS-based spheroid culture chip. Reproduced from [8] with permission from Fuji Technology Press

prised of multicavities (512 wells, 1.0 mm in diameter, 1.05 mm pitch, and 1060 μm in depth) on a 25 × 25 mm section of the cell culture area.

16.4.4 Chondrocyte Culture Using a PDMS-Based Culture Chip

It has been reported that the size and shape of the cell aggregates are important for cell proliferation and differentiation. Numerous studies have reported that dense cell pellets consisting of an aggregate of MSCs or chondrocytes enhance the expression of cartilage-specific genes, secretion of extracellular matrix such as glycosaminoglycan (GAG) and type II collagen, and development of cartilage-like morphology [36].

The murine embryonic carcinoma cell line ATDC5 was previously shown to differentiate into chondrocytes from mesenchymal condensation in the presence of insulin [37]. ATDC5 cells at different concentrations (5×10^5, 1×10^6, 2×10^6, and 4×10^6 cells/chip) were added to the chips and grown in chondrogenic medium. Cells spontaneously dropped into the cavities, and a single spheroid formed in each within 3 h. In addition, the spheroid diameters increased depending on the initial cell density. Cells were collected from the culture chip by washing them out with PBS or culture medium without enzymatic treatment at the end of the culture period. The culture chip produced well-rounded spheroids with narrow size distributions (Fig. 16.7). The secretion of sulfated GAG by ATDC5 cells tended to increase with time in all samples (Fig. 16.8). Moreover, sulfated GAG secretion in the spheroid culture system was higher than that in the monolayer culture and increased as the initial cell density increased. After 14 days of culture, the cells seeded at a density of 4×10^6 cells/chip had the greatest secretion (4.0–fold higher than that in the monolayer culture). A higher cell seeding density significantly improved the differentiation of ATDC5 cells in the spheroid culture chip. Further studies are required to elucidate the mechanism that facilitates the chondrogenic phenotype of ATDC5 cells by assessing the culture chip at the molecular level, including changes in the gene expression level during chondrogenic differentiation. Some groups have

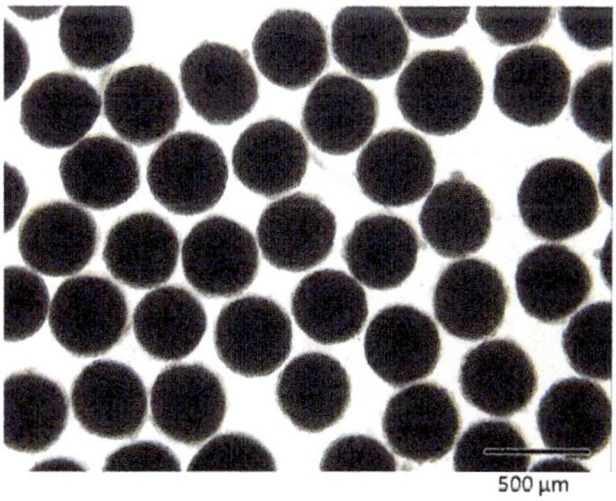

Fig. 16.7 Light microscopic image of ATDC5 spheroids collected from the culture chip on day 14 of culture. Reproduced from [8] with permission from Fuji Technology Press

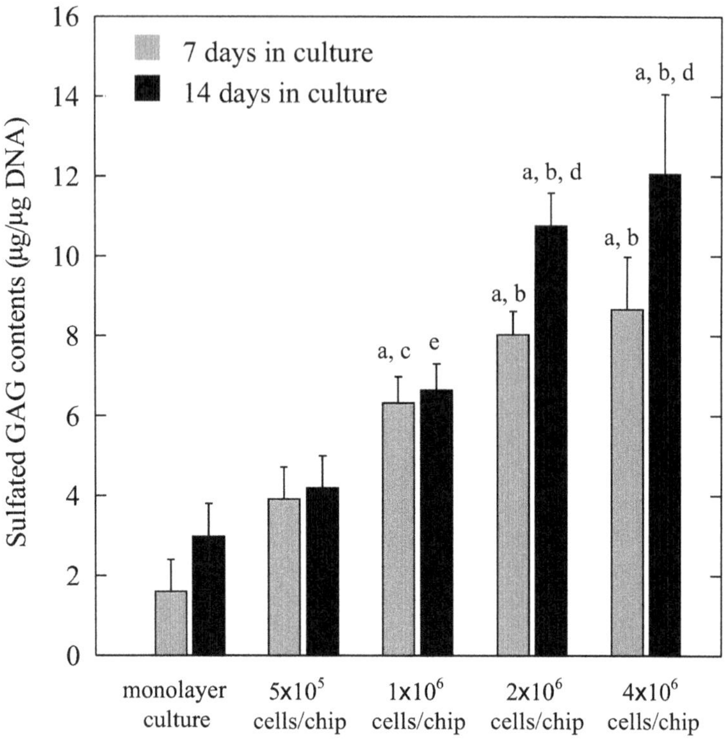

Fig. 16.8 Changes in sulfated GAG content of ATDC5 cells normalized by DNA. [a]$p<0.01$ compared to monolayer culture. [b]$p<0.01$ compared to 5×10^5 cells/chip. [c]$p<0.05$ compared to 5×10^5 cells/chip. [d]$p<0.01$ compared to 1×10^6 cells/chip. ep <0.05 compared to monolayer culture. Reproduced from [8] with permission from Fuji Technology Press

reported on the regeneration of cartilage using chondrocyte spheroids injected without scaffold material [38–40] or with hydrogel [41]. Our system could be useful for the mass production of transplantable spheroids, and we believe that this culture system can be potentially used for a wide variety of tissue engineering strategies.

16.5 Conclusion

The present studies underscore the effectiveness of the scaffold material, and in particular that of OCP crystals, to enhance the activity of calcified tissue-relating cells such as osteoblasts. Furthermore, the device used for cellular construct preparation effectively formed 3D spheroids and facilitates cellular activity, especially chondrocytes. Further studies are underway to establish the effect of the interaction between the scaffold materials and the cellular construct on cellular function, such as differentiation, and may be used to efficiently repair tissue.

Acknowledgements This study was supported in part by Grants-in Aid (23106010, 23390450, 17076001, and 19390490) from the Ministry of Education, Science, Sports and Culture of Japan and the Suzuken Memorial Foundation.

References

1. Ozawa H, Hoshi K, Amizuka N (2008) Current concepts of bone mineralization. J Oral Biosci 50(1):1–14
2. Mironov V, Visconti RP, Kasyanov V, Forgacs G, Drake CJ, Markwald RR (2009) Organ printing: tissue spheroids as building blocks. Biomaterials 30(12):2164–2174
3. Wagoner JAJ, Herschler BA (2011) A review of the mechanical behavior of CaP and CaP/polymer composites for applications in bone replacement and repair. Acta Biomater 7(1):16–30
4. Kawai T, Anada T, Masuda T, Honda Y, Sakai Y, Kato Y et al (2011) The effect of synthetic octacalcium phosphate in a collagen scaffold on the osteogenicity of mesenchymal stem cells. Eur Cell Mater 22:124–136
5. Kotobuki N, Hirose M, Takakura Y, Ohgushi H (2004) Cultured autologous human cells for hard tissue regeneration: preparation and characterization of mesenchymal stem cells from bone marrow. Artif Organs 28(1):33–39
6. Suzuki O (2010) Octacalcium phosphate: osteoconductivity and crystal chemistry. Acta Biomater 6(9):3379–3387
7. Suzuki O (2013) Octacalcium phosphate (OCP)-based bone substitute materials. Jpn Dent Sci Rev 49:58–71
8. Anada T, Suzuki O (2013) Size regulation of chondrocyte spheroids using a PDMS-based cell culture chip. J Robot Mechatron 25(4):644–649
9. Handa T, Anada T, Honda Y, Yamazaki H, Kobayashi K, Kanda N et al (2012) The effect of an octacalcium phosphate co-precipitated gelatin composite on the repair of critical-sized rat calvarial defects. Acta Biomater 8(3):1190–1200
10. Miura K, Anada T, Honda Y, Shiwaku Y, Kawai T, Echigo S et al (2013) Characterization and bioactivity of nano-submicro octacalcium phosphate/gelatin composite. Appl Surf Sci 282(0):138–145
11. Anada T, Kumagai T, Honda Y, Masuda T, Kamijo R, Kamakura S et al (2008) Dose-dependent osteogenic effect of octacalcium phosphate on mouse bone marrow stromal cells. Tissue Eng Part A 14(6):965–978
12. Shibuya I, Yoshimura K, Miyamoto Y, Yamada A, Takami M, Suzawa T et al (2013) Octacalcium phosphate suppresses chondrogenic differentiation of ATDC5 cells. Cell Tissue Res 352(2):401–412
13. Suzuki O, Kamakura S, Katagiri T, Nakamura M, Zhao B, Honda Y et al (2006) Bone formation enhanced by implanted octacalcium phosphate involving conversion into Ca-deficient hydroxyapatite. Biomaterials 27(13):2671–2681
14. Takami M, Mochizuki A, Yamada A, Tachi K, Zhao B, Miyamoto Y et al (2009) Osteoclast differentiation induced by synthetic octacalcium phosphate through RANKL expression in osteoblasts. Tissue Eng Part A 15(12):3991–4000
15. Wang X, Suzawa T, Miyauchi T, Zhao B, Yasuhara R, Anada T et al (2013) Synthetic octacalcium phosphate-enhanced reparative dentine formation via induction of odontoblast differentiation. J Tissue Eng Regen Med. doi:10.1002/term.1669
16. Fuji T, Anada T, Honda Y, Shiwaku Y, Koike H, Kamakura S et al (2009) Octacalcium phosphate-precipitated alginate scaffold for bone regeneration. Tissue Eng Part A 15(11):3525–3535

17. Suzuki K, Honda Y, Anada T, Handa T, Miyatake N, Takahashi A et al (2012) Stimulatory capacity of an octacalcium phosphate/gelatin composite on bone regeneration in a rabbit tivia defect model. Phosphorus Res Bull 26:53–58
18. Murakami Y, Honda Y, Anada T, Shimauchi H, Suzuki O (2010) Comparative study on bone regeneration by synthetic octacalcium phosphate with various granule sizes. Acta Biomater 6(4):1542–1548
19. Honda Y, Anada T, Kamakura S, Morimoto S, Kuriyagawa T, Suzuki O (2009) The effect of microstructure of octacalcium phosphate on the bone regenerative property. Tissue Eng Part A 15(8):1965–1973
20. Pampaloni F, Reynaud EG, Stelzer EH (2007) The third dimension bridges the gap between cell culture and live tissue. Nat Rev Mol Cell Biol 8(10):839–845
21. Fukuda J, Nakazawa K (2005) Orderly arrangement of hepatocyte spheroids on a microfabricated chip. Tissue Eng 11:1254–1262
22. Tong JZ, Lagausie PD, Furlan V, Cresteil T, Bernard O, Alvarez F (1992) Long-term culture of adult rat hepatocyte spheroids. Exp Cell Res 200:326–332
23. Sakai Y, Nakazawa K (2007) Technique for the control ofspheroid diameter using microfabricated chips. Acta Biomater 3:1033–1040
24. Kale S, Biermann S, Edwards C, Tarnowski C, Morris M, Long MW (2000) Three-dimensional cellular development is essential for ex vivo formation of human bone. Nat Biotechnol 18:954–958
25. Kronenberg HM (2003) Developmental regulation of the growth plate. Nature 423(6937):332–336
26. Santini MT, Rainaldi G (1999) Three-dimensional spheroid model in tumor biology. Pathobiology J Immunopathol Mol Cell Biol 67(3):148–157
27 Friedrich J, Seidel C, Ebner R, Kunz-Schughart LA (2009) Spheroid-based drug screen: considerations and practical approach. Nat Protoc 4:309–324
28. Kunz-Schughart LA, Kreutz M, Knuechel R (1998) Multicellular spheroids: a three-dimensional in vitro culture system to study tumour biology. Int J Exp Pathol 79(1):1–23
29. Yamada K, Kamihira M, Hamamoto R, Iijima S (1998) Efficient induction of hepatocyte spheroids in a suspension culture using a water-soluble synthetic polymer as an artificial matrix. J Biochem 123(6):1017–1023
30. Kelm JM, Timmins NE, Brown CJ, Fussenegger M, Nielsen LK (2003) Method for generation of homogeneous multicellular tumor spheroids applicable to a wide variety of cell types. Biotechnol Bioeng 83(2):173–180
31. Markovitz-Bishitz Y, Tauber Y, Afrimzon E, Zurgil N, Sobolev M, Shafran Y et al (2010) A polymer microstructure array for the formation, culturing, and high throughput drug screening of breast cancer spheroids. Biomaterials 31(32):8436–8444
32. Kikuchi K, Sumaru K, Edahiro J, Ooshima Y, Sugiura S, Takagi T et al (2009) Stepwise assembly of micropatterned co-cultures using photoresponsive culture surfaces and its application to hepatic tissue arrays. Biotechnol Bioeng 103(3):552–561
33. Wang W, Itaka K, Ohba S, Nishiyama N, Chung UI, Yamasaki Y et al (2009) 3D spheroid culture system on micropatterned substrates for improved differentiation efficiency of multipotent mesenchymal stem cells. Biomaterials 30(14):2705–2715
34. Anada T, Masuda T, Honda Y, Fukuda J, Arai F, Fukuda T et al (2010) Three-dimensional cell culture device utilizing thin membrane deformation by decompression. Sens Actuators B 147(1):376–379
35. Sugiura S, Edahiro J, Sumaru K, Kanamori T (2008) Surface modification of polydimethylsiloxane with photo-grafted poly(ethylene glycol) for micropatterned protein adsorption and cell adhesion. Colloids and surfaces B. Biointerfaces 63(2):301–305
36. Takagi M, Umetsu Y, Fujiwara M, Wakitani S (2007) High inoculation cell density could accelerate the differentiation of human bone marrow mesenchymal stem cells to chondrocyte cells. J Biosci Bioeng 103(1):98–100

37. Shukunami C, Ishizeki K, Atsumi T, Ohta Y, Suzuki F, Hiraki Y (1997) Cellular hypertrophy and calcification of embryonal carcinoma-derived chondrogenic cell line ATDC5 in vitro. J Bone Miner Res 12(8):1174–1188
38. Gelse K, Brem M, Klinger P, Hess A, Swoboda B, Hennig F et al (2009) Paracrine effect of transplanted rib chondrocyte spheroids supports formation of secondary cartilage repair tissue. J Orthop Res 27(9):1216–1225
39. Lee JI, Sato M, Kim HW, Mochida J (2011) Transplantatation of scaffold-free spheroids composed of synovium-derived cells and chondrocytes for the treatment of cartilage defects of the knee. Eur Cell Mater 22:275–290 (discussion 90)
40. Schubert T, Anders S, Neumann E, Scholmerich J, Hofstadter F, Grifka J et al (2009) Long-term effects of chondrospheres on cartilage lesions in an autologous chondrocyte implantation model as investigated in the SCID mouse model. Int J Mol Med 23(4):455–460
41. Yoon HH, Bhang SH, Shin JY, Shin J, Kim BS (2012) Enhanced cartilage formation via three-dimensional cell engineering of human adipose-derived stem cells. Tissue Eng Part A 18(19–20):1949–1956

Chapter 17
The Visualization of Human Organogenesis from Stem Cells by Recapitulating Multicellular Interactions

Ran-Ran Zhang, Hiroyuki Koike and Takanori Takebe

Abstract Due to the severe shortage of organ donors, over 100,000 patients waiting for transplants per year die of end-stage organ failure in the United States alone. Laboratory-grown functional organs derived from pluripotent stem cells, such as induced pluripotent stem cells (iPSCs), may be one promising solution to this organ shortage. However, it has been considered impractical to recapitulate in stem cell cultures the spatiotemporal control of the multicellular interactions that occur during embryogenesis to generate a complex and vascularized organ, such as the liver. Recently, we reported that a three-dimensional iPSC-derived rudimentary organ (liver bud) could be grown from mixed human liver progenitors in a specific three-dimensional culture by mimicking early organogenic cellular interactions. The transplantation of this iPSC-derived liver bud was effective in rescuing lethal liver failure, a proof of principle that highlights the enormous therapeutic potential of this approach (*i.e.*, *in vitro*-grown organ-bud transplantation). The present manuscript provides an overview of this new paradigm of organ regenerative medicine. This chapter also discusses a new live imaging modality for intravitally assessing the degree of maturation to dissect the cellular processes that occur during human organogenesis.

Keywords Induced pluripotent stem cells (iPSCs) · Liver bud · Organogenesis

17.1 Introduction

Organ transplantation is the only curative method for treating end-stage organ failure. This groundbreaking concept has given new hope to patients since the first successful kidney transplantation was reported in the 1950s; this initial success was followed by a series of technical achievements that allowed us to expand the field of transplantation to the liver, pancreas, heart and lungs [1]. However,

T. Takebe (✉) · R.-R. Zhang · H. Koike
Department of Regenerative Medicine, Yokohama City University Graduate School of Medicine, Kanazawa-ku 3-9, Yokohama, Kanagawa 236–0004, Japan
e-mail: ttakebe@yokohama-cu.ac.jp

T. Arai et al. (eds.), *Hyper Bio Assembler for 3D Cellular Systems*,
DOI 10.1007/978-4-431-55297-0_17

over the past 10 years, there has been an increased demand for organ transplantation throughout the world due to the increased incidence of organ failure. There remains a shortage of organs relative to the existing demand; for example, in the United States, 20,000 patients are actively waiting for a donor liver for transplantation to treat end-stage liver disease, and this high demand for organs will eventually result in the death of thousands of waiting patients [2]. Recent progress in stem cell biology has highlighted the use of stem cell-based therapeutics such as induced pluripotent stem cells (iPSCs) [3], particularly for growing three-dimensional organs in culture [4]. In this chapter, we summarize the potential for generating functional three-dimensional organs from iPSCs. The present manuscript also review a unique method for live imaging analysis that can be used to dissect the previously uncharacterized roles of multiple cellular communication events in human organ development.

17.2 Current Paradigm of Regenerative Medicine (Cell Differentiation)

Current stem cell therapy primarily targets only minor diseases that can be treated with the transplantation of cells that are derived by the cell-type specification of stem cells [5]. In fact, numerous publications describing the use of iPSCs include protocols involving combinations of inductive or reprogramming factors that can induce the differentiation of cells and result in gene expression and functional profiles that resemble those of terminally differentiated cells, such as hepatocytes [6], insulin-producing cells [7] or dopaminergic neurons [8]. However, no definitive protocol has successfully generated fully functional cells that possess a function comparable to that of primary human parenchymal cells. The conventional unidirectional cell differentiation approach ignores the three-dimensional and temporal multicellular communication that physiologically coincides with proper cell differentiation during organogenesis. Therefore, to derive functional and therapeutically effective cells from pluripotent stem cells, it is essential to develop a novel three-dimensional culture platform that allows the cells to undergo developmental interactions leading to proper organogenesis.

Furthermore, this conventional approach, also known as "cell-based therapy", offers a rather limited therapeutic outcome: several clinical trials have reported the disappointing outcomes of cell transplantation methods and the potentially adverse events associated with these methods. Given the limitations of cell-based approaches, the development of an innovative technology that enables the reconstitution of three-dimensional organs from pluripotent stem cells is urgently required to solve the severe donor shortage crisis and address the high medical costs for the increasing number of patients waiting for organs [9].

17.3 The Importance of Multi-Cellular Communication During Normal Organogenesis

Organogenesis is initiated with the specification of the three primary germ layers: ectoderm, mesoderm and endoderm. During vertebrate development, organ formation is achieved by the differentiation of less-specialized cells into cell types that are more specialized via the dynamic intercellular communications that occur among the germ layers to mutually control cell fate decisions and patterning in every tissue. However, the complex cellular processes that are involved in three-dimensional organ development in culture remain poorly understood, and methods for growing complicated and vascularized organs are completely lacking. More specifically, the complexity and fluidity of the developmental regulation of the spatial and temporal changes that occur during organogenesis prevent the unidirectional cell differentiation approach from recapitulating the biophysical stimuli, the biochemical and molecular signals and the spatial organization that mediate organ development [10].

17.4 Example: Liver Bud Development

Pioneering studies have implied that intercellular communication plays an important role in directing liver development. In the mouse, liver organogenesis is initiated at embryonic day 8.5 via hepatic specification, which is driven by an interaction between the ventral portion of the foregut endoderm and the constituents of the adjacent immature endothelial cells (Fig. 17.1, upper). The specified hepatic endodermal cells extend into the surrounding septum transversum mesenchyme (STM) and form the hepatic diverticulum, which is the origin of the liver bud. The region of the hepatic diverticulum that is closest to the mesenchyme delaminates into the STM with the endothelium, and the liver bud is then generated. The STM cells and endothelial cells induce endodermal proliferation [11].

At approximately embryonic day 9.5, mesenchymal cells such as cardiac mesoderm and septum transversum mesenchyme have been shown to play an important role in the establishment of the liver bud, at minimum, producing the signaling molecules bone morphogenetic protein (BMP) and fibroblast growth factor (FGF) [12]. The phenotype of mouse embryos that are deficient for BMP4 is similar to the phenotype produced by the knockout of vascular endothelial growth factor receptor (VEGFR) in endothelial cells: a multi-layered liver epithelium forms, indicating defective liver bud outgrowth. Because, the absence of endothelial cells results in the failure of the liver epithelial cells to migrate into the surrounding mesenchyme, followed by the failure of liver bud formation. These findings suggest that stromal-cell-dependent paracrine factors such as BMP and FGF are essential for liver bud formation [13].

At a later stage of liver bud formation, the endothelial cells migrate to form a branched structure, and blood flow then perfuse to the immature endothelial

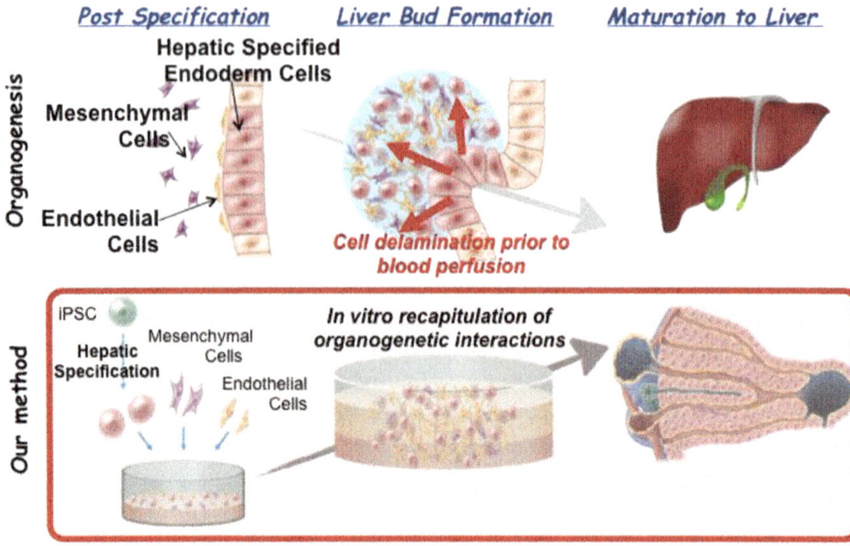

Fig. 17.1 Generation of liver bud from pluripotent stem cells by mimicking early organogenesis. (Modified from [14, 17])

network in the liver bud at embryonic day 10.5. This perfusion stimulates the maturation of the blood network and the epithelium in the liver bud. The hepatic endodermal cells in the liver bud undergo a morphological and functional transition from hepatoblasts to mature hepatocyte [11].

Minimally, the mechanisms mentioned above are essential to recapitulate the early process of hepatocyte differentiation from iPSCs, but this concept is completely lacking in conventional iPSCs differentiation systems. Specifically, the induction of three-dimensional liver bud formation by multiple cellular interactions and the subsequent blood perfusion into the liver bud, the two essential components of liver organogenesis *in vivo*, were rarely examined in the conventional system. Therefore, we established a culture system that recapitulates these early organogenic processes and succeeded in generating functional human liver tissue from iPSCs [14, 15].

17.5 iPSC-based Liver Bud Self-Organization

As described previously, the intercellular interactions among endothelial cells, mesenchymal cells and hepatically committed endodermal cells are necessary for liver bud formation *in vivo* (Fig. 17.1, lower) [11]. Therefore, we first derived hepatic endodermal cells from human iPSCs. Using a stepwise differentiation protocol designed to mimic liver development; human iPSCs were first differentiated into definitive endoderm using Activin A and further treated with FGF2 and BMP4 to

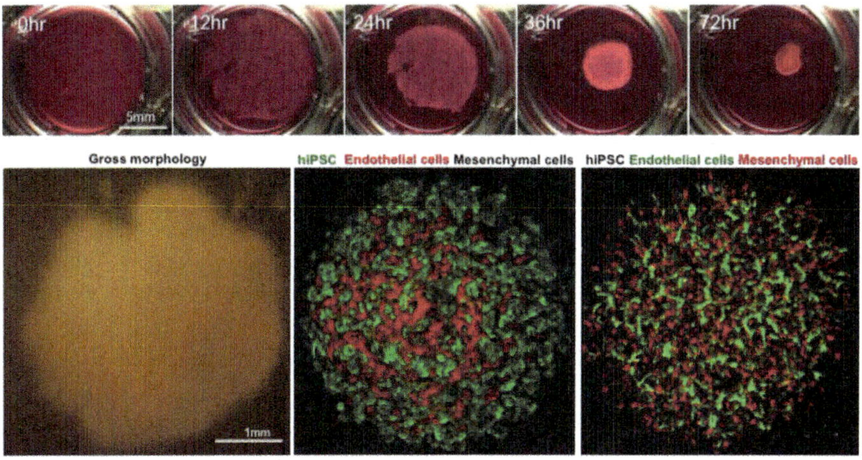

Fig. 17.2 Self-formation of three-dimensional liver bud from human iPSC *in vitro*. (Modified from [14, 15])

generate hepatic endoderm cells. This step was followed by a combination of hepatocyte growth factor (HGF), Oncostain M (OSM) and dexamethasone (Dex) to induce maturation of the hepatic progenitor cells and generate cells with hepatocyte-like properties [6]. Based on the principle of early hepatogenesis, we selected hepatic endoderm cells to examine liver bud formation after BMP4 and FGF2 exposure. We recently developed a specific culture condition in which cells successfully self-organized into a three-dimensional vascularized liver bud in a dish. This liver bud self-organization was achieved by recapitulating the liver developmental process as follows: human iPSC-derived hepatic endoderm cells were mixed with human umbilical vein endothelial cells (HUVECs) and human mesenchymal stem cells (MSCs) in a petri dish *in vitro*; the mixed cells self-aggregated to form a three-dimensional structure, approximately 4 to 5 mm in diameter (Fig. 17.2, upper). These aggregated mass self-organized into a tissue that had an inner-branched endothelial structure and liver developmental gene expression properties identical to those of the early liver bud prior to blood perfusion in the physiological development process (Fig. 17.2, lower) [14].

17.6 Proof-of-Concept Demonstration of Liver Bud Transplantation

To determine whether blood perfusion would stimulate the maturation of the liver bud into functional liver tissue *in vivo*, *in vitro*-generated liver buds were transplanted into immunodeficient mice. These liver buds were rapidly permeated with host vasculature within 48 h after transplantation. These rudimentary organs also produced greater amounts of human proteins and exhibited the capacity for drug metabolism than cells derived from reported method, suggesting that these liver buds underwent marked hepatic maturation compared with conventional human iPSCs that have been directed to differentiate into hepatic endoderm cells. Finally, we demonstrated that the transplanted human iPSC-derived liver buds rescued a lethal liver failure model using immunodeficient mice. These results highlight the enormous therapeutic potential of *in vitro*-grown organ bud transplantation in the treatment of organ failure [14]. The achievement of human liver bud generation and transplantation is noteworthy. We could use the normal supportive cells in the iPSC-derived liver cell culture, and this efficiently promoted a significant maturation of the transplanted tissue function. Moreover, this new approach may also help in predicting human drug metabolite profiles in preclinical animal models to promote drug development. The pre-screening of lead compounds may thus enable the early-stage detection of undesired side effects before starting clinical trials, rather than the less desirable identification of these effects after the drug enters routine clinical practice. Ultimately, this novel concept raises the distinct possibility that we may be able to generate organs from the skin cells of a patient dying of liver failure. This proof-of-concept demonstrates that organ bud transplantation offers an alternative approach for treating organ failure by generating a three-dimensional and vascularized organ and provides a promising direction in regenerative medicine.

17.7 Live Imaging Analysis of Transplanted Tissues (with a Cranial Window)

In parallel with the advances in methods for reconstituting organs, we also developed a real-time system for monitoring the transplanted tissues inside living animals (Fig. 17.3). Intravital imaging is a valuable tool for the characterization of dynamic cellular processes, including vascularization. An in-depth characterization by repetitive live imaging will further promote our understanding of the communication among the various cell types and the maturity of the transplants based on their morphology. A variety of *in vivo* optical imaging techniques currently exist; these include diffuse optical imaging and tomography, which are capable of identifying sources of absorption, scattering and fluorescence contrast. However, these approaches generally provide a low resolution with no long-term repetitive access [16]. To overcome these limitations, we recently demonstrated that a cranial window is one promising technique for evaluating the successful engraftment and subsequent maturation of endodermal organs *in vivo* by performing continuous intravital imaging.

This cranial window is created by performing a craniotomy and replacing the skull with a cover glass; using this approach, we are able to directly monitor variations in the transplanted tissue in living mice for a long term with high resolution. The cranial window model is the most commonly used live imaging

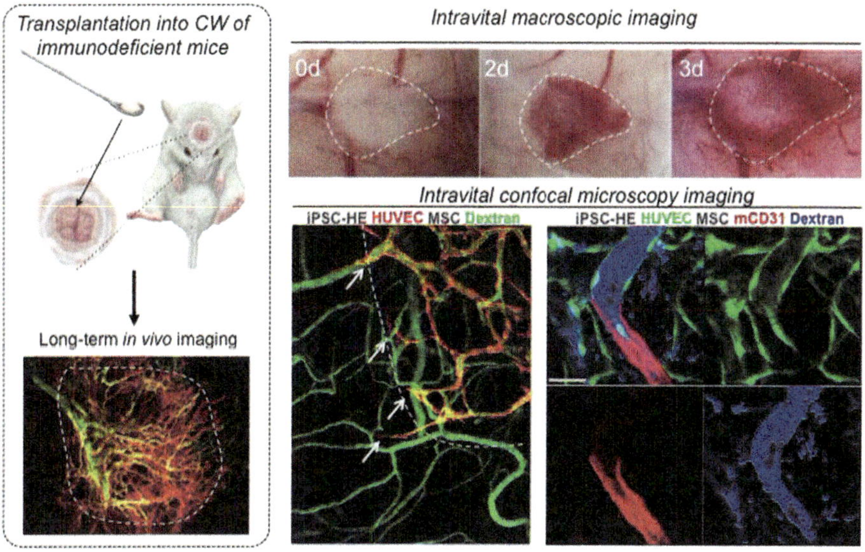

Fig. 17.3 Transplantation of liver buds inside mouse cranium and intravital visualization of visualarized liver buds inside mouse cranium. (Modified from [15])

modality in the study of brain physiology via repetitive intravital fluorescence microscopy, and we were the first group to demonstrate the applicability of this approach to the endodermal and mesodermal organs, including the liver, pancreas and kidneys, by transplanting embryonic tissue explants. The ability to visualize the vasculogenesis and angiogenesis of the transplanted tissues enables the collection of data regarding processes such as vascular remodeling, which cannot be directly measured by any of the previous techniques. High-resolution imaging at the single cell level enables us to quantify chronological changes in morphology to determine the degree of maturation. Furthermore, monitoring a single animal at multiple time points during dynamic processes minimizes animal-to-animal variability and thus decreases experimental variance [17, 18]. In fact, intravital confocal microscopy was used to track transplanted human fetal liver cells under a cranial window, revealing that the human vasculature directly connected with the recipient circulation within 48 h *in vivo* (Fig. 17.3). Several specific parameters of neovascularization, including the total vessel length, branching, vascular leakage and vessel diameter, were continuously quantified to determine the vessel maturation into an organ-specific vasculature. Notably, we found that we could predict the state of differentiation based on the morphology of the transplanted human hepatic cells via intravital imaging combined with functional analysis, a change similar to that used to follow the natural murine developmental process. In our experience, at least 3–4 months of monitoring in the same animal is possible using this model [14].

Our cranial window transplant model thereby provides a unique live imaging system to study the precise spatiotemporal relationships among the multiple cell types *in vivo*. Combined with our recently developed iPSC-derived organ bud-based approach, the cranial window method will be especially useful in studying the developmental processes of human embryonic sources, as they are currently not well characterized. A further understanding of the molecular and/or cellular mechanisms that underlie the process of organogenesis will allow us to expand our knowledge of human developmental biology and will contribute to a revision of the regenerative medicine strategy by more precisely recapitulating the dynamic physiological cellular processes during organogenesis.

17.8 Conclusion

Although studies have highlighted the use of stem cell-based therapeutics, including iPSCs, the cellular processes involved in the generation of three-dimensional organs in culture remain poorly understood, and methods for the growth of complicated and vascularized organs are completely lacking. With the hope of eliminating this deficit in cultured organs, our recently developed approach is very promising and, using a liver model, for the first time enables the growth of an iPSC-derived rudimentary organ ("organ bud") from a mixture of somatic cell-derivatives. Given this proof-of-principle demonstration, we will further optimize the transplantation procedures for iPSC-derived organ buds to evaluate the therapeutic efficacy and safety of human iPSC-derived organ bud transplantation by various in-depth cellular, biochemical and genomic assays aimed at future clinical use. Our results have the potential to advance the future understanding of organogenic cellular interactions in the study of human developmental biology. Collectively, the studies described here will provide novel therapeutic approaches to producing an unlimited amount of functional organs from pluripotent stem cells for the treatment of patients suffering from end-stage organ failure, opportunities to potentially address the critical shortage of donated organs and insight into the potential metabolic profiles and adverse effects of candidate lead compounds in the pharmaceutical industry to promote the development of new drugs [19].

References

1. Robson NZMH, Razack AH, Dublin N (2010) Review paper: organ transplants: ethical, social, and religious issues in a multicultural society. Asia Pac J Public Health 22(3):271–278
2. Punch JD, Hayes DH, LaPorte FB, McBride V, Seely MS (2007) Organ donation and utilization in the United States, 1996–2005. Am J Transplant 7:1327–1338
3. Tsuji O et al (2010) Therapeutic potential of appropriately evaluated safe-induced pluripotent stem cells for spinal cord injury. Proc Natl Acad Sci 107(28):12704–12709
4. Rashid T, Takebe T, Nakauchi H (2014) Novel strategies for liver therapy using stem cells. Gut 64(1):1–4

5. Sasai Y (2013) Next-generation regenerative medicine: organogenesis from stem cells in 3D culture. Cell Stem Cell 12:520–530
6. Si-Tayeb K et al (2010) Highly efficient generation of human hepatocyte–like cells from induced pluripotent stem cells. Hepatology 51(1):297–305
7. Zhang D et al (2009) Highly efficient differentiation of human ES cells and iPS cells into mature pancreatic insulin-producing cells. Cell research, 19(4):429-438.
8. Swistowski A et al (2010) Efficient generation of functional dopaminergic neurons from human induced pluripotent stem cells under defined conditions. Stem Cells 28(10):1893–1904
9. Takebe T et al (2012) Self-organization of human hepatic organoid by recapitulating organogenesis in vitro. Transplant Proc 44(4):1018–1020
10. Zorn AM, Wells JM (2009) Vertebrate endoderm development and organ formation. Ann Rev Cell Dev Biol 25:221
11. Matsumoto K et al (2001) Liver organogenesis promoted by endothelial cells prior to vascular function. Science 294(5542):559–563
12. Lammert E, Cleaver O, Melton D (2003) Role of endothelial cells in early pancreas and liver development. Mech Dev 120(1):59–64
13. Zorn AM (2008) Liver development. StemBook PMID: 20614624 [PubMed] 1–26.
14. Takebe T et al (2013) Vascularized and functional human liver from an iPSC-derived organ bud transplant. Nature 499(7459):481–484
15. Takebe T et al (2014) Generation of a vascularized and functional human liver from an iPSC-derived organ bud transplant. Nat Protoc 9(2):396–409
16. Isshiki M, Okabe S (2014) Evaluation of cranial window types for in vivo two-photon imaging of brain microstructures. Microscopy 63(1):53–63
17. Takebe T et al (2014) Engineering of human hepatic tissue with functional vascular networks. Organogenesis 10(2):0–1.
18. Koike N et al (2004) Tissue engineering: creation of long-lasting blood vessels. Nature 428(6979):138–139
19. Takebe T, Taniguchi H (2014) Human iPSC-derived miniature organs: a tool for drug studies. Clinical Pharmacology Therapeutics. 96(3): 310-313

Chapter 18
Bionic Simulator Based on Organ-Explant-Chip

Taisuke Masuda, Hirofumi Owaki, Tomohiro Kawahara and Fumihito Arai

Abstract In the drug discovery and regenerative medicine research, in vitro models based on cells or tissues are widely desired for attaining reliable data and high throughput as well as ethical concerns. This model: "bionic simulator" is required to simulate the biological function in the context of whole living organs. In this chapter, to achieve an organ-explant-chip for evaluating the function of an explanted biological organ by the mechanical system, we focused on the connection technique because we need to reconstruct the explanted organs or tissues as part of the biological simulator by connecting them with the external environment, other organs, and tissues. By using this microfluidic device, we have succeeded in constructing a hybrid circulatory system between artificial tubes and the explanted heart. Moreover, we have succeeded in measuring the response of the explanted heart with a variety of solutions. We confirmed from the response of the explanted heart to agonist drugs that the explanted organ maintained normal function.

Keywords Microvascular Connection · Microfluidic Device · Biological Simulator · Embryonic Chick Heart · Explant Culture

18.1 Introduction

Since Langendorff first described his perfusion system [1], isolated mammalian hearts have been used in numerous studies that have led to many important discoveries in the fields of cardiac physiology, pharmacology, and surgery. Isolated hearts are unique because they allow for the study of myocardial characteristics separate from the environment of the intact animal. The Langendorff perfusion system was

T. Masuda (✉) · H. Owaki · T. Kawahara · F. Arai
Department of Micro-Nano Systems Engineering, Graduate School of Engineering, Nagoya University, Nagoya, Japan
e-mail: masuda@mech.nagoya-u.ac.jp

T. Masuda
Furo-cho, Chikusa-ku, 464-8603 Nagoya, Japan

T. Arai et al. (eds.), *Hyper Bio Assembler for 3D Cellular Systems*,
DOI 10.1007/978-4-431-55297-0_18

further applied to evaluate the effect of different cardiovascular drugs on the coronary vasculature, muscle contraction, and heart rate. Today, a variety of cardiovascular researchers still use this vital technique to investigate the heart [2–4]. In this method, isolated heart preparations are performed using small animals such as the rat, rabbit, and guinea pig, although large animal preparations have been described for primarily [5, 6]. Selection of an experimental model depends on many factors, including cost, ease of data collection, relevance to the human condition, and experimental questions to be answered. Cost and ease of data collection are the primary reasons for using hearts from small animals [7]. However, it is difficult to apply the conventional Langendorff perfusion system to organs from small animals, and the circulation system is crucial for long-term monitoring.

Microfluidic chips are often described as miniature versions of their micro/nanoscale counterparts of laboratory equipment [8]. In addition to "lab-on-a-chip" technologies, microfluidic chips have been widely used to provide small volumes and fluid connections for biomedical and drug discovery applications. Microfluidic chips offer unique advantages, such as sample handling, reagent mixing, separation, and detection in a closed space. Moreover, microfluidic chips offer an approach that can circumvent many of the limitations of in vitro cell culture. Li et al. developed a microfluidic chip with improved cell retention properties and used this to monitor calcium mobilization in single cardiomyocytes, in real-time, as a part of a drug screening application [9]. A cultivation device based on the microfluidic chip with the circulation system has contributed to the long-term representation of the complex cell–cell and cell–stroma interactions.

18.2 Heart-Explant-Chip

In the conventional Langendorff perfusion system, technical skill is especially required to set up this experimental environment, which can be an obstacle for researchers in other fields to introduce this invaluable experimental method [10]. To achieve the microfluidic chip-based perfusion system, on-chip circulation was performed using an explanted heart, in which a connection was made between the cardiovascular system and an artificial system. A microfluidic chip-based perfusion system has been proposed to maintain a viable and functioning heart from a small animal. We call this prototype chip the "Organ-Explant-Chip". Herein, we first introduce the method for connecting the cardiovascular system to the artificial channel and then demonstrate chemical stimulation using an explanted embryonic chick heart. Figure 18.1 shows the concept of the "Heart-Explant-Chip" based on the Organ-Explant-Chip. An explanted heart from an organism is maintained by perfusion culturing in a microfluidic chip, while monitoring and evaluating the function of the explanted organ in real-time with mechanical, electrical, or chemical stimuli. There are three main advantages of the proposed approach: (1) living organs can be used, (2) the function of organs can be observed in detail and in an in vitro environment,

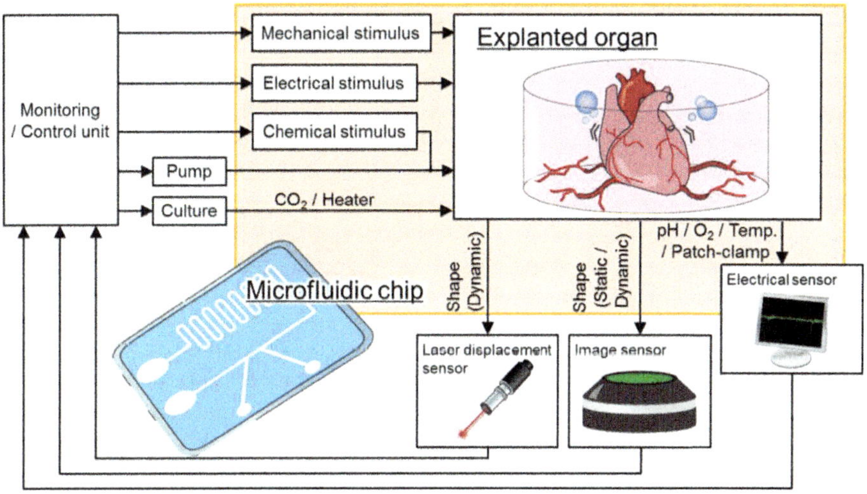

Fig. 18.1 The concept of the Heart-Explant-Chip based on the Organ-Explant-Chip. An explanted heart from an organism is maintained by perfusion culturing in a microfluidic chip, while monitoring and evaluating the function of the explanted organ in real-time with mechanical, electrical, or chemical stimuli. The advantages of proposed approach are that we can use living 3D organs for bionic simulator and that we can observe the function of organs in detail in vitro by taking out the organ from the body

and (3) a single organ separated from the biological system as well as the interaction between other organs can be functionally evaluated.

18.2.1 Isolation of Embryonic Chick Heart

Recently, drug discovery research and gene movement operations using chick embryos, zebrafish, and African clawed frogs as alternative animals have become important topics, especially in Europe where intense ethical constraints remain [11, 12]. Here, we chose the chick embryo as an animal model for the following reasons: (1) the circulatory system of the chick embryo has a structure similar to that of the human circulatory system and it is widely used in developmental biology and cardiovascular development research [13], and (2) sperm eggs are easily acquired and inexpensive, and it is easy to generate the chick embryo in a short time without special culture medium (i.e., only using the yolk in the egg as the nutrient). Use of the chick embryo is suitable for application to the Heart-Explant-Chip in terms of cost, scale of the experimental environment, and ethical issues. In this experiment, a fertilized chicken egg incubated for 11 days (Hamburger and Hamilton stage [HH] 37, Fig. 18.2a) [14] was used because the tissue extraction procedure is easy and the degree of organ growth is the minimum necessary for the Heart-Explant-Chip. After extracting the chick embryo from the egg as showing in Fig. 18.2b, only the heart

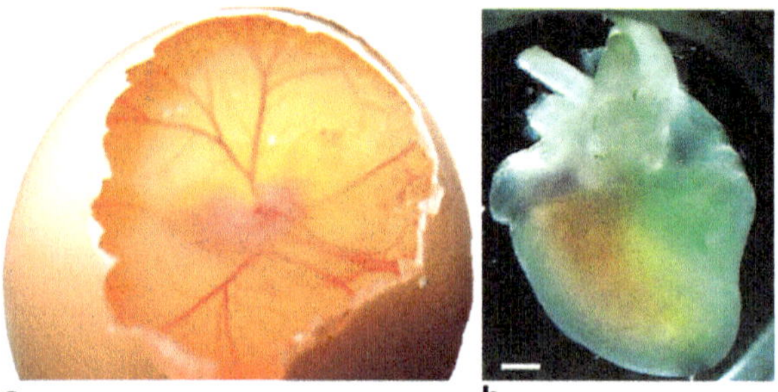
a b

Fig. 18.2 Experimental model: embryonic chick heart, **a** A developing chick embryo inside the egg, **b** confocal fluorescence image of an explanted heart from a chick embryo. To transparentize the embryonic chick heart, we prepared the fixed heart and incubated it in SCALEVIEW-A2 (Olympus, Tokyo, Japan). A bar is 1 mm

and the blood vessels leading to the heart (ascending aorta (AO), pulmonary vein (PV), vena cava (VC), and pulmonary artery (PA)) were isolated and maintained in DMEM (Sigma-Aldrich, Arklow, Ireland) supplemented with 10% FBS and antibiotics (100 U/ml of 200 mM L-Glutamine and 100 U/ml penicillin-G). All procedures were approved by the Animal Research Committee of Nagoya University.

18.2.2 Approach of Connecting Cardiovascular and Artificial Tube

To achieve the Heart-Explant-Chip, the following three techniques are required: (1) connection of the organ–artifact system for supplying nutrients, oxygen, and drugs as well as for incorporation into a simulator system, (2) control of the state of the organ and environment for long-term culturing and creating an arbitrary environment (i.e., pH, temperature, and oxygen level), and (3) evaluation of the response of the explanted organ to the environment of the biological simulator for understanding its biological function. We chose to first focus on the connecting technique because we need to reconstruct the explanted organs or tissues as part of the biological simulator by connecting them with the external environment, other organs, and tissues. Here, we describe an innovative method for connecting blood vessels and artificial tubes. Figure 18.3 shows the proposed vascular connection approach using the suction-induced vascular fixation (SVF) method. In general, microvascular (< 1 mm in diameter) anastomosis is difficult and there are skill-dependent and time-consuming problems because blood vessels are flexible structure with multi-DOF and hard to operate. In particular, there are several blood vessels led to the heart of chick embryo, which are concentrated in one place. Therefore, it is necessary to simplify the task of vascular anastomosis technique. SVF method is a way of fixing the position and orientation of the blood vessels automatically with a microfluidic device by negative pressure for easy vascular connection. The SVF method uses negative pressure to fix the position and orientation of the blood vessels automatically on a microfluidic device for easy vascular connection. The functions of the microfluidic device are summarized as follows.

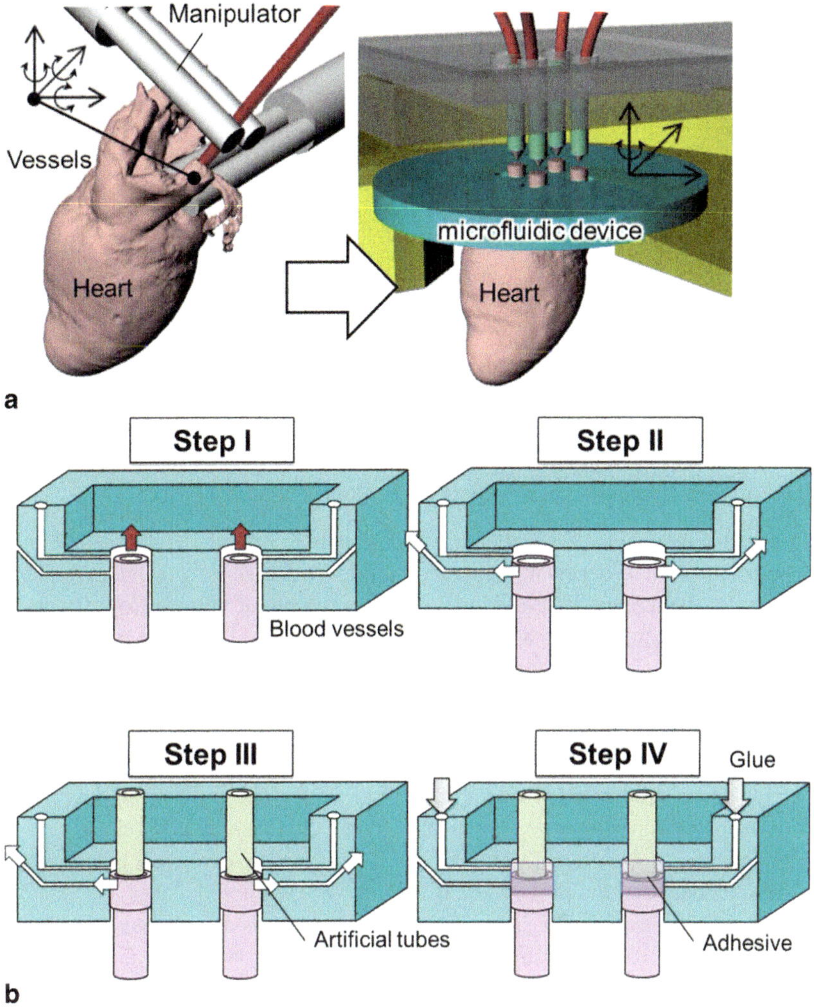

Fig. 18.3 Suction-induced vascular fixation (*SVF*) method, **a** concept of the SVF method, and **b** the SVF procedure using the microfluidic device. The SVF method uses negative pressure to fix the position and orientation of the blood vessels automatically on a microfluidic device for easy vascular connection

- Positioning: blood vessels can be positioned by suctioning the blood vessels through the holes patterned on a microfluidic device from the vertical direction (Step I). Moreover, multiple blood vessels can be aligned concurrently by negative pressure.
- Securing: by suctioning the blood vessels from the channel of the device patterned in the circumferential direction, blood vessels can be expanded in diameter and fixed (Step II), which facilitates insertion of the artificial microtubes

Table 18.1 Parameters of the microfluidic device

Description	Value (μm)
Diameter of the holes patterned on the device (d1)	750
Distance between the holes (d2)	1500
Thickness of the layer I	250
Thickness of the layer II	250
Thickness of the layer III	1500
Width of the microchannel	100

into the blood vessels. The multiple degrees of freedom (DOF) of the flexible blood vessels can also be reduced.

- Administering: after inserting the artificial tubes into the blood vessels (Step III), the connection portion is secured and the drugs are administered externally by pouring the glue from the channels of the device (Step IV).

We present a new vascular connection technique (SVF method) that uses a microfluidic device with a suction mechanism. In order to determine the parameters of the microfluidic device, we first attempt to suction the blood vessels from the vertical direction through the device with variety of shapes (Step I). In this experiment, we used an isolated heart from the fertilized chicken egg incubated for 11 days (HH stage 37). Because of the variability in individual chick embryo, an outer diameter of cardiovascular (D1) and a distance between two cardiovascular (D2) are defined as approximately 750 and 1000 μm, respectively. Considering the short length of the blood vessels extracted from the developing chick embryo, we set the thickness of the microfluidic device (t) as 1.5 mm. Based on these assumptions; we varied two parameters of the microfluidic device; the diameter of the each holes (d1), and the distance between the holes (d2). If the size of each hole is large, several blood vessels are suctioned into the same hole, while the vessels do not pass through small holes. Moreover, if the distance between holes is too long, several blood vessels cannot be suctioned into each holes, and too short a distance will cause interference between the artificial tubes. From the experimental results, we set the parameters of the microfluidic device as shown in Table 18.1. Using the fabricated microfluidic device, we can operate four blood vessels connected to the extracted embryonic chick heart through the holes of the device. Using the suction, we can correspond to the specific size of the chick embryo. Furthermore, the shape of the device is easily changed by microfabrication, which makes it possible to use hearts from chicks during the major embryonic stages.

18.2.3 Construction of the Hybrid Circulatory System

Figure 18.4a shows the constructed hybrid circulatory system between the extracted biological organ and artificial system. By constructing a circulatory system between the explanted heart and the artificial microtubes, we could quantitatively evaluate

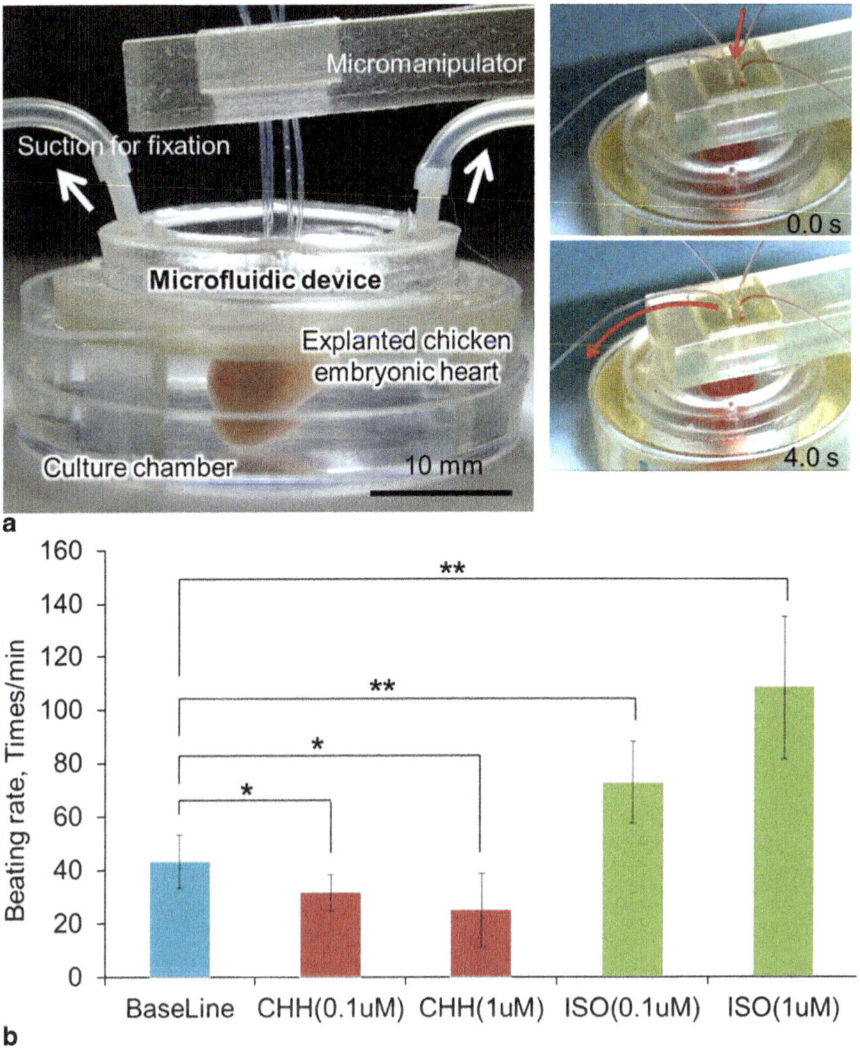

Fig. 18.4 Perfusion culturing and chemical stimulation of the explanted heart. **a** Constructed hybrid circulatory system between explanted embryonic chick heart and artificial tubes, and **b** response to the heart agonist drugs, isoproterenol (*ISO*) and carbachol (*CHH*). The heart rates were significantly different between groups, respectively. $N=6$, $^{*}P<0.05$, $^{**}P<0.01$

the response of the organ to externally supplied reagents or nutrients. First, we circulated the culture medium (DMEM+ 10% FBS) at different temperatures (37 °C, 4 °C). We then circulated the culture medium including the heart agonist drugs, isoproterenol (ISO; Tokyo Chemical Industry, Tokyo, Japan) and carbachol (CHH; Wako, Osaka, Japan). The evaluation of heart function by using agonist drugs is a traditional method in cardiac surgery and clinical tests [15, 16]. In order to evaluate the effects of ISO and CHH on the explanted heart, we prepared 0.1 and 1 μM concentrations of the drugs. In all experiments, the explanted embryonic chick heart was embedded in phosphate buffered saline (PBS), and culture medium was supplied (10 μl/s) using a mechanical pump. During the experiment, the behavior of the explanted heart was observed by digital HD video camera (HDR-CX180, Sony, Tokyo, Japan) and confocal microscope (MVX10-DSU, Olympus, Tokyo, Japan).

By connecting the external system and an explanted heart, we can easily use the organ as a biological simulator to control the solutions introduced into the organ. The beating heart showed change by the difference in the temperature of the introduced culture medium, and it was reversibly changed. In this result, this system can control the environment and monitor the reaction of the explanted heart. In general, cardiomyocytes consume a lot of nutrients and oxygen to maintain continuous beating; the cardiac rate and systolic performance for an extracted heart will increase after supplying culture medium. This change was more significant than the perfusion culturing only outside of the explanted heart. Figure 18.4b shows the response of the heart to ISO and CHH. These well-known heart agonist drugs clearly changed the heart rate of the explanted heart. A proposed approach suggested that explanted organ had maintained the normal heart function.

The heart isolated from the developing chick embryo was used as an experimental model in the present study. Until now, the heart of chick embryo has rarely been used in the Langendorff system. The circulatory system of a chick embryo, however, has a structure similar to that of the human circulatory system, and is attracting attention as an alternative for animal experimentation, especially in developmental biology. While a chick embryo was used in the present study, the proposed method can be applied to other animals more commonly used, such as mouse, rat, and rabbit. As a basic experiment for the evaluation of cardiac function, we investigated the response of the explanted embryonic chick heart to isoproterenol (ISO) and carbachol (CHH), two heart agonist drugs. The experimental results show that an explanted embryonic chick heart is useful as a simulator for evaluating cardiac function. Furthermore, the explanted heart will maintain a heart function during a maximum of 6 days using our circulatory system.

18.3 Challenges to Bionic Simulator Based on Organ-Explant-Chip

The reason why the Langendorff methodology remains largely unchanged to this day is the simplicity of the technique [1, 17, 18]. The heart is perfused by cannulating from the aorta, and by flowing the perfusion buffer retrograde to the aorta, the aortic valve is closed under pressure and, consequently, the perfusate enters the coronary arteries. This simple and effective method has been wildly used over the last 100 years in cardiovascular and pharmacological research, in spite of a few shortcomings. In conducting actual experiments, however, the technical skill required to set up the experimental environment becomes an obstacle for researchers in other fields to introduce this invaluable experimental method. In the Langendorff method, a vascular connection technique is necessary to perfuse from the aorta. Simplification of the connection procedure is one of the issues that should be addressed, especially when using a heart isolated from a small animal, such as a mouse. Thus, we proposed a simple microvascular connection method. Our approach, called the suction-induced vascular fixation (SVF) method, which uses a microfluidic device with suction mechanism, not only facilitates the connection method but also realizes vascular connection automatically by collaborating with the robotic technology in the future.

The proposed Heart-Explant-Chip, which uses a viable heart explanted onto a microfluidic chip, can be used for providing the knowledge combining the in vivo and vitro investigations. Along these lines, we believe that a Heart-Explant-Chip will increase the reality and the flexibility of future biological simulators for advanced biological and cardiovascular research. A biological simulator having biomaterials such as cells, tissues, and organs is widely desired for understanding the biological function in in vitro environments. To realize a bionic simulator, both the Langendorff methods and organ-on-a-chip [19, 20] are innovative technologies that will be widely used in the future. There is, however, a large difference in the type of data obtained with the experiments. The Langendorff technique provides an abundance of information about the heart isolated from animals, but it is difficult to analyze the mutual interaction between other organs and conduct long-term observations. On the other hand, the organ-on-a-chip is suitable for co-culturing of the cells and long-term observation in a microfluidic chip. However, functional analysis of organs beyond the tissue level is difficult to realize at present. We believe that our proposed Organ-Explant-Chip may be a solution. Moreover, we believe that the connection technique can be applied not only for organs and artificial tubes but also for organ–organ connections. Furthermore, by combining 3D cell culture techniques, such as "bottom-up" approaches, the proposed method will contribute significantly to regenerative medicine and drug discovery research.

18.4 Conclusion

We propose a simple and rapid method for creating a hybrid circulatory system using an embryonic chick heart to achieve a biological simulator. We used a layer stacking technique to fabricate a microfluidic device with a suction-induced vascular fixation mechanism. We confirmed experimentally that the vessels of the embryonic chick heart can be concurrently connected to the artificial tubes and that solution could be circulated through the heart from the artificial tubes connected to the vessels. We believe that the connection technique can be applied not only for organs and artificial tubes but also for organ–organ connections.

Acknowledgment This research was supported by a Grant-in-Aid from the Ministry of Education, Culture, Sports, Science, and Technology for Scientific Research (24106506).

References

1. Langendorff O (1895) Untersuchungen am überlebenden Säugethierherzen. Arch Gesamte Physiol 61:291–332
2. Chinchoy E, Soule CL et al (2000) Isolated four-chamber working swine heart model. Ann Thorac Surg 70:1607–1614
3. Neubauer S, Ingwall JS (1991) The isolated, buffer-perfused ferret heart—a new model for the study of cardiac physiology and metabolism. Lab Anim 25:348–353
4. Vick JA, Herman EH (1971) Isolated dog or monkey heart preparation for studying cardioactive compounds. Pharmacology 6:290–299
5. Broadley KJ (1979) Langendorff heart preparation—reappraisal of its role as a research and teaching model for coronary vasoactive drugs. J Pharmacol Method 2:143–156
6. Suga H, Sagawa K (1974) Instantaneous pressure-volume relationships and their ratio in excised, supported canine left-ventricle. Circ Res 35:117–134
7. Tada H, Thompson CI et al (2000) Myocardial glucose uptake is regulated by nitric oxide via endothelial nitric oxide synthase in Langendorff mouse heart. Circ Res 86:270–274
8. Duffy DC, McDonald JC, Schueller OJA, Whitesides GM (1998) Rapid prototyping of microfluidic systems in poly(dimethylsiloxane). Anal Chem 70:4974–4984
9. Li XJ, Li PCH (2005) Microfluidic selection and retention of a single cardiac myocyte, on-chip dye loading, cell contraction by chemical stimulation, and quantitative fluorescent analysis of intracellular calcium. Anal Chem 77:4315–4322
10. Skrzypiec-Spring M, Grotthus B, Szeląg A, Schulz R (2007) Isolated heart perfusion according to Langendorff-Still viable in the new millennium. J Pharmacol Toxicol Methods 55:113–126
11. Szita N, Boccazzi P et al (2005) Development of a multiplexed microbioreactor system for high-throughput bioprocessing. Lab Chip 5:819–826
12. van der Worp HB, Howells DW et al (2010) Can animal models of disease reliably inform human studies? Plos Med 7:e1000245
13. Happel CM, Klose C et al (2010) Non-destructive, high-resolution 3-dimensional visualization of a cardiac defect in the chick embryo resembling complex heart defect in humans using micro-computed tomography double outlet right ventricle with left juxtaposition of atrial appendages. Circulation 122:E561–E564
14. Bellairs R, Osmond M (2005) The atlas of chick development, 2nd edn. Elsevier, Boston
15. Protas LL (1992) Positive and negative inotropic effects of carbachol on the embryonic chick atrium. Experientia 48:1007–1009
16. Satoh H, Sperelakis N (1993) Hyperpolarization-activated inward current in embryonic chick cardiac myocytes—developmental-changes and modulation by isoproterenol and carbachol. Eur J Pharmacol 240:283–290
17. Ravelli F, Allessie M (1997) Effects of atrial dilatation on refractory period and vulnerability to atrial fibrillation in the isolated Langendorff-perfused rabbit heart. Circulation 96:1686–1695
18. Zimmer HG (2000) Modifications of the isolated frog heart preparation in Carl Ludwig's Leipzig Physiological Institute: relevance for cardiovascular research. Can J Cardiol 16:61–69
19. Huh D, Hamilton GA, Ingber DE (2011) From 3D cell culture to organs-on-chips. Trends Cell Biol 21:745–754
20. Huh D, Matthews BD et al (2010) Reconstituting organ-level lung functions on a chip. Science 328:1662–1668

Chapter 19
Tempo-Spatial Dynamics of Cellular Mechanics

Takeomi Mizutani and Ryosuke Tanaka

Abstract The mechanical properties of the cytoskeleton are important for many cellular events. Recent studies have focused on the non-uniformity of these properties because of their contribution to the local regulation of activities such as mechanochemical transduction and force generation. In the present study, we developed a method to measure local deformation along the longitudinal direction of cytoskeletal fibers using fluorescent microscopy and digital image analysis. We evaluated local deformation of actin fibers and α-actinin fibers on the actin fibers. Thus, we observed (i) significant shortening in a region and little change in the other region; and that (ii) in the significantly shortened region, actin fiber shortening was more severe than α-actinin fiber shortening. In addition, we evaluated local deformation along the longitudinal direction of curved microtubules. We observed that (i) a significant shortened region and a lengthened region emerged simultaneously; (ii) these non-uniform deformation events occurred a short distance away from the site of curving. These non-uniform deformations of the cytoskeleton are vital for understanding the mechanisms of local cellular regulation.

Keywords Cytoskeleton · Mechanical property · Image processing · Finite element analysis

19.1 Introduction

Evaluation of deformation of extracellular matrix and intracellular cytoskeletal networks is important to discuss single-cellular and multi-cellular movement during 3-dimentional morphogenesis. In this section, we introduce a method to evaluate deformation along a fiber using fluorescent microscopy and digital image analysis, and show an application of this method to cytoskeletal networks (actin fiber and microtubule).

T. Mizutani (✉) · R. Tanaka
Faculty of Advanced Life Sciences, Hokkaido University, North 10 West 8, Kita-ku, 060-0810 Sapporo, Japan
e-mail: mizutani@sci.hokudai.ac.jp

T. Arai et al. (eds.), *Hyper Bio Assembler for 3D Cellular Systems*,
DOI 10.1007/978-4-431-55297-0_19

Cytoskeletal elements such as actin filaments and microtubules play important roles in the maintenance of cell shape, cell movement, cell division and mechanical sensing ([1–3] and references therein). Mechanical factors, such as viscoelastic moduli, stress, and strain, on cytoskeletons are vital for elucidating the underlying mechanisms of these functions. In particular, to discuss the biochemical reactions induced by mechanical factors, local information on the mechanical is required. Therefore, the methods for measuring mechanical factors are a primary focus in current research.

The analysis of local deformation of the cytoskeleton in living cells has primarily focused on the actin cytoskeleton. To evaluate local deformation, combined fluorescent microscopy and image processing methods have been widely used. Peterson et al. observed that green fluorescent protein (GFP)-tagged α-actinin, which is an actin filament binding protein known to distribute in a punctate manner along actin stress fibers, changed its periodicity in response to actomyosin contraction [4]. Hebert et al. developed a method to trail fluorescent-tagged objects in living cells [5], and Adachi et al. developed a method for measuring the strain field in a 2D actin filament network [6]. Using fluorescent-tagged actin and kymographic analysis along actin stress fibers, Smith et al. discovered that local deformation of actin stress fibers induced actin binding protein accumulation at the site of deformation, allowing actin binding proteins to interact with other proteins [7]. Thus, a question arises regarding whether the deformation of the distribution of α-actinin is similar to that of actin fibers. To address this question, simultaneous observation of fluorescent-tagged actin and α-actinin and local deformation analysis for both fibers are required.

The mechanical behaviors of microtubules have been investigated in vivo and in vitro (see review article [8] and references therein). Polymerization of α- and β-tubulin heterodimers results in the creation of a hollow tube structure with an outer diameter of 25 nm. The heterodimers can be added at or dissociate from either end. The end with a greater tendency to add heterodimers, which is located along the periphery of cells, is named the "plus end." According to in vitro mechanical measurements, the stiffness of microtubules along the longitudinal direction is uniform [9]. Microtubule dynamics has also been investigated in vivo by using fluorescent microscopy and image processing. Bicek et al. proposed a method to measure dynamic changes in microtubule curvature [10] and observed that microtubule bending was caused by local microtubule displacement [11]. Improvements of their technique would permit the clarification of local microtubule deformation and allow the results to be compared with those of computational analysis of microtubule deformation in response to compression or elongation, assuming the microtubule to be a simple elastic body.

This study aimed to measure local deformation along stress fibers and microtubules. We proposed a method that can evaluate fluorescence intensity along a targeted fiber and detect similar regions from a set of time-course images. We measured and compared the local deformation of actin fibers and α-actinin fibers along the actin fibers. Moreover, we observed non-uniform deformation along the longitudinal direction of microtubules. The non-uniform deformation of microtubules was compared using finite element analysis (FEA). The non-uniformity of the mechanical properties of the cytoskeleton is described.

19.2 Materials and Methods

Fibroblasts (NIH-3T3) were purchased from the RIKEN Cell Bank (Tsukuba, Japan) and cultured in low-glucose Dulbecco's modified Eagle's medium (DMEM; Sigma–Aldrich, MO, USA) supplemented with 10% fetal bovine serum (FBS) and 1% antibiotics (Sigma–Aldrich) in 5% CO_2 at 37 °C. The cells were trypsinized and cultured at a density of 1×10^2 cells per mm^2 on a glass dish. Transfection of the cells with fluorescent-tagged cytoskeletal protein-expressing plasmids was performed using Lipofectamine 2000 reagent (Invitrogen, CA, USA). Twelve hours after transfection, the cells were recultured on a new glass dish at a density of 1×10^2–5×10^2 cells per mm^2. One day after transfection, the culture medium was replaced with HEPES-buffered DMEM (pH 7.2–7.3) containing 10% FBS and 1% antibiotic solution and covered with liquid paraffin (Wako Pure Chemical Industries, Ltd., Osaka, Japan) to avoid culture medium evaporation during live cell imaging.

We used 2 confocal laser microscope systems: DIGITAL ECLIPSE C1 (Nikon, Tokyo, Japan), equipped with an argon laser (488 nm) and a helium-neon laser (594 nm), for Azami Green (AG)-α-actinin and Raspberry-actin; and TCS SP5 (Leica Microsystems K.K., Tokyo, Japan), equipped with an argon laser (488 nm), for GFP-tubulin. The combination of 488-nm and 594-nm lasers was chosen to minimize the amount of light leaking from the α-actinin fluorescent signal to the β-actin signal image. As shown in Fig. 19.1, no correlation was observed between the intensity profiles of AG-α-actinin and Raspberry-actin along a filament (the correlation may have been negative rather than nonexistent), indicating that AG-α-actinin signals do not leak into the Raspberry-actin image, and vice versa. TCS SP5 was suitable for the observation of rapid microtubule dynamics. A set of time-course images was obtained as follows. Four *Z* plane images (a *Z* stack) separated by an equal 0.5-μm space were captured every 5 min for α-actinin and β-actin and every 5 s for α-tubulin. The *Z* plane images were merged (i.e., in our experiment, 1.5 μm of *Z* information was mixed); thus, we obtained 2D time course images. The size of each image was 512 pixels × 512 pixels.

Our method to define and track individual cytoskeletal fibers is based on acquiring time-course images of fibers and analyzing the fluorescence intensity profile along the fibers. If a region on a fiber is differently displaced during the time-course imaging, we could evaluate the local strain (or the incorporation and dissociation of local filamentous components) on the filament. The algorithm to determine the coordinates of targeted fibers was based on previous research [10, 11], except for drawing a user-defined initial curve along the fiber of interest at the beginning of

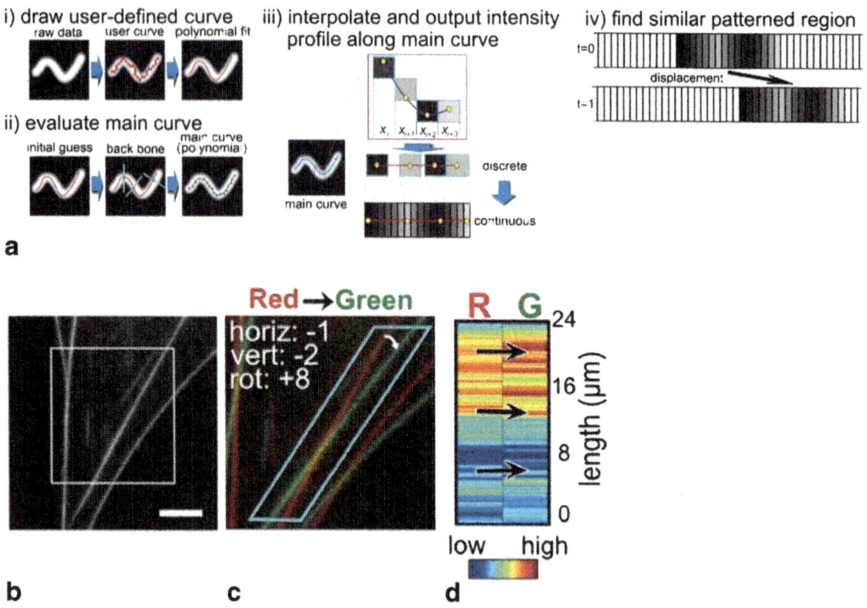

Fig. 19.1 Cartoon of the image processing protocol and the validity of the processing for the parallel displacement and rotation of the samples. **a** Flow of the image processing. The program consists of 4 parts (i–iv). (i) A user-defined curve of the fiber of interest is drawn. The user-defined curve is modified to be a polynomial. (ii) The central positions (*backbone*) of the fiber are evaluated by Gaussian fit. Here, the polynomial curve is used for the initial estimates of the Gaussian fit. The evaluated central positions are fitted to meet a polynomial (*main curve*). (iii) Each pixel on the main curve is projected to be the actual distance along the curve. After the projection, the intensity between pixels is linearly interpolated. (iv) To find identical regions on the fiber from the time-course images, similar intensity profiles are searched. **b** Fluorescent protein-tagged actin-expressing cells were cultured on a glass dish, fixed, and observed. The square region was magnified, as shown in **c**. Bar = 10 μm. **c** The cell culture glass dish was dislocated (combination of parallel displacement and rotation), and then the distribution of actin was observed. *Red* and *green* indicate the distribution before and after dislocation, respectively. The dislocation parameters of the dish are as follows: horizontal displacement to the *left*, 1 μm; vertical displacement to the bottom, 2 μm; rotation angle to clockwise, 10°. **d** The normalized intensity profiles before (R column) and after (G column) dislocation along the fiber were arranged after eliminating parallel displacement

the analysis (Fig. 19.1). The user-defined curve was modified to be a polynomial by the least-squares method. The polynomial curve was used to define the x-axis and y-axis of the image so that the absolute value of the angle between the alignment of the curve and x-axis was less than 45°; if the angle exceeded 45°, then the image was rotated 90°. Additionally, the polynomial curve was used to determine the initial estimates for the following Gaussian fit to determine the backbone coordinates ($xcenter_i$, $ycenter_i$) of the fiber:

$$I(x_{ij}, y_{ij}) = I_{bg} + I_0 e^{-\{(x_{ij}-xcenter_i)^2+(y_{ij}-ycenter_i)^2\}/\sigma^2} \tag{19.1}$$

where $I(x_{ij}, y_{ij})$ is the intensity value; x_{ij} and y_{ij} are the discrete coordinates along normal line of i th fit; and I_{bg}, I_0, $xcenter_i$, $ycenter_i$, and σ are fit parameters. The estimated coordinates ($xcenter_i$, $ycenter_i$) were fitted to the polynomial curve (named the main curve). The main curve was used as the initial estimate in the subsequent image (i.e., time-course tracking of the fiber). The fluorescence intensity profile along the main curve was determined by linear interpolation between discrete pixel data on the main curve. In this manner, we obtained the intensity profile $I(s, t)$ along the fiber, where s is a defined coordinate along the fiber (from the starting point to the end of the user-defined curve) and t is time. To determine an identical point on the fiber between t and $t+1$, the least-squares method was applied to minimize the following function:

$$Min(d) = \int_{cent-\frac{L}{2}}^{cent+\frac{L}{2}} \{I(s,t) - I(s+d,t+1)\}^2 \, ds \qquad (19.2)$$

where *cent* is the coordinate of the targeted point that we wish to identify and L is the range of the searching region. The estimated d was defined as the displacement of the identical point. The intensity profile along a fiber at any time was normalized by its average intensity. In the case of microtubule analysis, the time course of the intensity profiles was 7 pixels × 3 pixels (distance axis × time axis)-median filtered.

To ensure that our method is applicable for the displacement and/or rotation of the targeted fiber, Raspberry-actin–expressing cells were fixed with 4 % formaldehyde (Sigma–Aldrich) and observed before and after the application of sample stage locomotion. Normalized intensity profiles along fibers were very similar, and defined points were identified by our method.

We assessed the accuracy of the analysis. From a set of time-course images, identical points were determined using the aforementioned procedures. In this process, noise and errors, which were derived from our microscope system, made the identification difficult. To evaluate the accuracy of the system, fluorescent protein-expressing cells were fixed with 4 % formaldehyde and observed over time (Fig. 19.2). We identified certain points and statistically analyzed the time change of their distances. In the absence of noise and errors, each distance should be stable. However, each distance data changed to some extent. We evaluated the standard deviation (S.D.) of the distance, for which the distance data from other points were treated as different datasets. The S.D.s of the distance were independent of their length and were nearly identical, indicating that the accuracy of this analysis depended on the S.D. (i.e., accuracy depended on the evaluation of the position). We assumed that each distance dataset included errors in the t-distribution with $n-1$ degrees of freedom. Therefore, we speculated that the significant alteration in the live imaging data can be determined by the following equation:

$$|L(0) - L(t)| > S.D. \times tvalue(0.1) \qquad (19.3)$$

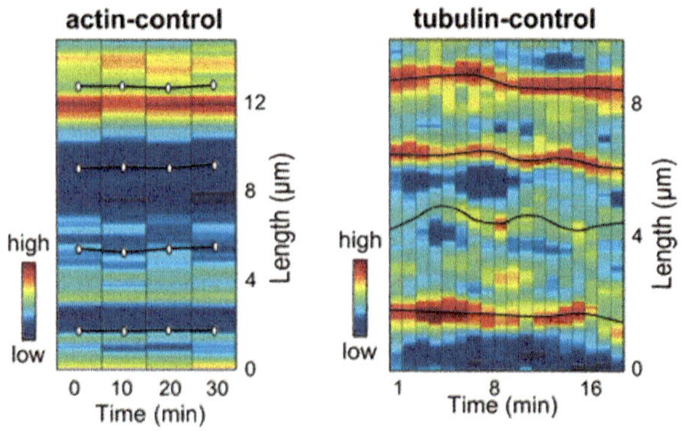

Fig. 19.2 Stability of analysis for the cytoskeletal fiber deformation. Raspberry-actin or Green fluorescent protein (*GFP*)-tubulin expressing cells were fixed and, the cytoskeletal fibers were observed at fixed time intervals by confocal microscope. Fluorescent intensity along single actin fiber or microtubule was analyzed. The intensity was normalized and shown as a *blue-red* color profile. Four points on the profiles were identified by the manner described in Eq. (19.2) and these points were connected with *black lines*

where $L(0)$ and $L(t)$ are the distances between the identical points, *S.D.* is standard deviation calculated from the fixed fiber data and the time course, and *t value* (0.1) is the *t* value of the 2-sided 90 % confidence interval.

FEA was conducted using FlexPDE 6 (PDE Solutions Inc., WA, USA). To generate a 2D elastic body mesh model of microtubules, the geometry of the region was determined by the main curve (see "Image processing" section) and a microtubule width of 25 nm [12, 13]. The approximate elasticity parameters of microtubules (Young's modulus, 1×10^9 N/m^2; shear modulus, 1×10^3 N/m^2; Poisson's ratio, 0.4) followed those of a previous study [14]. Stress was applied at each terminus or in the middle of the mesh model, and the direction was tangential.

19.3 Results and Discussion

First, we evaluated the deformation of actin and α-actinin fibers. Living cells co-expressing Raspberry-actin and AG-α-actinin every 5 min under 10 nM calyculin A treatment (Fig. 19.3). Calyculin A is known to increase cellular contractile force via the change in actomyosin activity [15]. The intensity profiles along the fibers were analyzed to examine the time-dependent variation in the length of the 2 defined regions (Fig. 19.3c and d). The initial length of both regions was 3.5 µm. We observed (i) significant shortening in a region, whereas little change was observed in the other region; and that (ii) in the region with significant shortening, actin fiber shortening was more severe than α-actinin fiber shortening. The shortening velocities were approximately 0.01 µm/min for α-actinin and 0.03 µm/min for actin.

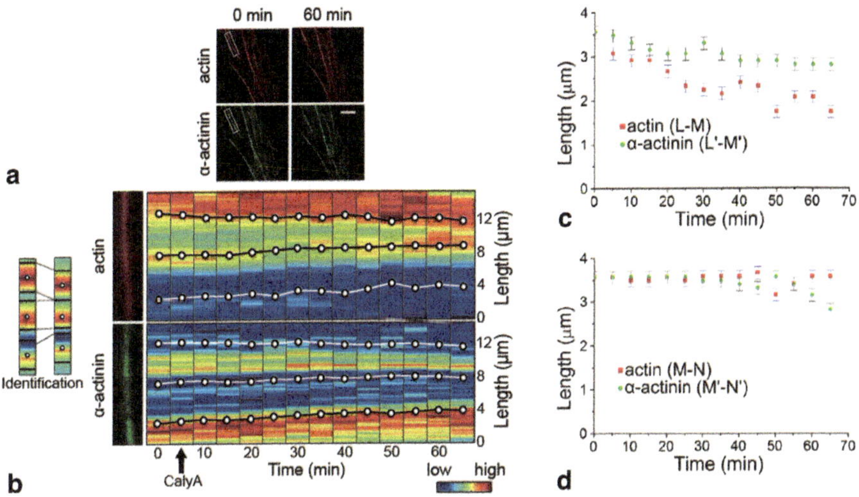

Fig. 19.3 Evaluation of the local deformation of actin stress fibers. Cells co-expressing Raspberry-tagged actin and Azami Green-tagged α-actinin were observed every 5 min. Cells were treated with calyculin A at 5 min. **a** Representative actin and α-actinin images are shown. The white box was magnified, as shown in (B). Bar = 5 μm. **b** The normalized intensity profiles of actin and α-actinin along the actin fiber are shown in the kymograph. Initially, 3 points (L, M, and N in actin; L′, M′, and N′ in α-actinin) were chosen to equally divide the total length of the fiber. Each identical point was selected using our created software. **c**, **d** The distance between identical points was plotted as a function of time

These non-uniform deformations were observed in every sample (3 samples). These data provided unexpected aspects of actin stress fiber dynamics.

If the mechanical properties along the longitudinal axis of an actin fiber are uniform, then the strain from each region will be constant. However, we observed heterogeneous deformation along actin fibers, which was previously discussed by Deguchi et al. ([16] and the references in the review article). Heterogeneous forces generated by actomyosin contraction along an actin stress fiber may produce heterogeneous deformation. Additionally, we presume that heterogeneous deformation is due to the heterogeneous elasticity distribution along an actin stress fiber. Katoh et al. observed the detailed structure of extracted actin stress fibers by using electron microscopy and reported the existence of small structures on actin stress fibers [17]. The binding of these structures may locally reinforce actin fibers and induce heterogeneous deformation when force is exerted. We also observed that actin fiber shortening was more severe than α-actinin fiber shortening. In many studies, changes in the periodicity of the α-actinin band pattern (sarcomere length) are used to measure the deformation of actin stress fibers. However, our data suggest that changes in the length of α-actinin do not completely reflect the deformation of actin stress fiber. Mansour et al. applied 10 % strain to longitudinally aligned cardiomyocytes and observed an approximate 6 % change in sarcomere length [18]. Their results support our findings. Although the molecular dynamics explaining the deviation

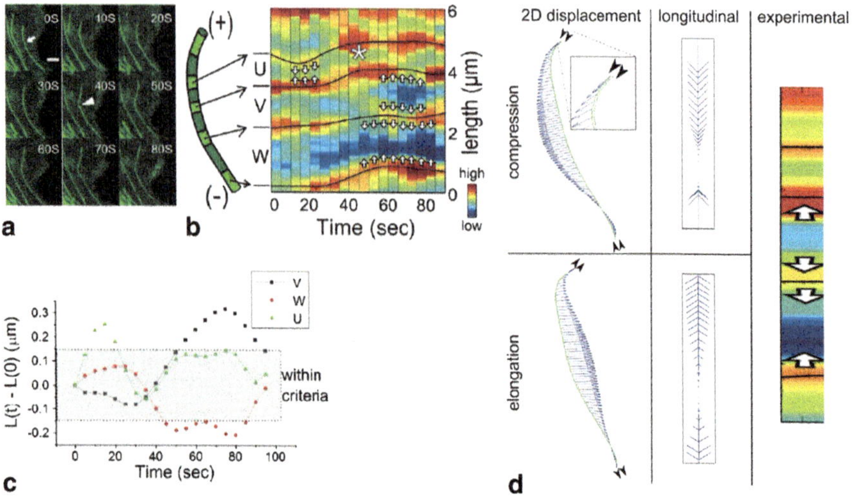

Fig. 19.4 Evaluation of the local deformation on microtubules and finite element analysis. *Green* fluorescent protein (*GFP*)-tagged tubulin-expressing cells were observed every 5 s. **a** A growing and bending microtubule is shown. At t=40 s, the microtubule bended strongly (*arrowhead*). The targeted microtubule (*arrow*) was analyzed, as shown in **b**. Bar=2 μm. **b** Normalized and median-filtered intensity profiles of a microtubule along its longitudinal direction is shown in the kymograph. The 4 brightest points were manually traced (*black curves*). The regions sandwiched between the curves were named U, V, and W. Significantly deformed regions relative to the initial length were denoted by inner or outer arrows. * denotes the point of bending. **c** The length of each region was subtracted from its initial value and plotted as a function of time. The gray zone denotes criteria. **d** Simplified finite element analysis of a microtubule visualized as an elastic body. The arrowheads indicate the direction of the exerted stress. The displacement vectors are shown as *arrows*

remain unclear, we speculate that transient α-actinin departure from the actin fiber may explain the differences between α-actinin fiber and actin fiber deformation.

We measured local deformation along curved microtubules. The estimated criteria value of the significant change was ±0.15 μm. We also observed dynamic changes in the GFP-tubulin distribution in living cells (Fig. 19.4a). The targeted microtubule elongated to the upper side of the images (white arrow in Fig. 19.4a). The growing end was defined as the plus end, and other end was defined as the minus end. The bending of microtubules increased 40 s after the start of the observation (indicated as * in Fig. 19.4a and white arrowhead in Fig. 19.4b). We evaluated local deformation along the longitudinal direction of the microtubule (Fig. 19.4b). Four identical points were traced, and 3 regions were defined. For the 3 regions, significant changes in their lengths compared to their respective initial values are shown as arrows. We observed that (i) a shortened region and a lengthened region emerged simultaneously; (ii) the region around the site of bending did not exhibit significant changes in its length, and the shortened and lengthened regions were both located at least 1 μm from the site of curving. These non-uniform deformations were observed in every sample (3 samples). To explain the observed non-uniform deformation, we performed a very simplified FEA of an elastic microtubule model. The same degree of stress (elongation or compression) was exerted at each end of the mesh model in the tangential direction (Fig. 19.4d). The displacement vector map under the compression condition predicted the following bending region. Displacement component along the longitudinal direction of the model was extracted and shown as arrows. Neither elongation nor compression condition explained deformation as seen in our experiment.

Non-uniform deformation along the longitudinal direction of microtubules has not been previously reported. To explain our observation, the introduction of local stress may be adequate. Microtubules are known to interact with actin filaments and to buckle [19]. Although our observed microtubule buckling did not accompany longitudinal deformation of the microtubule, actomyosin contraction is a likely candidate as the local deformation generator. In addition, microtubule motor proteins such as kinesin and dynein may contribute to the local deformation. Further experiment under actomyosin or microtubule motor proteins force-inhibited conditions would indicate the mechanism.

19.4 Conclusion

In this study, we proposed a method to evaluate deformation of cytoskeletal networks using fluorescent microscopy and digital image analysis. Local deformation of actin fibers and microtubules was not uniform, suggesting that heterogeneity of mechanical property of these cytoskeletal fibers. Cell shape and direction of cell movement may originate in the cytoskeletal heterogeneity. The next step of our study is to evaluate deformation of cytoskeletal networks in 3-dimentional circumstances. The current and future works will help understanding of 3-dimentional single cell movement and morphogenesis in multi-cellular systems.

References

1. Hall A (2009) The cytoskeleton and cancer. Cancer Metastasis Rev 28(1–2):5–14
2. Sun SX, Walcott S, Wolgemuth CW (2010) Cytoskeletal cross-linking and bundling in motor-independent contraction. Curr Biol 20(15):R649–R654
3. Vogel V, Sheetz MP (2009) Cell fate regulation by coupling mechanical cycles to biochemical signaling pathways. Curr Opin Cell Biol 21(1):38–46
4. Peterson LJ, Rajfur Z, Maddox AS, Freel CD, Chen Y, Edlund M, Otey C, Burridge K (2004) Simultaneous stretching and contraction of stress fibers in vivo. Mol Biol Cell 15(7):3497–3508
5. Hebert B, Costantino S, Wiseman PW (2005) Spatiotemporal Image Correlation Spectroscopy (STICS) theory, verification, and application to protein velocity mapping in living CHO cells. Biophys J 88(5):3601–3614
6. Adachi T, Okeyo KO, Shitagawa Y, Hojo M (2009) Strain field in actin filament network in lamellipodia of migrating cells: implication for network reorganization. J Biomech 42(3):297–302

7. Smith MA, Blankman E, Gardel ML, Luettjohann L, Waterman CM, Beckerle MC (2010) A zyxin-mediated mechanism for actin stress fiber maintenance and repair. Dev Cell 19(3):365–376
8. Hawkins T, Mirigian M, Yasar MS, Ross JL (2010) Mechanics of microtubules. J Biomech 43(1):23–30
9. Schaap IAT, Carrasco C, de Pablo PJ, MacKintosh FC, Schmidt CF (2006) Elastic response, buckling, and instability of microtubules under radial indentation. Biophys J 91(4):1521–1531
10. Bicek AD, Tuezel E, Kroll DM, Odde DJ (2007) Analysis of microtubule curvature. Cell Mech 83:237–268
11. Bicek AD, Tuzel E, Demtchouk A, Uppalapati M, Hancock WO, Kroll DM, Odde DJ (2009) Anterograde microtubule transport drives microtubule bending in llc-pk1 epithelial cells. Mol Biol Cell 20(12):2943–2953
12. Li HL, DeRosier DJ, Nicholson WV, Nogales E, Downing KH (2002) Microtubule structure at 8 angstrom resolution. Structure 10(10):1317–1328
13. Evans L, Mitchison T, Kirschner M (1985) Influence of the centrosome on the structure of nucleated microtubules. J Cell Biol 100(4):1185–1191
14. Tuszynski JA, Luchko T, Portet S, Dixon JM (2005) Anisotropic elastic properties of microtubules. Eur Phys J E 17(1):29–35
15. Chartier L, Rankin LL, Allen RE, Kato Y, Fusetani N, Karaki H, Watabe S, Hartshorne DJ (1991) Calyculin-a increases the level of protein-phosphorylation and changes the shape of 3t3 fibroblasts. Cell Motil Cytoskeleton 18(1):26–40
16. Deguchi S, Sato M (2009) Biomechanical properties of actin stress fibers of non-motile cells. Biorheology 46(2):93–105
17. Katoh K, Kano Y, Masuda M, Onishi H, Fujiwara K (1998) Isolation and contraction of the stress fiber. Mol Biol Cell 9(7):1919–1938
18. Mansour H, de Tombe PP, Samarel AM, Russell B (2004) Restoration of resting sarcomere length after uniaxial static strain is regulated by protein kinase C epsilon and focal adhesion kinase. Circ Res 94(5):642–649
19. Brangwynne CP, MacKintosh FC, Kumar S, Geisse NA, Talbot J, Mahadevan L, Parker KK, Ingber DE, Weitz DA (2006) Microtubules can bear enhanced compressive loads in living cells because of lateral reinforcement. J Cell Biol 173(5):733–741

Chapter 20
Four-Dimensional Analysis for a Tumor Invasion

Masato Tamura and Hirofumi Matsui

Abstract Four-dimensional analysis provides the understandings for a tumor invasion. The analysis is of three-dimensional with times. In this part, we describe the cellular mobility focused on reactive oxygen species from mitochondria. Mitochondrial ROS is involved in cancer properties such as growth, metastasis and invasion. On the other hand, carcinogenesis and malignancy are involved in environmental factors such as foods. Thus, we evaluated whether sodium chloride and ethanol induce mitochondrial ROS or not. These results indicated that mitochondrial ROS enhances cancer cellular invasion, meanwhile, sodium chloride and ethanol induced a production of mitochondrial ROS in cells. Taking these results into accounts, sodium chloride and ethanol probably enhances tumor invasion. In addition, cellular morphology is alternated when cells invades into pericellular matrix. If cellular characters such as an ability of invasion can be suggested by cellular observation, classification of cancer cellular malignancy is easy. Therefore, we have developing a novel method of cell picking from three-dimensional culture.

Keywords Four-dimensional analysis · Invasion · Reactive oxygen species

20.1 Introduction

Cancer cells show properties of tumorigenesis, anchorage-independent growth and invasion. Understandings of these mechanisms should contribute a developing cancer treatment and prevention because the properties coincided with cancer malignancy. We have been investigating a tumor invasion involved in reactive oxygen species (ROS). ROS regulates a tumor growth and invasion and metastasis through control of cell signaling, resulting in a cancer malignancy. In vitro study revealed that ROS concentration in cancer cells was greater than that in normal cells. Many reports reported about the relation between cancer malignancy and ROS. For example, ROS such as hydroxyl radical and superoxide anion activated a tumor growth and invasion and metastasis, meanwhile, ROS scavenger

H. Matsui (✉) · M. Tamura
Faculty of Medicine, University of Tsukuba, Tsukuba, Ibaraki, Japan
e-mail: hmatsui@md.tsukuba.ac.jp

T. Arai et al. (eds.), *Hyper Bio Assembler for 3D Cellular Systems*,
DOI 10.1007/978-4-431-55297-0_20

suppresses these properties. In our life, environmental factors such as ethanol and sodium chloride become an oxidative stressor by an excess uptake. Herein, we describe an in vitro four-dimensional analysis with normal and cancer cells, concluding an enhancement of tumor invasion by ROS. We also show a possibility of oxidative stressors in ethanol, sodium chloride and acetaldehyde. In final session, we describe the progress study about a novel cell separation method in three-dimensional culture.

20.2 Four-Dimensional Analysis and Cancer Specific Properties

In this article, four-dimensional means an analysis in three dimensional cultures with time. Why we need an analysis of four dimensional is related with cancer properties. As 20.1, cancer cells show three properties which are tumorigenesis and anchorage-independent growth and invasion. These properties show only in three-dimensional culture. For example, these properties could be confirmed by a cell culture in 0.33 % soft agar. In this condition, normal cells cannot growth and form a colony, but cancer cells can. This experimental method, which is called colony-forming assay, is used for an evaluation of cellular growth in radiobiology, and it also used for determination of anchorage-independent growth in cancer biology. The phenotype is dependent on an extracellular matrix (ECM). Recent study is applied collagen and laminin and others as ECM. Matrigel is a promising ECM in an invasion study because it is an extract from tumor tissue. We evaluated invasion ability in normal and cancer cells. In this study, rat gastric mucosal cells (RGM) and cancerous-muted RGM cells (RGK) were used as normal and cancer cells, respectively. The result clearly distinguished cells with or without invasion ability (Fig. 20.1). We confirmed that the invasion is involved in ROS.

20.3 ROS and Cancer

ROS regulates cell signaling through activation/ inactivation of transcriptional factors, resulting in cancer malignancy [1–7]. In one instance, ROS activate mitogen activated protein kinase (MAPK) through an activation of protein kinase C (PKC) [8]. This signaling induces cancer growth and migration. The ROS is mainly generated from cellular membrane and endoplasmic reticulum and mitochondria (Fig. 20.2). When cancer cells invade into tissue, cancer cells have to melt pericellular matrix by enzyme activity of proteases such as matrix metalloproteinase [5, 9, 10]. ROS from cell membrane regulates invasion through an activation of NADPH oxidase (NOX) because a formation of invadopodia requires superoxide [11].

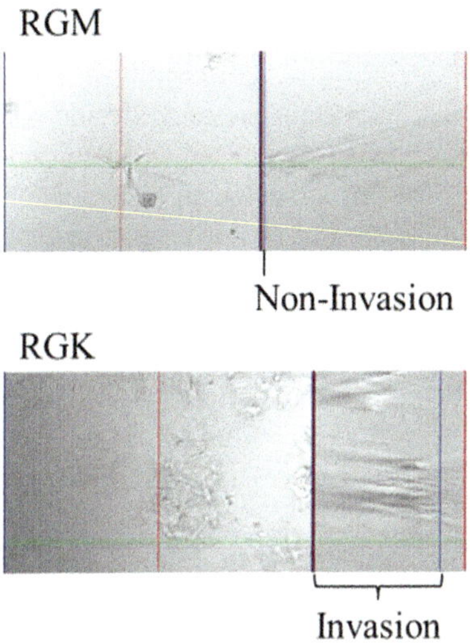

Fig. 20.1 The difference of invasion abilities between RGM and RGK in matrigel. Cells were dispersed on the matrigel, and then incubated for 96 h. These images were made from the images at different depth by captured using confocal microscopy (The length between slice and next slice was 6.02 μm). Normal cells (*RGM*) cannot invade into matrigel, but cancerous cells (*RGK*) can invade with forming invadopodia

Fig. 20.2 Major ROS generators in cell

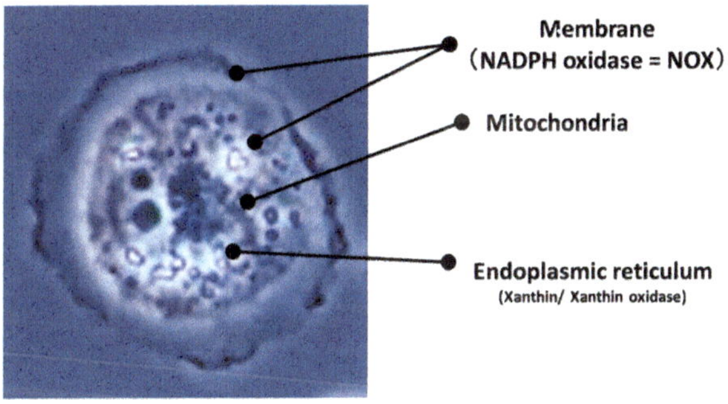

Mitochondria in cancer cells have the mutation which allows mitochondria to generate the excess of ROS [12, 13]. In fact, we detected that the amounts of cellular ROS in cancer cells were greater than that in normal cells [14]. Mitochondrial DNA encodes the proteins related with the electron transport system, and the proteins are located at the mitochondrial inner membrane [13]. Usually, a small percentage of electrons leak and trapped by oxygen in the electron transport system, and then it becomes superoxide. Since complex I and III within the electron transport system are muted in cancer cells, the production of ROS from mitochondria should be increased.

In this session, we focus on the ROS from mitochondria (mitochondrial ROS).

20.4 Mitochondrial ROS in Cancer Cells

Cancer invasion is the most important factor to decide patient's prognosis. Meanwhile tumor without an invasion is diagnosed as a benign lesion, an invaded tumor into normal tissue is diagnosed as a malignant lesion. Therefore, the malignancy is decided by the depth of invasion. For example, gastric cancer is classified by TNM system: the depth of tumor infiltration (T); node involvement (N); the presence of distant metastases (M) [15]. Index of cancer diagnosis indicated that decreasing of invasion lead to improve the prognosis.

As to 20.2, mitochondrion is a main source of ROS. However, the relation between mitochondrial ROS and cancer invasion is not well investigated although mitochondrial ROS is known to derive cancer cellular progression and metastasis [16–18].

Manganese superoxide dismutase (MnSOD) is particularly localized at mitochondria to scavenge their $O_2^{\cdot}$ [19, 20]. In other words, MnSOD is the specific scavenger for mitochondrial ROS from mitochondrial electron transport [21]. Overexpression of MnSOD after its gene transfection suppressed tumor proliferation and metastasis [17]. Mitochondrial ROS is thus likely to important for organizing tumor properties.

For a compared study of normal and cancer mobility, the pair both of normal and cancer cells is useful. We established two cell lines; a rat normal gastric mucosa cell-lines (RGM) [22] and a cancerous RGM cell-lines (RGK) [23]. These cell lines are useful to compare cellular properties between normal cells and cancerous cells. In addition, we also established MnSOD-expressing cell lines, because these cell-lines used to be regarded as good experimental model to clarify whether intracellular ROS were derived from mitochondria or not [13, 21, 24–28]. The fluorescence image after ROS stained supports these reports (Fig. 20.3). Using them, we have reported the roles of mitochondrial ROS to involve gastric epithelial cellular injuries after treatments with NSAIDs, bisphosphonate and exposure to acidic environments in previous studies [29, 30]. Following section, we describe that mitochondrial ROS involved in tumor cellular invasion.

20.5 Mitochondrial ROS Enhances Cancer Invasion

This section is described that the relation between mitochondrial ROS and tumor cellular invasion according to our report [14, 31]. Contents are the results of living cellular ROS spectra of electron paramagnetic resonance (EPR), cellular ruffling frequencies, wound healing assay and invasion assay using RGM, RGK and MnSOD-overexpressing stable clone RGK cell lines (RGK MnSOD). EPR study shows that cellular ROS is order as increasing RGM, RGK-MnSOD and RGK. Cellular ruffling frequencies of RGK showed greater than that of RGM, and the frequencies of RGK was suppressed by MnSOD-overexpressing. The results of wound

Fig. 20.3 ROS-stained fluorescence images of RGM, RGK and RGK MnSOD-expressing cells. Scale bars: 50 μm

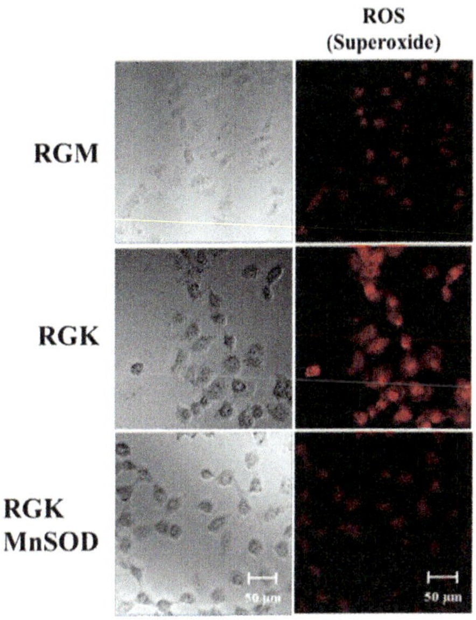

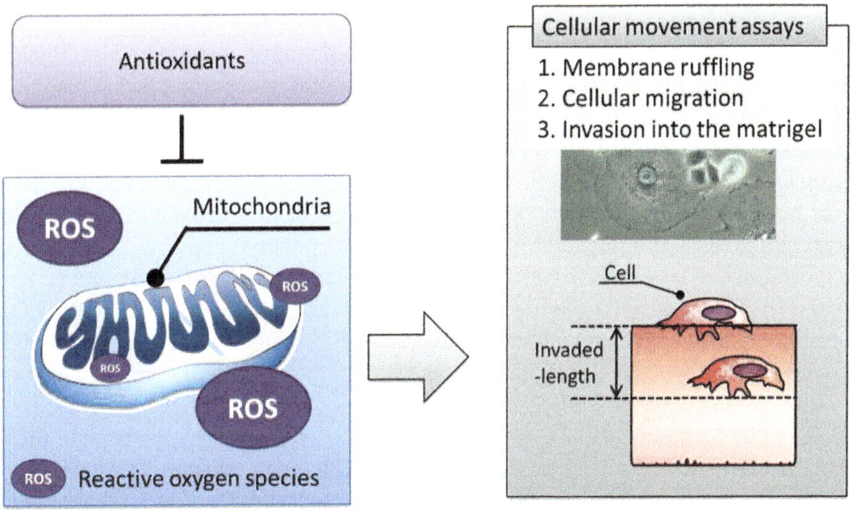

Fig. 20.4 Mitochondrial ROS enhances membrane ruffling and cellular migration and invasion

healing assay and invasion were same as cellular ruffling frequencies. These results indicated that mitochondrial ROS involved in tumor invasion (Fig. 20.4).

As to 20.2, NOX-ROS pathway is a common understanding in invasion study. In the pathway, ROS enhances cellular invasion through an activation of Rho family. Rho family is a family of small molecules G proteins including Cdc42, Rac1 and RhoA. These functions are regulation of organclle development and cellular migration by controlling actin remodeling or other molecules. If mitochondrial ROS also activates Rho family, the understandings is easy because the mechanism should same as NOX-ROS pathway. However, mitochondrial ROS did not activate Rac1/cdc42 in preliminary experiment. This data suggest mitochondrial ROS seem to be a dependent from NOX-ROS pathway. What molecules enhanced cellular invasion by mitochondrial ROS is still under investigation.

20.6 Oxidative Stressors in Environmental Factor

Theoretically, carcinogenesis and malignancy is involved in environmental factors such as foods. Many reports support this suggestion by statistics; however, a certification by experimental study is few. Since oxidative stress is deeply related with cancer (see to 20.2), we evaluated whether foods or its metabolites becomes oxidative stressor or not. Additionally, we targeted mitochondrial ROS because of our indication (see to 20.3). As environmental factors, we used sodium chloride and ethanol.

Food with high concentration salt is an aggressive factor and carcinogen for gastric mucosae [32]. It has reported to destroy a mucosal barrier and leads to inflammations such as diffuse erosions and degenerations. An inflammation is known as a factor of cancer malignancy. Salt-inducible cell injuries were often investigated in botany because sodium chloride induces stresses affecting crop yield all over the world. This is called “Salt stress”. Salt stress is oxidative stress with superoxide production in plant [33, 34]. In animal cells, Salt stress disturbs cellular metabolism and iron homeostasis [35], however, the relation between salt stress and ROS was not been well known. Previous study indicated that hypertonic salt treatments were an inducer of mitochondrial ROS in gastric epithelial cells [36]. While cells exposed under less one M NaCl environment died immediately, cells under less 650 mM NaCl survived for a few hours with ROS production. In addition, the ROS production was decreased by MnSOD-overexpression. These results suggested high concentration of sodium chloride is a necrotizing factor, and moderate concentration of sodium chloride is an oxidative stressor.

Alcohol/ethanol is also known as an aggressive factor for gastrointestinal tract. The alcohol such as a beer or wine is contained 4–20% ethanol. Previous study showed that cells were survived for a hew hors under less than 10% ethanol with production of mitochondrial ROS, while more than 15% ethanol exposure derived cell death immediately within 1 h [37]. Same as salty environment, more than 15% ethanol was a necrotizing factor, while moderate condition of ethanol was an oxidative stressor (Fig. 20.5). Alcohol is a casual factor in 60 types of both diseases and component causes in 200 others. Ethanol is metabolized in microsomal ethanol oxidizing system (MEOS). MEOS requires cytochrome P450 family (CYP2E1) for generating oxidized NADPH in cytoplasm [38, 39]. CYP2E1 accelerates the expression of cyclooxygenase-2 (COX-2) in liver [40], and it produces prostaglandins. Through the metabolism, ethanol becomes to acetaldehyde.

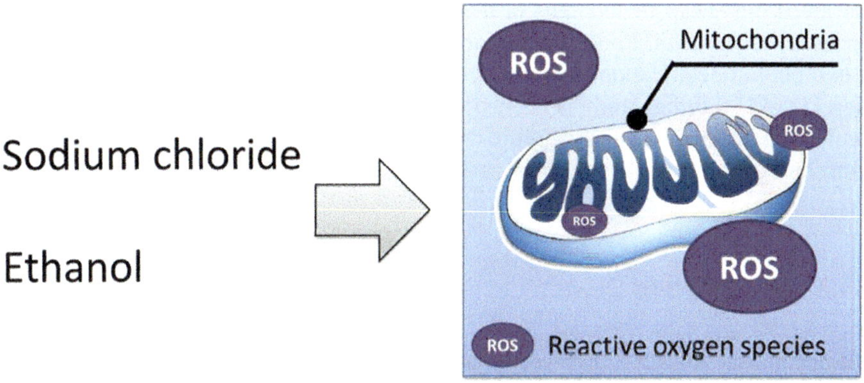

Fig. 20.5 Sodium chloride and ethanol are an oxidative stressor

More recently, we reported that acetaldehyde also involved in mitochondrial ROS [41]. As mentioned above, mitochondrial ROS derives the tumor invasion (see to 20.3). Our previous study suggested that sodium chloride and ethanol and acetaldehyde could be an enhancer of tumor cellular invasion. For evaluating further details of cell, we need a novel method for a cell picking from three-dimensional culture.

20.7 Cell Separation from Three-Dimensional Culture

For developing the method of cell separation from three-dimensional culture, we and collaborators used photodegradable hydrogels with photocleavable crosslinker [42]. Optical cell manipulation techniques form the cell pattern using the modified chemicals, which causes structural alterations by optical energy, on the surface of the dish. Some modified chemicals contain arginine-glycine-aspartate (RGD) cell-adhesive peptide.[43–45] Photo irradiation cleaves the chemicals or alternate the chemical structure, as a result, the chemicals controls whether the cell adhesion or not by the existence of RGD peptide. While the non-cell-adhesive peptide polymer such as 2-methacryloyloxyethyl phosphorylcholine (MPC) polymer was used for cell patterning [46]. In other studies also regulated the cell adhesive site by photo-irradiation. J. Nakanishi et al. controlled the dimention of cellular migration by the photo-selectived adherence site on the substrate closed to the cell [47, 48]. Q. Guo et al. regulated the dimention of cellular migration with a photoacitvatable-Rac [49]. Collaborators also developed photo/thermal responsive culture surface with pNSp-NIPAAm which can control a cell adhesion site by photo irradiation and temperature [50]. These reports indicated cells are very sensitive the pericellular matrix. Recently, collaborators synthesized a novel cross-linker NHS-PC-4arm PEG, and cells on photodegradable hydrogels could detach from the dish by photo irradiation [51]. Used this photodegradable gels, we have successfully picked and retrieved the encapsulated cells in the photo irradiated area with any damages in terms of cell growth. Photo irradiation can control the depth of degrading gels by a two photon degradation [52] or by the dose of photo irradiation after gelatinization [53, 54]. Further study is needed to determine the stiffness of hydrogels for a prolonged three-dimensional culture due to analyse a cellular characters by cellular morphology.

20.8 Conclusion

Mitochondrial ROS are involved in the tumor cellular invasion which should be produced by sodium chloride, ethanol and acetaldehyde. Mitochondrial ROS seem to be a dependent from NOX-ROS pathway in cellular membrane. For distinguishing the characterization of cells by its morphology in three-dimensional culture, an evaluation of pericellular matrix will be needed. In the evaluation, ROS should become a key for understanding cellular characters from cell morphology in three-dimensional culture. The four-dimensional analysis is also needed for evaluating cellular invasion and other functions.

References

1. Nishikawa M (2008) Reactive oxygen species in tumor metastasis. Cancer Lett 266(1):53–59
2. Kamata H, Honda S-I, Maeda S, Chang L, Hirata H, Karin M (2005) Reactive oxygen species promote TNFalpha-induced death and sustained JNK activation by inhibiting MAP kinase phosphatases. Cell 120(5):649–661
3. Lokeshwar BL, Selzer MG, Zhu B-Q, Block NL, Golub LM (2002) Inhibition of cell proliferation, invasion, tumor growth and metastasis by an oral non-antimicrobial tetracycline analog (COL-3) in a metastatic prostate cancer model. Int J Cancer 98(2):297–309
4. Kim E-Y, Seo J-M, Cho K-J, Kim J-H (2010) Ras-induced invasion and metastasis are regulated by a leukotriene B4 receptor BLT2-linked pathway. Oncogene 29(8):1167–1178
5. Kumar B, Koul S, Khandrika L, Meacham RB, Koul HK (2008) Oxidative stress is inherent in prostate cancer cells and is required for aggressive phenotype. Cancer Res 68(6):1777–1785
6. Bernstein H, Bernstein C, Payne CM, Dvorakova K, Garewal H (2005) Bile acids as carcinogens in human gastrointestinal cancers. Mutat Res 589(1):47–65
7. Wu W-S, Wu J, Hu C (2008) Signal cross talks for sustained MAPK activation and cell migration: the potential role of reactive oxygen species. Cancer Metastasis Rev 27(2):303–314
8. Wu W-S (2006) The signaling mechanism of ROS in tumor progression. Cancer Metastasis Rev 25(4):695–705
9. Westermarck J, Kähäri VM (1999) Regulation of matrix metalloproteinase expression in tumor invasion. FASEB J 13(8):781–792
10. Curran S, Murray GI (1999) Matrix metalloproteinases in tumour invasion and metastasis. J Pathol 189(3):300–308
11. Shinohara M, Adachi Y, Mitsushita J, Kuwabara M, Nagasawa A, Harada S, Furuta S, Zhang Y, Seheli K, Miyazaki H, Kamata T (2010) Reactive oxygen generated by NADPH oxidase 1 (Nox1) contributes to cell invasion by regulating matrix metalloprotease-9 production and cell migration. J Biol Chem 285(7):4481–4488

12. Majima HJ, Toyokuni S (2012) Mitochondria and free radical studies on health, disease and pollution. Free Radic Res 46(8):925–926
13. Indo HP, Davidson M, Yen H-C, Suenaga S, Tomita K, Nishii T, Higuchi M, Koga Y, Ozawa T, Majima HJ (2007) Evidence of ROS generation by mitochondria in cells with impaired electron transport chain and mitochondrial DNA damage. Mitochondrion 7(1–2):106–118
14. Tamura M, Matsui H, Tomita T, Sadakata H, Indo HP, Majima HJ, Kaneko T, Hyodo I (2013) Mitochondrial reactive oxygen species accelerate gastric cancer cell invasion. J Clin Biochem Nutr 54(1):12–17
15. Dikken JL, Velde CJH, Gönen M, Verheij M, Brennan M, Coit DG (2012) The New American Joint Committee on Cancer/International Union Against Cancer Staging System for Adenocarcinoma of the Stomach: Increased Complexity without Clear Improvement in Predictive Accuracy Ann Surg Oncol 19:2443–2451
16. Zhong W, Oberley LW, Oberley TD, Clair DKSt (1997) Suppression of the malignant phenotype of human glioma cells by overexpression of manganese superoxide dismutase. Oncogene 14(4):481–490
17. Sotgia F, Martinez-Outschoorn UE, Lisanti MP (Jan 2011) Mitochondrial oxidative stress drives tumor progression and metastasis: should we use antioxidants as a key component of cancer treatment and prevention? BMC Med 9(1):62
18. Indo HP, Davidson M, Yen H-C, Suenaga S, Tomita K, Nishii T, Higuchi M, Koga Y, Ozawa T, Majima HJ (2007) Evidence of ROS generation by mitochondria in cells with impaired electron transport chain and mitochondrial DNA damage. Mitochondrion 7(1–2):106–118
19. a Okado-Matsumoto and I. Fridovich (2001) Subcellular distribution of superoxide dismutases (SOD) in rat liver: Cu, Zn-SOD in mitochondria. J Biol Chem 276(42):38388–38393
20. Zhang HJ, Yan T, Oberley TD, Oberley LW (1999) Comparison of effects of two polymorphic variants of manganese superoxide dismutase on human breast MCF-7 cancer cell phenotype. Cancer Res 59(24):6276–6283
21. Indo HP, Inanami O, Koumura T, Suenaga S, Yen H-C, Kakinuma S, Matsumoto K-I, Nakanishi I, St Clair W, St Clair DK, Matsui H, Cornette R, Gusev O, Okuda T, Nakagawa Y, Ozawa T, Majima HJ (2012) Roles of mitochondria-generated reactive oxygen species on X-ray-induced apoptosis in a human hepatocellular carcinoma cell line, HLE. Free Radic Res 46(8):1029–1043
22. Kobayashi I, Kawano S, Tsuji S, Matsui H, Nakama A, Sawaoka H, Masuda E, Takei Y, Nagano K, Fusamoto H, Ohno T, Fukutomi H, Kamada T (1996) RGM1, a cell line derived from normal gastric mucosa of rat. In Vitro Cell Dev Biol Anim 32(5):259–261
23. Shimokawa O, Matsui H, Nagano Y, Kaneko T, Shibahara T, Nakahara A, Hyodo I, Yanaka A, Majima HJ, Nakamura Y, Matsuzaki Y (2008) Neoplastic transformation and induction of H+,K+-adenosine triphosphatase by N-methyl-N'-nitro-N-nitrosoguanidine in the gastric epithelial RGM-1 cell line. In Vitro Cell Dev Biol Anim 44(1–2):26–30
24. Motoori S, Majima HJ, Ebara M, Kato H, Hirai F, Kakinuma S, Yamaguchi C, Ozawa T, Nagano T, Tsujii H, Saisho H (2001) Overexpression of mitochondrial manganese superoxide dismutase protects against radiation-induced cell death in the human hepatocellular carcinoma cell line HLE. Cancer Res 61(14):5382–5388
25. Majima HJ, Indo HP, Suenaga S, Kaneko T, Matsui H, Yen HC, Ozawa T (2011) Mitochondria as source of free radicals. Free Radic Biol 29:12–22
26. Majima HJ, Oberley TD, Furukawa K, Mattson MP, Yen HC, Szweda LI, Clair DKSt (1998) Prevention of mitochondrial injury by manganese superoxide dismutase reveals a primary mechanism for alkaline-induced cell death. J Biol Chem 273(14):8217–8224
27. Majima HJ, Indo HP, Suenaga S, Matsui H, Yen H-C, Ozawa T (2011) Mitochondria as possible pharmaceutical targets for the effects of vitamin E and its homologues in oxidative stress-related diseases. Curr Pharm Des 17(21):2190–2195
28. Rai K, Matsui H, Kaneko T, Nagano Y, Shimokawa O, Udo J, Hirayama A, Hyodo I, Indo HP, Majima HJ (2011) Lansoprazole inhibits mitochondrial superoxide production and cellular lipid peroxidation induced by indomethacin in RGM1 cells. J Clin Biochem Nutr 49(1):25–30

29. Nagano Y, Matsui H, Tamura M, Shimokawa O, Nakamura Y, Kaneko T, Hyodo I (Jan 2012) NSAIDs and acidic environment induce gastric mucosal cellular mitochondrial dysfunction. Digestion 85(2):131–135
30. Nagano Y, Matsui H, Shimokawa O, Hirayama A, Nakamura Y, Tamura M, Rai K, Kaneko T, Hyodo I (2012) Bisphosphonate-induced gastrointestinal mucosal injury is mediated by mitochondrial superoxide production and lipid peroxidation. J Clin Biochem Nutr 51(3):196–203
31. Tamura M, Mutoh M, Fujii G, Matsui H (2013) Involvement of mitochondrial reactive oxygen species in gastric carcinogenesis. J Gastrointest Dig Syst 3(4):150
32. Sonnenberg A (1986) Dietary salt and gastric ulcer. Gut 27(10):1138–1142
33. Katsuhara M, Otsuka T, Ezaki B (2005) Salt stress-induced lipid peroxidation is reduced by glutathione -transferase, but this reduction of lipid peroxides is not enough for a recovery of root growth in. Plant Sci 169(2):369–373
34. Liu R, Garvin JL, Ren Y, Pagano PJ, O. a Carretero (2007) Depolarization of the macula densa induces superoxide production via NAD(P)H oxidase. Am J Physiol Renal Physiol 292(6):F1867–72
35. Griffiths ME, Orians CM (2003) Salt spray differentially affects water status, necrosis, and growth in coastal sandplain heathland species. Am J Bot 90(8):1188–1196
36. Tamura M, Matsui H, Nagano YN, Kaneko T, Indo HP, Majima HJ, Hyodo I (2013) Salt is an oxidative stressor for gastric epithelial cells. J Physiol Pharmacol 64(1):89–94
37. Tamura M, Matsui H, Kaneko T, Hyodo I (2013) Alcohol is an oxidative stressor for gastric epithelial cells: detection of superoxide in living cells. J Clin Biochem Nutr 53(2):75–80
38. Julkunen RJ, Di Padova C, Lieber CS (1985) First pass metabolism of ethanol-a gastrointestinal barrier against the systemic toxicity of ethanol. Life Sci 37(6):567–573
39. Pronko P, Bardina L, Satanovskaya V, Kuzmich A, Zimatkin S (2002) Effect of chronic alcohol consumption on the ethanol- and acetaldehyde-metabolizing systems in the rat gastrointestinal tract. Alcohol Alcohol 37(3):229–235
40. Uzma N, Kumar BS, Priyadarsini KI (2011) Hepatoprotective, immunomodulatory, and anti-inflammatory activities of selenocystine in experimental liver injury of rats. Biol Trace Elem Res 142(3):723–734
41. Tamura M, Ito H, Matsui H, Hyodo I (2014) Acetaldehyde is an oxidative stressor for gastric epithelial cells. J Clin Biochem Nutr 55(1):26–31
42. Tamura M, Yanagawa F, Sugiura S, Toshiyuki T, Kimio S, Hirofumi M, Toshiyuki K (2014) Optical cell separation from three-dimensional environment in photodegradable hydrogels for pure culture techniques. Sci Rep 4:4793
43. Ohmuro-Matsuyama Y, Tatsu Y (2008) Photocontrolled cell adhesion on a surface functionalized with a caged arginine-glycine-aspartate peptide. Angew Chem Int Ed Engl 47(39):7527–7529
44. Petersen S, Alonso JM, Specht A, Duodu P, Goeldner M, del Campo A (Jan 2008) Phototriggering of cell adhesion by caged cyclic RGD peptides. Angew Chem Int Ed Engl 47(17):3192–3195
45. Liu D, Xie Y, Shao H, Jiang X (Jan 2009) Using azobenzene-embedded self-assembled monolayers to photochemically control cell adhesion reversibly. Angew Chem Int Ed Engl 48(24):4406–4408
46. Jang K, Sato K, Mawatari K, Konno T, Ishihara K, Kitamori T (2009) Surface modification by 2-methacryloyloxyethyl phosphorylcholine coupled to a photolabile linker for cell micropatterning. Biomaterials 30(7):1413–1420
47. Nakanishi J, Kikuchi Y, Inoue S, Yamaguchi K, Takarada T, Maeda M (2007) Spatiotemporal control of migration of single cells on a photoactivatable cell microarray. J Am Chem Soc 129(21):6694–6695
48. Kaneko S, Nakayama H, Yoshino Y, Fushimi D, Yamaguchi K, Horiike Y, Nakanishi J (2011) Photocontrol of cell adhesion on amino-bearing surfaces by reversible conjugation of poly(ethylene glycol) via a photocleavable linker. Phys Chem Chem Phys 13(9):4051–4059

49. Guo Q, Wang X, Tibbitt MW, Anseth KS, Montell DJ, Elisseeff JH (2012) Light activated cell migration in synthetic extracellular matrices. Biomaterials 33(32):8040–8046
50. Edahiro J-I, Sumaru K, Tada Y, Ohi K, Takagi T, Kameda M, Shinbo T, Kanamori T, Yoshimi Y (2005) In situ control of cell adhesion using photoresponsive culture surface. Biomacromolecules 6(2):970–974
51. Sugiura S, Takagi T, Yamaguchi M, Sumaru K, Kanamori T (2012) A photodegradable hydrogel sheet for microscale optical control of cell adhesion and detachment. Proc mTAS 671–673
52. Wong DY, Griffin DR, Reed J, Kasko AM (2010) Photodegradable hydrogels to generate positive and negative features over multiple length scales. Macromolecules 43(6):2824–2831
53. Kloxin AM, Tibbitt MW, Kasko AM, Fairbairn JA, Anseth KS (2010) Tunable hydrogels for external manipulation of cellular microenvironments through controlled photodegradation. Adv Mater 22(1):61–66
54. Kloxin AM, Kasko AM, Salinas CN, Anseth KS (April 2009) Photodegradable hydrogels for dynamic tuning of physical and chemical properties. Science 324(5923):59–63

Chapter 21
Three-Dimensional Mineralized Tissue Formation of Cultured Bone Marrow Stromal Cells

Takanori Kihara

Abstract Bone marrow stromal cells (BMSCs) can differentiate into osteocytes and form mineralized matrices in culture conditions. This chapter describes the structural characteristics of in vitro-cultured bone tissue, which includes mineralized matrices and differentiated cells, formed by rat BMSCs. In addition, I introduce a three-dimensional (3D) visualization method for cultured bone tissue using confocal laser scanning microscopy. This 3D visualization showed that the in vitro-constructed tissue contains layers of mineralized matrices with round cells in the matrix lacunae and osteoblastic cells on the mineralized matrices. Type I collagen fibers were well integrated within the mineralized matrices. Mineralized matrices near the culture dish surface did not contain deposited collagen fibers; however, in the upper part of the matrices, calcium was deposited as small particles along with collagen fibers. The structure of the in vitro-mineralized tissue formed by BMSCs is comparable to that of normal bone tissue formed in vivo. This 3D visualization method will aid in the in vitro construction of 3D bone tissue for tissue engineering applications.

Keywords Bone marrow stromal cells · Mineralized matrices · Confocal laser scanning microscope · Three-dimensional imaging

21.1 Introduction

Mesenchymal stem cells are tissue stem cells that can differentiate into not only mesenchymal lineage cells but also cells of other lineages [1–6]. In particular, bone marrow stromal cells (BMSCs), a cell population that includes mesenchymal stem cells, can differentiate into osteocytes, chondrocytes, adipocytes, skeletal muscle cells, cardiac muscle cells, neural cells, hepatocytes, and insulin-producing

T. Kihara (✉)
Department of Life and Environment Engineering,
Faculty of Environmental Engineering, The University of Kitakyushu
1-1 Hibikino, Wakamatsu, Kitakyushu, Fukuoka 808-0135, Japan
e-mail: takanori.kihara@gmail.com

T. Arai et al. (eds.), *Hyper Bio Assembler for 3D Cellular Systems*,
DOI 10.1007/978-4-431-55297-0_21

pancreatic beta cells. BMSCs are thought to act as repair cells in vitro [7]; thus, in clinical studies, they have been transplanted into patients with many types of disease [8].

Osteogenic differentiation is the most well-understood differentiation process of BMSCs. When BMSCs are implanted into ectopic sites and combined with ceramic blocks, they exhibit osteogenic potential, resulting in bone formation [9, 10]. BMSCs also differentiate into osteoblasts and undergo mineralization when they are cultured in the presence of ascorbic acid, β-glycerophosphate, and dexamethasone in two-dimensional (2D) common culture substrates [11]. Physicochemical and biochemical analyses of this in vitro mineralization show that this process is not the precipitation of just simple calcium and phosphorus, but is in fact biological apatite, which is present in natural bone and is associated with the formation of bone matrix protein [11, 12]. Bone development requires the concerted action of several microenvironmental signals, including cytokines, extracellular matrix, and cell-cell interactions [13]. Therefore, in vitro osteogenic differentiation of BMSCs can be used to study the cascade of bone development and regeneration.

In this chapter, the structural characteristics of in vitro-cultured bone tissue formed by BMSCs is described. In addition, a three-dimensional (3D) visualization method for cultured bone tissue using confocal laser scanning microscopy (CLSM) is introduced. This 3D visualization method clarifies the physical relationship between the mineralized matrix and cells. This method will aid in the in vitro construction of 3D bone tissue for use in tissue engineering applications.

21.2 Experimental Procedures

21.2.1 Cell Culture

Human osteosarcoma cells (HOS) and MC3T3-E1 cells were cultured in DMEM supplemented with 10% FBS and antibiotics. To induce mineralization of HOS cells, these cells were cultured in medium supplemented with 10 mM β-glycerophosphate. To induce mineralization of MC3T3-E1 cells, these cells were cultured in medium supplemented with 82 μg/mL ascorbic acid-2-phosphate, 10 mM β-glycerophosphate, and 10 nM dexamethasone.

21.2.2 Preparation of Rat BMSCs

Rat BMSCs were isolated and cultured as described previously [14]. Briefly, BMSCs were obtained from the femoral shafts of 7-week-old male Fischer 344 rats. Cells were obtained from at least two rats, and then mixed. The mixed cells were seeded in one 75-cm^2 tissue culture flask per femur. The culture medium used was MEM (with Earle's salts and L-Gln) containing 15% FBS and antibiotics. After the cells reached confluence, they were released from the substratum, and then seeded into culture plates at 1.0×10^4 cells/cm^2. Under osteogenic differentiation conditions, the cells were cultured in medium supplemented with 82 μg/mL ascorbic acid-2-phosphate, 10 mM β-glycerophosphate, and 10 nM dexamethasone.

21.2.3 FITC-Labeled Type I Collagen

FITC-labeled type I collagen was prepared according to the following procedure [15]. Type I collagen in 1 mM HCl was mixed with an equal volume of 0.5 M $NaHCO_3$-Na_2CO_3 (pH 9.5) and FITC (3.5 % collagen on a dry weight basis). After a 3-h incubation, the solution was neutralized with HCl, and then FITC-labeled collagen was precipitated with 70 % EtOH. The FITC-labeled collagen was then dissolved in 0.5 M acetic acid and dialyzed against 1 mM HCl.

21.2.4 Three-Dimensional Reconstruction Images

The 3D fluorescence images were reconstructed by layering a series of X-Y sections using TRI/3D-Bon software (Ratoc System Engineering, Japan) or ImageJ software (NIH, Bethesda, MD).

21.3 Three-Dimensional Reconstruction of In Vitro-Mineralized Tissues From Osteogenic Cell Lines

The most typical and characteristic property of osteogenic cells is its mineralized tissue formation capability. This property can be observed even in culture. HOS cells and MC3T3-E1 cells are known to induce in vitro mineralization. HOS cells, which are an osteoblastic cell line derived from malignant tumors of mesenchymal cells, produce a prominent osteoid matrix with mineralization [16]. MC3T3-E1 cells, which are a clonal, pre-osteoblastic cell line derived from newborn mouse calvaria, form an abundant collagen matrix and also show mineralization [17].

Figure 21.1 shows the matrix mineralization of cultured HOS and MC3T3-E1 cells. These two different osteogenic cell lines formed mineralized matrices with different structures. HOS cells formed many granular mineralization matrices. In contrast, MC3T3-E1 cells formed large, amorphous mineralized matrices. The 3D structures of these mineralized matrices were observed by CLSM (Fig. 21.2). CLSM is an effective tool for determining the 3D structures of cultured tissues. The mineralized matrices were labeled with a calcium-binding fluorescent molecule, calcein [14, 18]. The granular mineralized matrices formed by HOS cells were embedded in the upper side of the cell layer. The martices were approximately 6 μm in diameter. In contrast, the mineralized matrices formed by MC3T3-E1 cells were

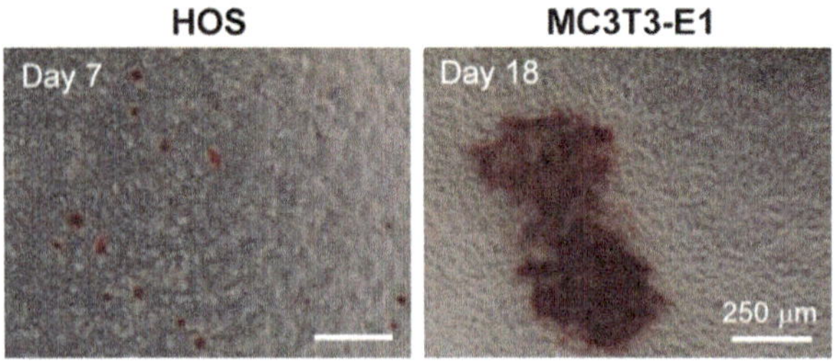

Fig. 21.1 Mineralization images of HOS and MC3T3-E1 cells. The mineralized cell layers were stained with Alizarin *red* S

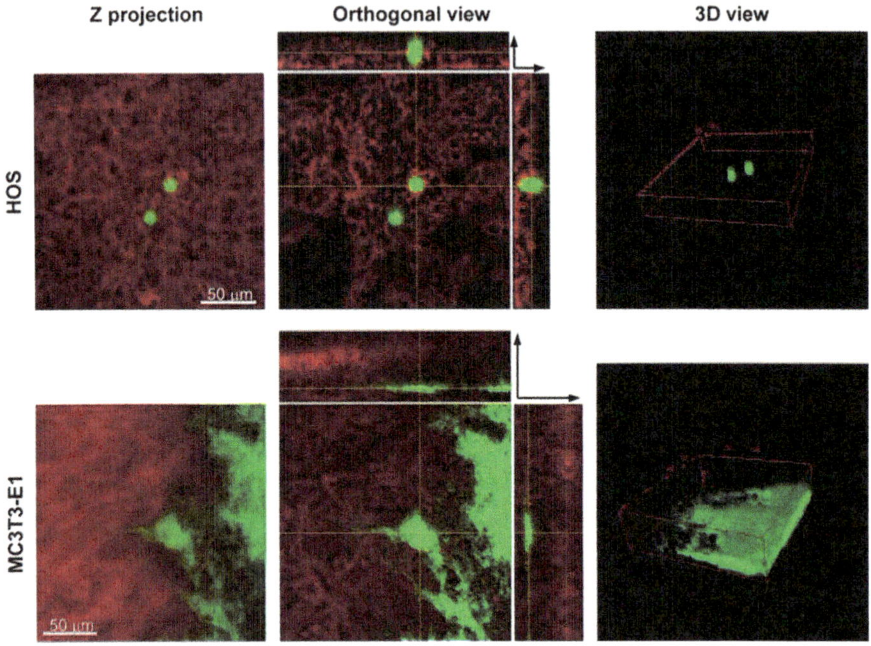

Fig. 21.2 Three-dimensional structures of mineralized matrices of HOS and MC3T3-E1 cells. The mineralized matrices were labeled with calcein (*green*) and the cells were stained with rhodamine-phalloidin (*red*). The mineralized cell layers were observed by CLSM

large and flat. A thick cell layer was overlaid on the mineralized matrices, and a single cell layer was located between the culture substrate and the mineralized matrices. The mineralized matrices formed by the MC3T3-E1 cells barely contacted the surface of the culture dish.

In summary, the structures of the in vitro-mineralized tissues, which include the mineralized matrices and cultured cells, varied according to cell. To construct bone tissues with various shapes and structures through tissue engineering approaches, the 3D structure of the mineralized tissues formed by the osteogenic cells should be considered.

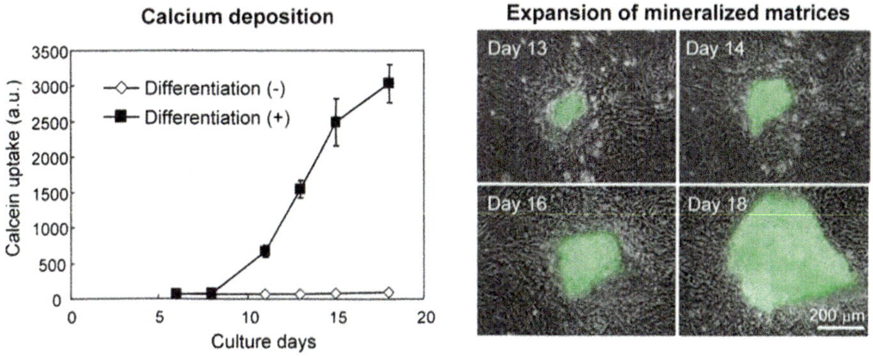

Fig. 21.3 Mineralization by rat BMSCs. Calcein depositions under osteogenic differentiation conditions were measured (*left*). Merged images of mineralized matrices (*green*) and phase contrast micrographs of the same region of the cell layer (*right*)

21.4 Three-Dimensional Reconstruction of BMSC-Derived In Vitro-Mineralized Tissues

Rat BMSCs cultured under osteogenic differentiation conditions formed solid, amorphous mineralized matrices (Fig. 21.3). The number and size of the mineralized matrices increased in a time-dependent manner. Then, the 3D structure of the solid mineralized matrices was observed by CLSM [14]. The tabular mineralized matrices contacted the surface of the culture dish, the surface of the mineralized matrices was covered with numerous cells (Fig. 21.4), and some round-shaped cells were present inside the mineralized matrices. These round cells were clearly distinct from the cells on the surface of the mineralized matrices. The cells on the matrices had stretched actin filaments, whereas the cells within the mineralized matrices did not. To clarify the structure of the matrices in detail, a 3D graphic was created using a stack of 2D slices collected by CLSM (Fig. 21.4). The top surface of the mineralized matrices was completely covered with clusters of cells. The cells within the matrix were entirely embedded in the lacunae of the matrix, and each cell was isolated from the others. When the cells in the 3D-reconstructed graphic were separated from the mineralized matrix, their dendritic morphology and short processes could be seen. The cross-sectional area of each dendritic cell in the matrix was approximately 57 μm^2.

CLSM analysis of natural bone showed that osteocytes are located in the mineralized matrix and that they have a dendritic morphology with many processes [19]. The cross-sectional area of the osteocyte lacunae in normal trabecular bone ranges from 50.6 to 53.8 mm^2 [20]. Based on these results, the round-shaped cells with short processes in the mineralized matrices formed by rat BMSCs are probably osteocytes that differentiated from osteoblasts that were originally localized on the matrix surface.

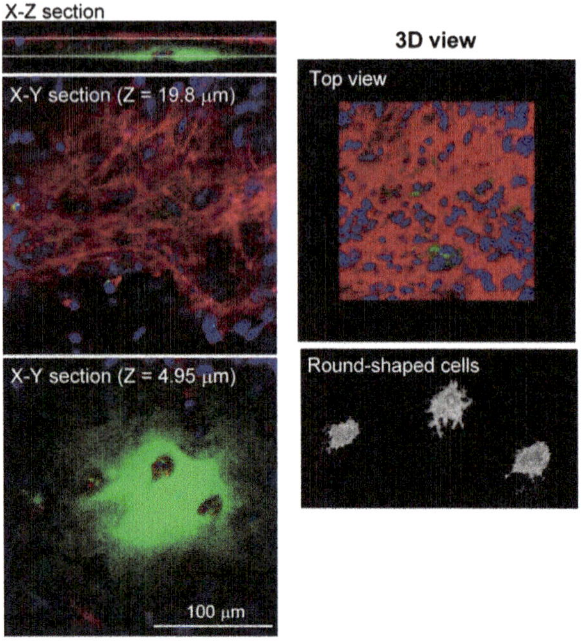

Fig. 21.4 Three-dimensional structures of mineralized matrices formed by BMSCs. The mineralized matrices were labeled with calcein (*green*), actin filaments of the cells were stained with rhodamine-phalloidin (*red*), and nuclei were stained with TOPRO-3 (*blue*)

21.5 Effect of Type I Collagen Fibrils on Mineralization by BMSCs

Type I collagen is a major organic component of bone matrix. A collagen coated dish or collagen gel increases the activity of osteoblasts and promotes the osteogenic differentiation of BMSCs [21–23]. Calcified nodules secreted by osteoblasts adhere to collagen fibers [24, 25].

The 3D relationship between type I collagen fibrils and mineralized matrices produced by BMSCs was examined [15]. Under physiological conditions (37 °C, neutral pH in saline), type I collagen self-assembles. We added type I collagen molecules to the cell culture medium to form fibrils. When BMSCs were cultured with type I collagen, the deposited calcium contents were higher than when the BMSCs were cultured without collagen (Fig. 21.5). Therefore, type I collagen facilitated matrix mineralization by BMSCs.

To investigate the location of the added type I collagen in the mineralized matrix, FITC-labeled type I collagen was added to the BMSCs culture (Fig. 21.5). After 2 days of culture, FITC fluorescence accumulated and was observed in the cell layer. The quantity of accumulated type I collagen increased with time, and after 14 days of culture, both high fluorescence and mineralized matrices were observed. Although the added FITC-labeled type I collagen was observed throughout the culture dish, mineralized matrices were also detected at many nodules along with intervening non-mineralized space. In the mineralization matrices, the FITC-labeled type I collagen also collocated.

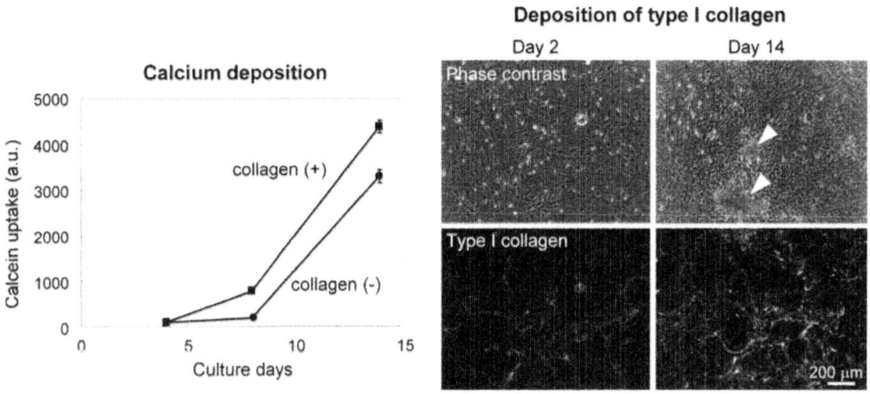

Fig. 21.5 Matrix mineralization by BMSCs cultured with type I collagen fibrils. Calcium depositions of the cells cultured with or without type I collagen were measured (*left*). Deposition of FITC-labeled type I collagen was observed by fluorescent microscope (*right*). At day 14, matrix mineralization was observed (*arrowheads*)

The 3D spatial distribution of FITC-labeled type I collagen and calcium, which was labeled with xylenol orange, in the mineralized matrices was observed by CLSM (Fig. 21.6). The mineralized matrices directly contacted the surface of the culture dish, and FITC-labeled type I collagen fibrils were detected in the matrices. Type I collagen fibrils were mainly observed near the surface of the mineralized matrices, and were barely seen in the bottom of the matrices. Thus, near the surface of the culture dish, the mineralized matrices did not contain the added type I collagen fibers. To clarify the details of the spatial relationship between the added type I collagen and calcium, fluorescence images were treated with an FFT filter. The FFT filter filters out large structures (shading correction) and small structures (smoothing) of specified sizes by Gaussian filtering in Fourier space. The calcium localized as aggregates of small particles that were deposited around the type I collagen in the mineralized matrices.

To summarize, near the culture dish surface, mineralized matrices did not contain deposited collagen fibers; however, in the upper part of the matrices, the calcium deposited as small particles along with the collagen fibers.

21.6 Process of In Vitro Bone Tissue Formation by BMSCs

The following mechanism is proposed for the formation of mineralized matrices by BMSCs. First, the BMSCs expand and cover the culture dish surface. Next, type I collagen fibrils, which are secreted by the BMSCs, are deposited on the culture dish by the cells. Then, the BMSCs differentiate into osteoblasts, and this process is enhanced by the deposited collagen fibers. The osteoblasts form numerous

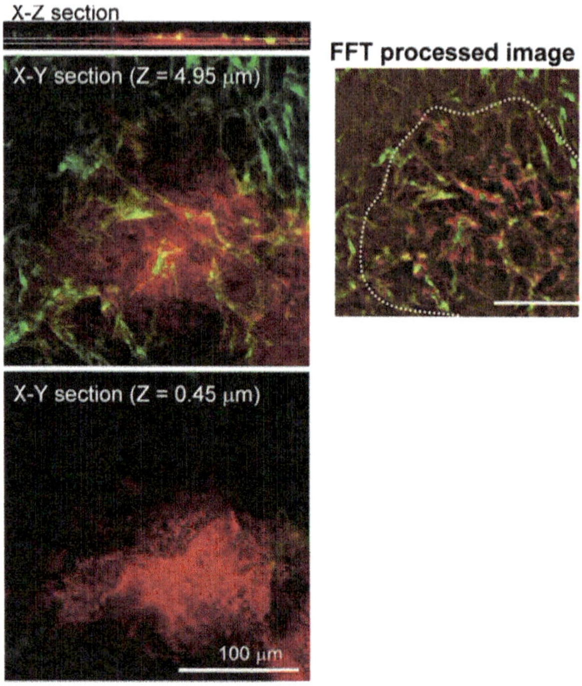

Fig. 21.6 Three-dimensional structure of mineralized matrices by BMSCs with type I collagen fibrils (*left*). The mineralized matrices were labeled with xylenol orange (*red*) and deposited type I collagen was prelabeled with FITC (*green*). The merged image of the calcium and collagen was processed with FFT filter (*right*). The region of the mineralized area is indicated by a *dotted white line*

mineralized matrices, and the initial mineralization area is independent of collagen deposition area. Afibrillar-mineralized matrices (i.e., the cement line) are constructed on the surface of the dish or culture substrate under the accumulated collagen fibrils in the osteogenic cell layer. The initial thin layer of non-collagenous mineralization functions as a template for collagenous mineralization formation followed by fully mineralized bone matrix maturation. The calcium particle aggregates surrounding the type I collagen fibrils and mineralized matrices expand throughout the type I collagen fibrils. In this matrix mineralization process, mineralized matrices lift the above osteogenic cell layer. Some osteoblasts remain in the mineralized matrices, probably in the collagenous mineralization area, and further differentiate into osteocytes.

21.7 Conclusion

The 3D structures of the in vitro-mineralized tissues varied according to the osteogenic cells used. The cultured mineralized matrices formed by rat BMSCs are comparable to normal bone tissue. This finding indicates that cultured bone tissue formed by osteogenic differentiated BMSCs can be used as a basal unit for modeling various 3D bone tissues in vitro, and the 3D imaging method reported here is a valuable tool for the evaluation of the assembled 3D bone tissues.

Acknowledgments The author thanks Ms. Yukari Furutani, Mr. Shunpei Takaoka, Mr. Taiki Uchiyama, and Ms. Chiya Umezu (The University of Kitakyushu) for their technical supports obtaining CLSM images of cultured mineralized matrices. The observations of the mineralized matrices were partly carried out with Nikon C2 confocal laser scanning microscope at the Instrumentation Center, The University of Kitakyushu. Part of this work was supported by a Grant-in-Aid for Scientific Research on Innovative Areas "Hyper Bio Assembler for 3D Cellular Innovation" (No. 2305) from the Ministry of Education, Culture, Sports, Science, and Technology of Japan.

References

1. Pittenger MF, Mackay AM, Beck SC, Jaiswal RK, Douglas R, Mosca JD, Moorman MA, Simonetti DW, Craig S, Marshak DR (1999) Multilineage potential of adult human mesenchymal stem cells. Science 284(5411):143–147
2. Beltrami AP, Cesselli D, Bergamin N, Marcon P, Rigo S, Puppato E, D'Aurizio F, Verardo R, Piazza S, Pignatelli A, Poz A, Baccarani U, Damiani D, Fanin R, Mariuzzi L, Finato N, Masolini P, Burelli S, Belluzzi O, Schneider C, Beltrami CA (2007) Multipotent cells can be generated in vitro from several adult human organs (heart, liver, and bone marrow). Blood 110(9):3438–3446
3. De Coppi P, Bartsch G Jr, Siddiqui MM, Xu T, Santos CC, Perin L, Mostoslavsky G, Serre AC, Snyder EY, Yoo JJ, Furth ME, Soker S, Atala A (2007) Isolation of amniotic stem cell lines with potential for therapy. Nat Biotechnol 25(1):100–106
4. Hu Y, Liao L, Wang Q, Ma L, Ma G, Jiang X, Zhao RC (2003) Isolation and identification of mesenchymal stem cells from human fetal pancreas. J Lab Clin Med 141(5):342–349
5. In 'tAPS, Scherjon SA, Kleijburg-van der Keur C, de Groot-Swings GM, Claas FH, Fibbe WE, Kanhai HH (2004) Isolation of mesenchymal stem cells of fetal or maternal origin from human placenta. Stem Cells 22(7):1338–1345
6. Liu TM, Martina M, Hutmacher DW, Hui JH, Lee EH, Lim B (2007) Identification of common pathways mediating differentiation of bone marrow- and adipose tissue-derived human mesenchymal stem cells into three mesenchymal lineages. Stem Cells 25(3):750–760
7. Kuroda Y, Wakao S, Kitada M, Murakami T, Nojima M, Dezawa M (2013) Isolation, culture and evaluation of multilineage-differentiating stress-enduring (Muse) cells. Nat Protoc 8(7):1391–1415. doi:10.1038/ nprot.2013.076
8. Wakao S, Akashi H, Kushida Y, Dezawa M (2014) Muse cells, newly found non-tumorigenic pluripotent stem cells, reside in human mesenchymal tissues. Pathol Int 64(1):1–9
9. Ohgushi H, Caplan AI (1999) Stem cell technology and bioceramics: from cell to gene engineering. J Biomed Mater Res 48(6):913–927
10. Ohgushi H, Goldberg VM, Caplan AI (1989) Heterotopic osteogenesis in porous ceramics induced by marrow cells. J Orthop Res 7(4):568–578
11. Maniatopoulos C, Sodek J, Melcher AH (1988) Bone formation in vitro by stromal cells obtained from bone marrow of young adult rats. Cell Tissue Res 254(2):317–330
12. Ohgushi H, Dohi Y, Katuda T, Tamai S, Tabata S, Suwa Y (1996) In vitro bone formation by rat marrow cell culture. J Biomed Mater Res 32(3):333–340
13. Lian JB, Stein GS (1992) Concepts of osteoblast growth and differentiation: basis for modulation of bone cell development and tissue formation. Crit Rev Oral Biol Med 3(3):269–305
14. Kihara T, Oshima A, Hirose M, Ohgushi H (2004) Three-dimensional visualization analysis of in vitro cultured bone fabricated by rat marrow mesenchymal stem cells. Biochem Biophys Res Commun 316(3):943–948
15. Kihara T, Hirose M, Oshima A, Ohgushi H (2006) Exogenous type I collagen facilitates osteogenic differentiation and acts as a substrate for mineralization of rat marrow mesenchymal stem cells in vitro. Biochem Biophys Res Commun 341(4):1029–1035

16. Sonobe H, Mizobuchi H, Manabe Y, Furihata M, Iwata J, Hikita T, Oka T, Ohtsuki Y, Goto T (1991) Morphological characterization of a newly established human osteosarcoma cell line, HS-Os-1, revealing its distinct osteoblastic nature. Virchows Arch B Cell Pathol Incl Mol Pathol 60(3):181–187
17. Al-Jallad HF, Nakano Y, Chen JL, McMillan E, Lefebvre C, Kaartinen MT (2006) Transglutaminase activity regulates osteoblast differentiation and matrix mineralization in MC3T3-E1 osteoblast cultures. Matrix Biol 25(3):135–148
18. Uchimura E, Machida H, Kotobuki N, Kihara T, Kitamura S, Ikeuchi M, Hirose M, Miyake J, Ohgushi H (2003) In-situ visualization and quantification of mineralization of cultured osteogenetic cells. Calcif Tissue Int 73(6):575–583
19. Kamioka H, Honjo T, Takano-Yamamoto T (2001) A three-dimensional distribution of osteocyte processes revealed by the combination of confocal laser scanning microscopy and differential interference contrast microscopy. Bone 28(2):145–149
20. Cane V, Marotti G, Volpi G, Zaffe D, Palazzini S, Remaggi F, Muglia MA (1982) Size and density of osteocyte lacunae in different regions of long bones. Calcif Tissue Int 34(6):558–563
21. Andrianarivo AG, Robinson JA, Mann KG, Tracy RP (1992) Growth on type I collagen promotes expression of the osteoblastic phenotype in human osteosarcoma MG-63 cells. J Cell Physiol 153(2):256–265
22. Lynch MP, Stein JL, Stein GS, Lian JB (1995) The influence of type I collagen on the development and maintenance of the osteoblast phenotype in primary and passaged rat calvarial osteoblasts: modification of expression of genes supporting cell growth, adhesion, and extracellular matrix mineralization. Exp Cell Res 216(1):35–45
23. Mizuno M, Fujisawa R, Kuboki Y (2000) Type I collagen-induced osteoblastic differentiation of bone-marrow cells mediated by collagen-alpha2beta1 integrin interaction. J Cell Physiol 184(2):207–213
24. Hoshi K, Kemmotsu S, Takeuchi Y, Amizuka N, Ozawa H (1999) The primary calcification in bones follows removal of decorin and fusion of collagen fibrils. J Bone Miner Res 14(2):273–280
25. Hoshi K, Ejiri S, Ozawa H (2001) Ultrastructural analysis of bone calcification by using energy-filtering transmission electron microscopy. Ital J Anat Embryol 106(2 Suppl 1):141–150

Chapter 22
Sociocytology Illuminated by Reconstructing Functional Tissue with Cell Sheet Based Technology

Kazuhiro Fukumori, Hironobu Takahashi, Jun Kobayashi, Masamichi Nakayama, Yoshikatsu Akiyama and Masayuki Yamato

Abstract To create cell-dense functional tissues, the authors have developed a new class of tissue reconstructing technology, called "cell-sheet engineering". In contrast to biodegradable scaffold-based three-dimensional (3D) tissue engineering, cell sheet-based bioassembler technology can reconstruct cell-dense tissues with extracellular matrix (ECM). For further mimicking living tissues and organs such as the liver and myotube, novel 3D bioassembler technologies are required for controlling the micro-scaled alignment of cells. This chapter summarizes the reconstruction of functional tissues with cell-sheet based technologies including (1) functional liver tissues using co-culture system of hepatocyte/nonparenchymal cells, (2) micropatterned co-culture by microcontact printing systems with fine alignment, and (3) multi-layered oriented myotube tissues using anisotropic cell sheets and gelatin gel coated devices. Eventually, the fabrications of functional tissues and organs with bioassembler technology would lead to illuminating "sociocytology".

Keywords Sociocytology · Cell sheet · Tissue engineering · Micropatterning · Micro contact printing · Liver · Myotube

22.1 Introduction

In the living body, cells coordinately assemble tissues such as nerve, muscle, epithelium, and connective tissues. Furthermore, the functional integration of tissues leads to the formation of organs such as heart, liver, and kidney. For exhibiting the mechanical robustness and flexibility of the human body, cells bind to neighboring cells through cell-cell junction and adhere to extracellular matrix (ECM). In this viewpoint, a methodology for reconstructing three dimensional (3D) tissues by cultivating cells on biodegradable scaffolds *in vitro*, called "tissue engineering", is a

M. Yamato (✉) · K. Fukumori · H. Takahashi · J. Kobayashi · M. Nakayama · Y. Akiyama
Institute of Advanced Biomedical Engineering and Science, Tokyo Women's Medical University, 8-1 Kawada-cho, 162-8666 Shinjuku, Tokyo, Japan
e-mail: yamato.masayuki@twmu.ac.jp

T. Arai et al. (eds.), *Hyper Bio Assembler for 3D Cellular Systems*,
DOI 10.1007/978-4-431-55297-0_22

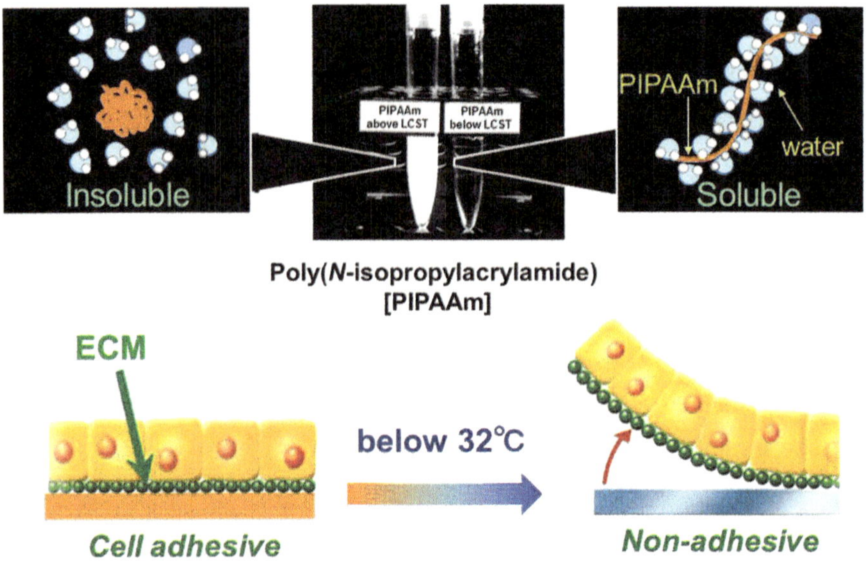

Fig. 22.1 Temperature-responsive cell culture surfaces. By simple lowing temperature, various types of cells are harvested as contiguous cell-sheet along with deposit extracellular matrix (*ECM*) from PIPAAm grafted cell culture surfaces

promising approach, because the biodegradable scaffolds serve as temporal matrix and facilitate cell growth and tissue organization [1]. Some successes are achieved for cell-sparse tissues such as heart valve [2], bone [3], and cartilage [4], as well as the reconstruct of human ear lobe on the back of a nude mouse [5]. However, the reconstruction of cell-dense and complicated tissue structures using the biodegradable scaffolds is fundamentally difficult, because the control of degradation rate of scaffolds is essential for replacing them with cells and ECMs. Although the degradation of scaffolds theoretically allows the proliferating cells to migrate into the spaces previously occupied by the degraded scaffolds, the large amounts of ECM components are often present within the scaffold structures.

For reconstructing cell-dense 3D tissues, cell sheet-based bioassembler technology using temperature-responsive culture dishes is a promising approach (Fig. 22.1) [6]. To create the temperature-responsive culture surfaces, temperature-responsive poly(*N*-isopropylacrylamide) (PIPAAm) is covalently modified onto commercially available tissue-culture-polystyrene (TCPS) surfaces [7, 8]. PIPAAm-grafted surfaces exhibit the reversible alteration of hydrophilic and hydrophobic surface properties in response to temperature changes across PIPAAm's lower critical solution temperature (LCST) at 32 °C [9]. In culture conditions at 37 °C, the surface is slightly hydrophobic, allowing cells to adhere and grow similarly to normal TCPS dishes. After reaching to confluence, temperature is lowered to 20 °C, allowing PIPAAm to become soluble in water. Under this condition, grafted PIPAAm swells and hydrates at PIPAAm-cell interface, and the cells are detached spontaneously, enabling them to be harvested as a contiguous and intact cell sheets, which contains

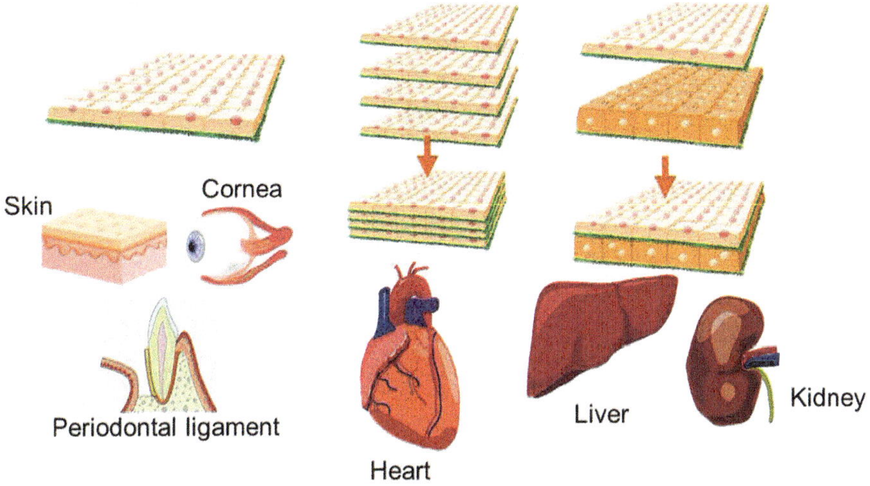

Fig. 22.2 Concept of cell sheet engineering. A single cell-sheet is applicable for skin, cornea, and periodontal ligament reconstruction, and multi-layered cell sheets used to reconstruct homogeneous or heterotypic three-dimensional tissues, such as myocardium, liver, and kidney. (Reprinted with permission from Ref. [14]. Copyright (2005) Elsevier.)

ECM deposited on its basal surface [10]. The deposited ECM component allows the cell sheets to adhere other surfaces including culture dish, other cell sheets, or host tissue. Therefore, the reconstruction of 3D tissues can be achieved by layering cell sheets (Fig. 22.2) [11]. For example, cardiac tissue is fabricated by layering cardiac cell sheets via the formation of gap junctions, leading to spontaneous, synchronized beating of the layered cardiac tissues [12]. In addition, an *in vitro* vascularized 3D cardiac tissue is formed on a perfusion bioreactor having collagen-based micro-channels. This method provides the connection of microfluidic channels with the vascularized cardiac tissue-like layers, which exceed 110 μm in thickness, well-beyond the diffusive transport limits for cells [13].

For further mimicking living tissues and organs, it is necessary to create a 3D tissue architecture with the heterotypic cells. In large organs such as heart, liver, and kidney, "sociocytology" is formed by interactions among various cell types for preserving differentiated cell functions. The connection of blood capillary vessels and nerve to the organs is required for cellular survival and functions, as well as the requirement of various infrastructures such as water supply and electrical cable for the operation of human society. Furthermore, tissues and organs possess their own unique structures such as multi-cellular, micrometer-scale organizations in the liver and kidney, aligned myocardium in heart. Therefore, the creation of highly functional tissues and organs is considered to be achieved by 3D bioassembler technologies that can be controlled micro-scaled alignment. In this chapter, the authors focus on the reconstruction of liver and myotube tissues using cell sheet-based bio-assembler technology. Eventually, the fabrication of mimicked tissues and organs would lead to illuminating "sociocytology".

22.2 Creation of Functinal Liver Tissues

22.2.1 Key Factors for Creating Functional Liver Tissues

Healthy liver is anatomically divided into liver lobules, minimal functional units (Fig. 22.3), and exhibits metabolic functions including detoxification and synthesis, the production of biochemicals for digestion, and the storage of nutrient in the liver. For reconstructing the functions of liver to treat various liver disease [15] and to investigate *in vitro* biological studies and drug screening [16], freshly isolated primary hepatocytes (HCs) are used. Once HCs are isolated from the native liver, however, these cells rapidly lose their viability and hepatic functions such as albumin secretion [17]. During the process of isolating HCs, the microscopic structure of liver lobule such as hepatic sinusoid (Fig. 22.3 *insert*) becomes impaired. Therefore, the reduction of the hepatic functions in the isolated HCs is considered to be caused by the loss of interaction with (1) neighboring non-HCs such as endothelial cells (ECs), (2) ECMs in space of Disse that is located between hepatic lamina and ECs, and (3) oxygen-rich blood from hepatic artery and nutrient-rich blood from portal vein. In other words, these factors mentioned above would be essential for creating functional liver tissues from the HCs.

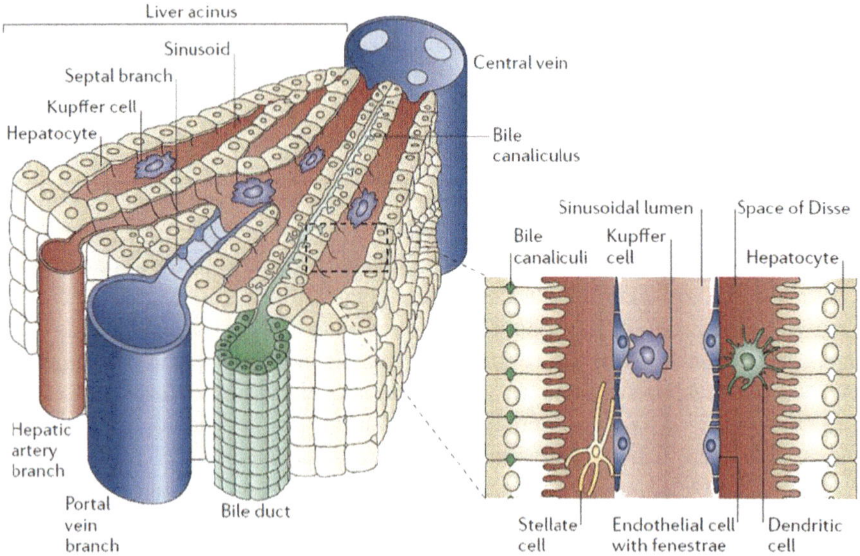

Fig. 22.3 Structure of liver microscopic structure of liver lobule, which is a minimal functional unit in the liver. (Reprinted with permission from Ref. [32]. Copyright (2006) Nature Publishing Group.)

22.2.2 Creation of Functional Liver Tissues by Cell Sheet-Based Bioassembler Technologies

The reconstitute of hepatic microstructure is an important issue for creating functional liver tissues. For improving *in vitro* hepatic function, several cell culture methods have been previously developed by using ECM and cell-cell communications, which resemble the surrounding environments of *in vivo* native liver tissues. *In vitro* phenotypic functions of HCs is enhanced on Matrigel [18] and within collagen hydrogel sandwiches [19]. Random co-culture of HCs with nonparenchymal cells such as fibroblasts and endothelial cells also improves the maintenance of hepatic functions [20]. Molecular machinery of hepatocyte-nonparenchymal cell interactions has never been fully understood. To date, the secretion of a proteoglycan, decorin [21, 22], soluble growth factors such as heparin growth factor (HGF) [23], and cell-cell contacts [24] have a pivotal role to maintain the hepatic functions. However, the creation of functional liver tissue is unable to be achieved only by mixing cells and ECMs. Spatial alignment of both cells and ECMs should be defined to resemble the microscopic structure of liver lobule in normal liver.

To precisely mimic the microenvironments of native tissues and organs, bioassembler technologies with multiple cells and ECMs are essential. A 2D micropatterning with hepatocyte/nonparenchymal cells can been achieved by using photolithography [25] and soft lithography [26]. 3D cellular micropatterning has also been investigated using several different cellular printing systems [27, 28]. However, the reconstruction of complicated tissues and organs remains a challenging issue, mainly due to a lack of transplantable 3D tissues with microscopically arranged, high-density cells.

An approach based on cell sheet-based bioassembler technology facilitates the fabrication of cell-dense 3D tissues, because 2D aligned cell sheets are transplantable and hierarchically stacked in a layered structure. Micropatterned temperature-responsive surfaces showing a dual phase-transition temperature enable both the culture and the recovery of patterned heterotypic cell sheets [29, 30]. Two kinds of temperature-responsive polymers showing different LCSTs, PIPAAm and *n*-butyl methacrylate (BMA)-co-grafted PIPAAm, are micropatterned using electron beam (EB)-generated polymerization of BMA onto PIPAAm-grafted surfaces through stainless stencil. Step-by-step cell seeding on the micropatterned surface showing a hydrophilic/hydrophobic property alteration allows the co-culture of rat primary HC and bovine carotid endothelial cell (EC) to be performed. First, HCs selectively adhere to the hydrophobized poly(IPAAm-BMA) co-grafted domains at 27 °C, whereas they are unable to adhere to the hydrophilic PIPAAm domains (Fig. 22.4a). When temperature is increased to 37 °C, the secondary seeding of ECs exclusively adhere to hydrophobized PIPAAm domains, resulting in micropatterned co-culture with HC (Fig. 22.4b). Hepatic functions, albumin secretion and urea synthesis, are up-regulated on micropatterned co-culture compared with those on monoculture

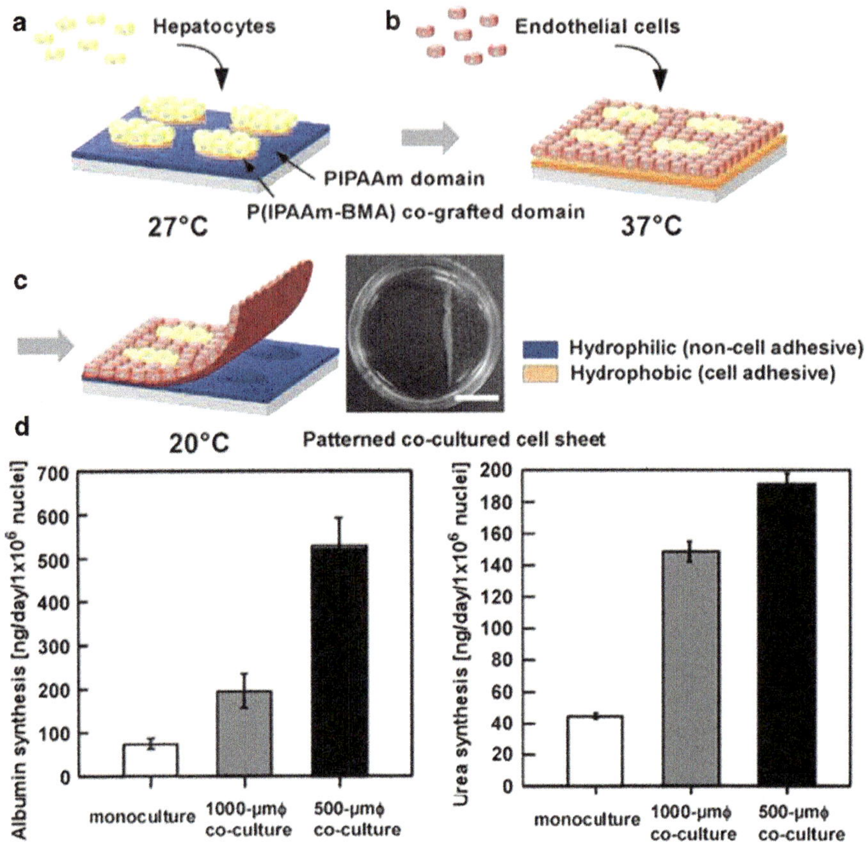

Fig. 22.4 Two-dimensional co-culture using dual temperature-responsive surfaces. **a** Hepatocytes (*HCs*) are selectively adhered to hydrophobized poly(IPAAm-BMA) co-grafted domains at 27 °C. **b** After increasing temperature to 37 °C, endothelial cells (*ECs*) are adhered to PIPAAm domains. **c** Co-culured HC/EC sheet is harvested from the surface with lowering temperature to 20 °C. **d** Albumin and urea synthesis are up-regulated on HC-EC co-culture compared with those on monoculture. (Reprinted with permission from Ref. [30]. Copyright (2006) Elsevier.)

(Fig. 22.4d) [30]. Eventually, co-cultured, micropatterned cell sheet is detached due to the hydration of both of the polymer-grafted domains when temperature is decreased to 20 °C (Fig. 22.4c) [29].

In addition to 2D alignment of multiple cells, stacking nonparenchymal/hepatic cell sheets produces 3D co-culture system [17, 31]. Cell sheet-based bioassembler technology enables a monolayer of ECs to be stratified onto confluent HCs, leading to the localization of EC sheet-derived ECMs between the layers of confluent HCs and ECs. Albumin secretion and urea synthesis are highly maintained at least during 4 weeks in double-layered HC-EC sheets, while HC monoculture exhibits

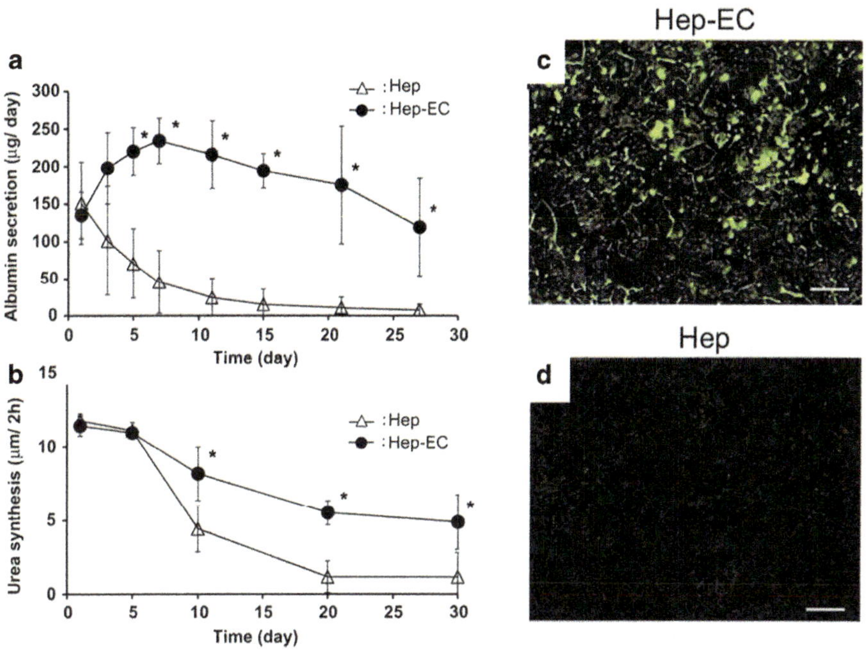

Fig. 22.5 Three-dimensional co-culture by layering nonparenchymal/hepatic cell sheets. Albumin secretion (**a**) and urea synthesis (**b**) are highly maintained on double-layered HC-EC sheets (*closed circles*) compared with those on HC monoculture (*open triangles*). Furthermore, the uptake of a fluorescent bile acid analogue, cholyl-lysyl-fluorescein, through the bile canaliculi is enhanced double-layered HC-EC sheets (**c**) cultured for 6 days, while little uptake is observed on HC monoculture (**d**). (Reprinted with permission from Ref. [17]. Copyright (2012) Elsevier.)

remarkable reductions in albumin secretion and urea synthesis (Fig. 22.5a, b). The expression of HC-specific genes is significantly higher in double-layered HC-EC sheets in comparison with HC monoculture after 2 weeks. Furthermore, the networks of bile canaliculi are created between HCs within double-layered HC-EC sheets after a 6-day cultivation (Fig. 22.5c, d). Barrier function of created bile canaliculi is maintained through tight junctions between HCs, indicating the formation of functional bile canaliculi. In the healthy liver, bile canaliculi are sealed with tight junctions to hinder bile leakage into surrounding space and blood circulation. Therefore, double-layered 3D co-cultured tissues by overlaying EC sheet on HC sheet possess a similarity to native liver tissues; hepatic-plate resembled microstructure with healthy functional bile canaliculi. These functional reconstructions are presumably considered to be attributed to mimicking a hepatic perisinusoidal space-like structure between HC and EC sheets with ECMs. Eventually, cell sheet-based bioassembler technologies facilitate creating tissue-mimicking multi-layer materials by layering the cell sheets.

22.3 Fabrication of Transplantable Micro-Patterned Thick Tissues on a Large Scale

22.3.1 Micro-Patterned Cell Sheet Fabricated with Micro Contact Printing (μCP)

As mentioned above, cell sheet-based bioassembler technology is considered to be a promising way for creating functional tissues. For constructing whole organs, however, some challenging issues still remain. First, the alignment of multiple cells is required with a position accuracy of a few micrometers. Co-culture of HC and EC on the dual temperature-responsive surface prepared by EB polymerization provides the cell sheet with a patterning size of more than 500 μm. Furthermore, micropattered cell sheets are required for each cellular layer in the organs, indicating the requirement of micropatterned surfaces for each layer. In the case of liver reconstruction, numbers of cell sheets containing hepatic lobules as a unit of liver tissue are required.

One of solutions for these issues is a protein patterning by micro contact printing (μCP) method, and the spontaneous organization of cells by the molecular interaction between cells and the protein [33]. Mircopattern substrates with cell-adhesive and non-adhesive regions using μCP prevent cells from growing to confluence [34]. To create a micropatterned and continuous cell sheet, PIPAAm-grafted tissue culture polystyrene dishes are printed with fibronectin (FN) using a patterned polydimethylsiloxane (PDMS) stamp. Vascular smooth muscle cells (VSMCs) are selectively attached onto the printed FN region on PIPAAm substrates in serum-free media. After the attachment and spreading of the cells, serum is added to promote the growth of adhered VSMCs toward non-FN-printed regions, resulting in the formation of aligned sheet within a few days (Fig. 22.6) [35]. In addition, after step-by-step cell

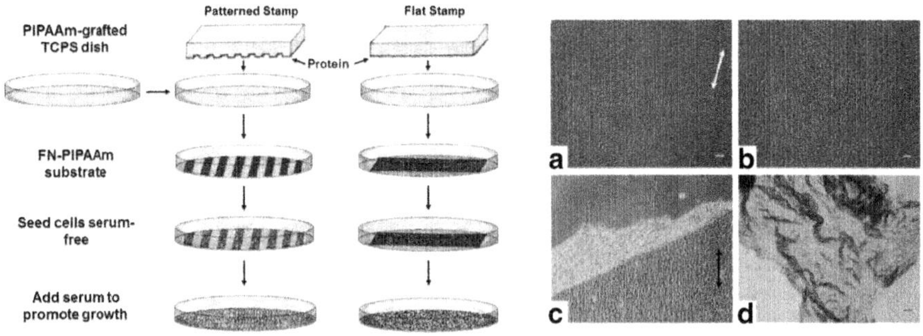

Fig. 22.6 Preparation of aligned vascular smooth muscle cell (*VSMC*) sheet with microcontact printing. Tissue culture polystyrene (*TCPS*) dishes modified with PIPAAm are printed with fibronectin (*FN*) using a micro patterned or a flat PDMS stamp. FN-PIPAAm substrate consists of FN stripes or a uniform FN region. At 37 °C, patterned (**a**) and non-patterned (**b**) cell sheets are fully attached to FN-PIPAAm substrates. At 20 °C, both patterned (**c**) and non-patterned (**d**) sheets are in the process of detaching. Double-headed arrows indicate direction of alignment. Scale bars: 100 μm. (Reprinted with permission from Ref. [35]. Copyright (2008) John Wiley and Sons.)

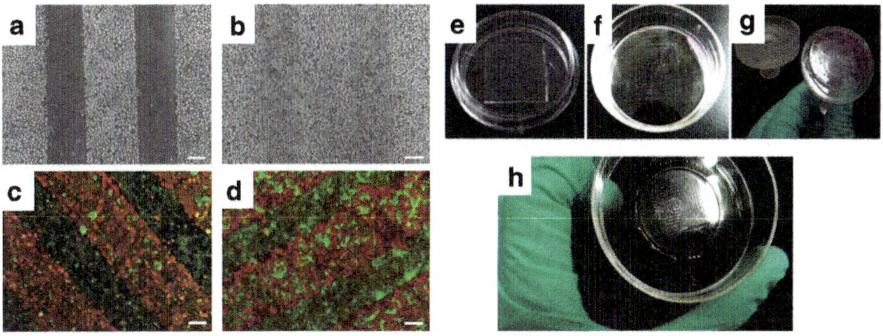

Fig. 22.7 Photographs of micro-patterned-co-culture of hepatocytes (*HCs*) and endothelial cells (*ECs*). **a** Phase contrast micrographs of HCs on fibronectin patterns in a serum free condition. **b** Phase contrast microscopic images of co-cultured HCs and ECs. **c** Fluorescent images of co-cultured HCs and ECs before cell sheet harvest. HCsare fluorescently labeled with AlexaFluor 594 (*red*) before seeding. ECs are labeled with AlexaFluor 488-conjugated isolectin GS-IB4 (*green*). **d** Fluorescent images of co-cultured HCs and ECs after cell-sheet transfer. Scale bars: 200 μm. Macroscopic views. **e** PDMS stamps. **f** Micropatterned HCs. **g** Harvest of micro-patterned cell sheets with gelatin-coated stamps. **h** Cell sheets transferred onto larger dishes. (Reprinted with permission from Ref. [36]. Copyright (2009) Elsevier.)

seeding using μCP method, HCs and ECs are co-cultured in a micro-patterned location (Fig. 22.7a, b). Double fluorescent staining with anti-albumin antibody (red fluorescent) specific for HCs and isolectin GS-IB4 (green fluorescent) specific for endothelial cells also reveals a cell micro-patterning with a few tens of micrometer (Fig. 22.7c) [36]. All of the cells are harvested as a single micro-patterned cell sheet by lowing temperature and transferred onto a new dish without the deformation of the patterns using a gelatin-coated cell-sheet manipulator (Fig. 22.7d, h). Therefore, micropattered, co-cultured cell sheets are obtained by μCP bioassembler technology with more accurate alignment than those by EB-induced polymerization. In addition, μCP method is relatively less time-consuming, because micropattered surfaces are obtained by using uniformly grafted PIPAAm cell culture surface and PDMS stamps without time-consuming polymerization process.

22.3.2 Micro-Structured Cell Sheets Fabricated with a Reproducible μCP System with Fine Alignment

In general, the procedure of μCP is manually performed as follows; applying ECM to the surface of PDMS stamp with a going-to-be-printed pattern, making the complete contact between the convex area of PDMS stamp and cell culture substrate with an appropriate contact force, and detaching PDMS stamp from the substrate after the contact. In the manual procedure, the reproducibility and quality of printed pattern become low, especially in the position accuracy, because the reproducibility and quality are strongly depends on the skill of technician performing μCP. When two or more different types of ECMs are printed to a designated position on the

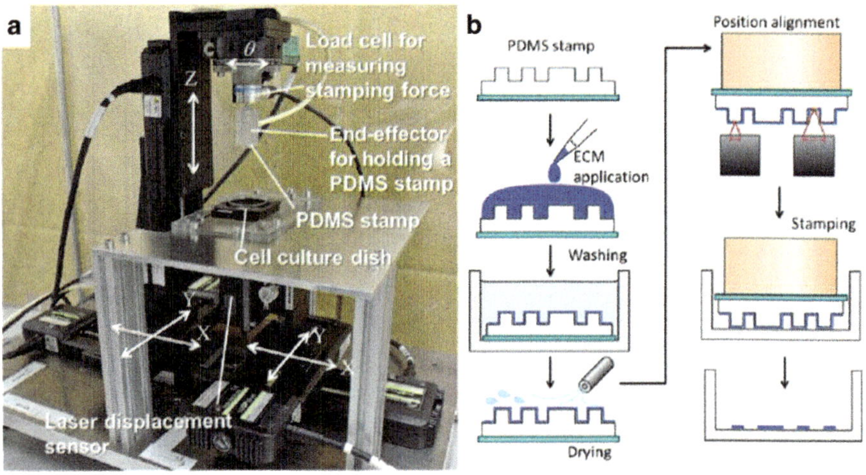

Fig. 22.8 Scheme of multiple micro contact printing with a fine alignment. **a** Fine alignment system consisting of x-y-z-θ stage, a laser sensor fixed on x-y stage, and a load cell. **b** Schematic illustrations show the simple procedure of multiple microcontact printing. (Reprinted with permission from Ref. [38]. Copyright (2013) IEEE.)

same dish, the manual procedure is quite difficult to give the high accuracy of position, indicating that a fine alignment system is strongly demanded. To solve this problem, a new bioassembler technology for a fine alignment of μCP is developed (Fig. 22.8) [37–39]. The procedure for aligning a PDMS stamp to the designated position is performed by following three steps. First, a cell culture dish is fixed on the position of the alignment system with a holder. Second, some markers for alignment on the PDMS stamp fixed on the tip of x-y-z-θ stage are detected with the sensor of stamp position and posture consisting laser displacement sensor and another x-y stage with scanning motion. After the error correction of PDMS stamp, the control of z-axis with the load cell gives a contact between the PDMS stamp and the surface of dish with an appropriate contact force. Two-cell-patterned co-culture is performed in a cell culture dish patterned with FN by the stamp stiffness calibrated printing as a manually procedure (Fig. 22.9). In addition, three different FN patterns (Alexa Fluor 350, 488, and 555 conjugated FN) are printed on the same dish (Fig. 22.10). These patterns are found to be identical to those of PDMS stamps and give the uniform intensity of fluorescent in position. This result indicates that FN printing is uniformly performed. As the result of measurement of pattern position, the position error is within 50 μm. Finally, as the demonstration of complex pattern, stamps with the combination of three different patterns are used. The merged image of each pattern is reconstructed into an original logotype (Fig. 22.11). This method is applicable for not only simple but also complex patterns and will lead to the development of rapid bioassembler technology by combining an automatic system such as robotics. Furthermore, this method is also useful for generating a multi-type-cellular tissue similar to a complex tissue formation and regeneration.

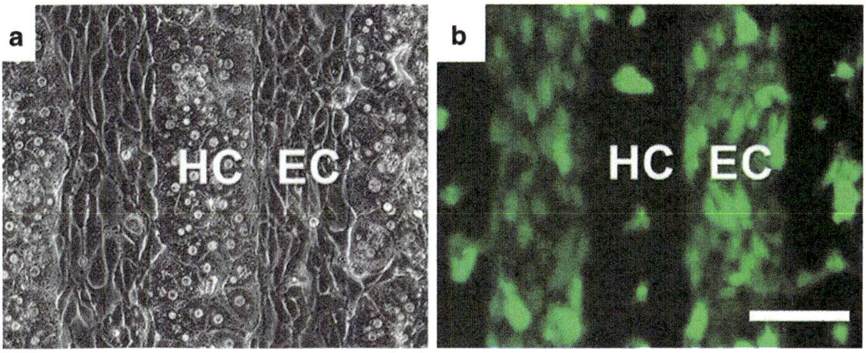

Fig. 22.9 Microscopic images of micro-patterned-co-culture of hepatocytes (*HC*) and endothelial cells (*EC*). **a** Phase contrast micrographs of co-cultured hepatocytes and endothelial cells. **b** Fluorescent micrographs of co-cultured hepatocytes and endothelial cells before cell sheet harvest. Endothelial cells were fluorescently labeled with cell tracker (*green*) before seeding. Scale bar: 100 μm

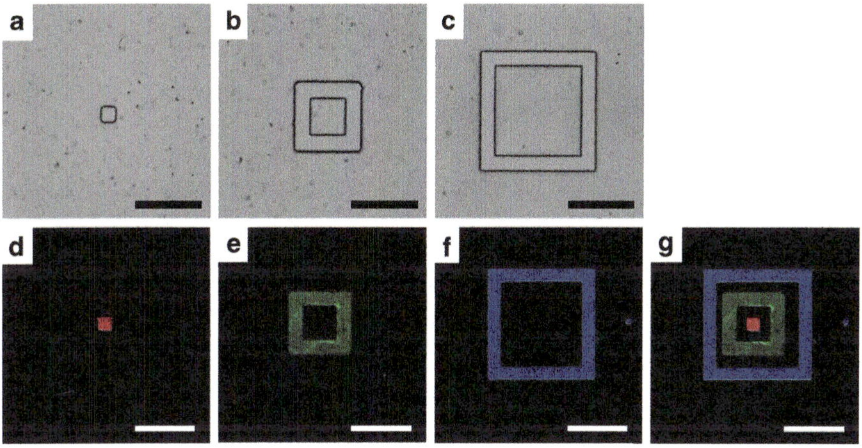

Fig. 22.10 Multiple micro contact printing with a fine alignment. **a**–**c** Phase contrast photographs of PDMS stamp. **d**–**f** Fluorescent photographs show various fibronectin printed patterns (labeled with Alexa 350, 488, 555, respectively). **g** Fluorescent photograph shows a combined pattern with differently labelled fibronectin. Scale bars: 500 μm. (Reprinted with permission from Ref. [38]. Copyright (2013) IEEE.)

Liver lobules contain not only parenchymal hepatocytes, but also capillary vessels and bile ducts, which provide blood and bile, respectively (Fig. 22.3). To our knowledge, no reports on the connection of both engineered capillary vessels and bile ducts in the liver tissues with host blood vessels and bile ducts. Therefore, further investigations of a new bioassembler technology to provide complex structure of liver lobules are desired.

Fig. 22.11 The merged fluorescent photograph of the original logotype of "Hyper Bio Assembler"

22.4 Creation of Aligned Myotube Tissues by Using Anisotropic Cell Sheets

Some parts of native tissues have well-organized cell/ECM orientations, which allow the tissue to have a tissue anisotropy producing important functional consequences (e.g., in skeletal muscle tissues and myocardial tissues) [40–43]. Therefore, the design of cell alignment is necessary to construct functional tissues that have the natively organized orientations of cells and ECM proteins. From this viewpoint, numerous microfabricated surfaces have been developed for regulating cell orientation [44–46]. For example, microgrooved PDMS substrates are widely used to control cell alignment. However, aligned cells are basically impossible to separate from these surfaces. Cell sheet-based bioassembler technology has a potential to achieve 3D tissue construction via cell sheet layering technique [47, 48]. The use of cell sheets, therefore, allows us to design anisotropy three-dimensionally by controlling cell alignment in cell sheets.

A well-aligned cell orientation is a key factor for producing mechanical functions in native skeletal muscle. In the development of mature skeletal muscle, muscle precursor cells, myoblasts, differentiate into myotubes, which orient themselves parallel to each other [49]. Therefore, the well-aligned orientation of myoblasts and myotubes is important to construct biomimetically muscle tissue. A micropatterned temperature-responsive surface has been developed to produce "anisotropic cell sheets" consisting of aligned cells. Hydrophilic poly(*N*-acryloylmorpholine) (PAcMo) is grafted spatio-selectively on a temperature-responsive PIPAAm surface through a conventional photolithographic process, resulting in a micropattered surface having block polymer, PIPAAm-*b*-PAcMo brush domains and PIPAAm brush domains (50/50 μm stripes) (Fig. 22.12) [50, 51]. Human skeletal muscle myoblasts, which can recognize a difference in cell-to-surface affinity between the two regions, are aligned in a parallel direction with the stripe patterns by one-pot cell seeding (Fig. 22.13) [52]. The aligned myoblasts can reach confluence with main-

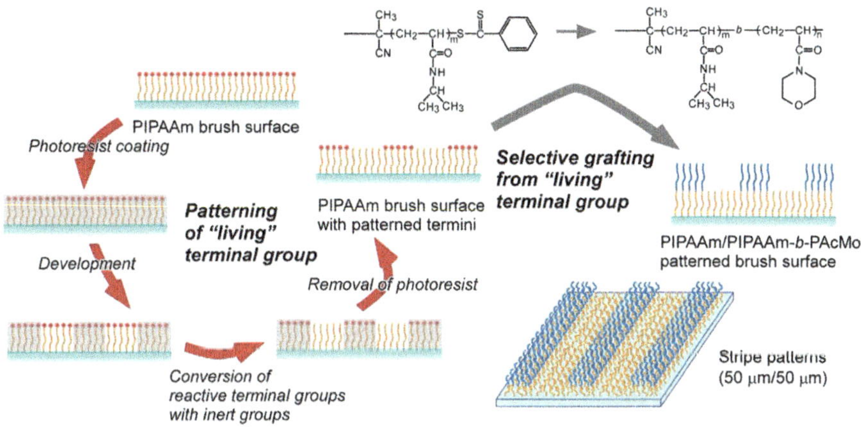

Fig. 22.12 Fabrication of patterned temperature-responsive surface through photolithographic process. A hydrophilic poly(*N*-acryloylmorpholine) (PAcMo) is selectively grafted from temperature-responsive PIPAAm brushes. *Red* termini indicate chain-transfer active ("living") terminal groups. PAcMo segments shown as the *blue* brushes are grafted only on the areas where the "living" terminal group remains. (Reprinted with permission from Ref. [50]. Copyright (2011) American Chemical Society.)

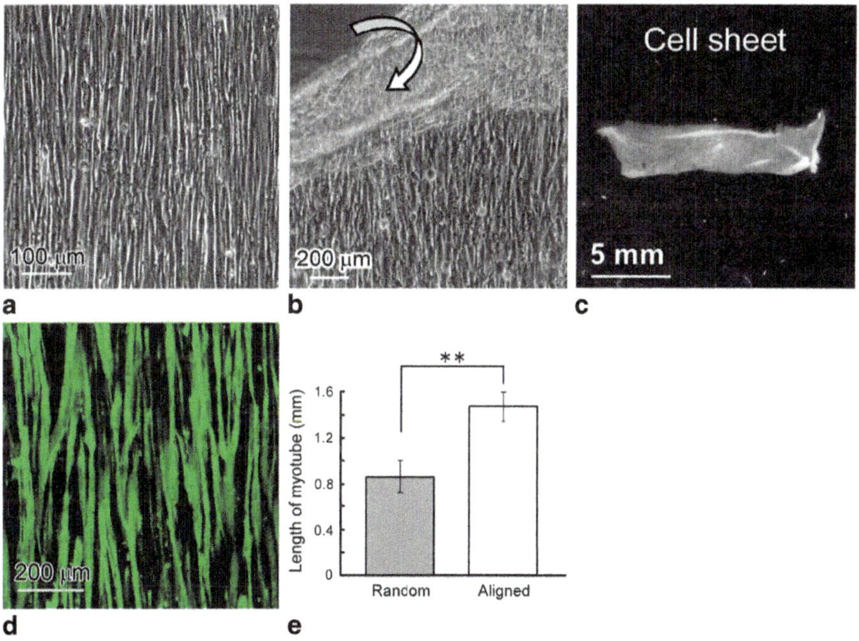

Fig. 22.13 Anisotropic properties of a cell sheet prepared using the patterned temperature-responsive surface. **a** Myoblasts are aligned on the surface, and **b** are harvested as a cell sheet by reducing culture temperature (20 °C). **c** The anisotropic cell sheet shows a unique shrinking that originated from the cell alignment (the original size of adherent cell sheet: 20 × 20 mm). **d** Aligned myoblasts differentiate into myotubes by incubation in differentiation media for 5 days. The image shows myosin heavy chain-positive myotubes. **e** Comparison of the length of myotubes in cell sheets composed of randomly oriented myoblasts (*Random*) and aligned myoblasts (*Aligned*) (**$p < 0.01$) ($n = 5$) [52]

taining the alignment and can be also harvested as an anisotropic cell sheet by reducing culture temperature (20 °C). Uniquely, whereas cell sheets shrink with keeping the original aspect of the cell monolayer by their cytoskeletal reorganization, the anisotropic cell sheet shows two distinctive shrinking rates in the vertical and parallel sides of the cell alignment. This behavior suggests that the cell sheet has a unique mechanical property. Moreover, by culturing anisotropic myoblast sheet in differentiation media (e.g., 2 % horse serum containing media), myotube constructs with aligned orientation can be produced. With maintaining the original cell alignment, the morphology of myoblasts is clearly changed microscopically, suggesting myotube formation. The control of aligned myoblasts influences not only the myotube orientation but also their length. Compared with randomly oriented myoblasts, the aligned myoblasts have a potential to forms longer myotubes. These characters indicate that this aligned myotube construct can be used as a component of 3D muscle tissue with a well-organized cell orientation.

More importantly, the transferable anisotropy of cell sheets allows us to uniquely produce 3D oriented myotube structures. In some native tissues, tissue anisotropy is organized three-dimensionally, and their oriented structures are essential for specific mechanical and biological functions. For example, native myocardial tissue consists of multiple layers of aligned cardiomyocytes that are oriented in various directions throughout the whole tissue [42, 43]. This well-organized and oriented structure is necessary for the tissue to produce a unique electrical propagation *in vivo*. To date, however, it has been still difficult to design a tissue that has different vertical cell orientations using conventional biomaterial scaffolds. Using the cell sheet layering technique, anisotropic cell sheets can be transferred onto desired sites and layered each other with keeping the controlled cell orientation. For example, when two aligned myotube constructs were layered by a gelatin gel-coated cell sheet manipulator, 3D perpendicularly orientated myotube constructs can be created and allowed to offer a potential to design various 3D oriented muscle tissues (Fig. 22.14). This strategy is simple and flexible for creating a variety of myotube architectures.

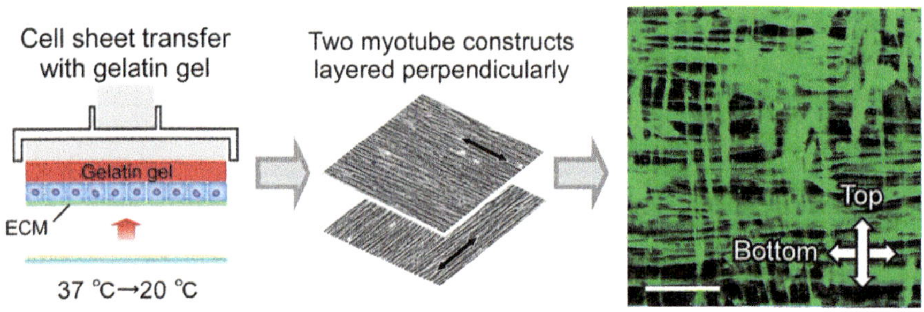

Fig. 22.14 Creation of multilayered myotube constructs through the cell sheet layering process. Aligned myoblasts are first differentiated to form myotubes, and then two myotube constructs are layered perpendicularly using a gelatin gel-coated plunger. A microscopic photographs shows a 3D oriented myotube structure as designed. Scale bar: 200 μm [52]

This cell sheet-based technology also allows us to find a unique myoblast behavior in 3D microenvironment of aligned cells and ECM. After being layered with an anisotropic myoblast sheet before differentiation to form myotube, randomly oriented myoblasts changed their orientation to be aligned with the layered anisotropic cell sheet. Consequently, myoblasts in two-layered myoblast sheets are all aligned in the same direction as the anisotropic cell sheet. As shown in Fig. 22.14, through the cell sheet layering process, myotube constructs can form an oriented structure with vertically different directions. On the other hand, when two anisotropic myoblast sheets are layered perpendicularly, myoblast orientation in the bottom sheet changed according to the anisotropy of the top sheet. This self-organization behavior can be observed only in the cell-dense scaffold-free tissue construct because scaffolds usually are unable to allow cells to contact directly each other. In multilayered cell sheets, myoblasts can recognize a cell-ECM microenvironment around them and re-oriented flexibly according to the tissue anisotropy (Fig. 22.15). Although the mechanism has never been elucidated yet, this self-organization process is probably mediated by some factors regulating cell/ECM orientation in native muscle tissues. As described above, myotubes can form oriented structures with vertical different directions through the cell sheet layering process. In addi-

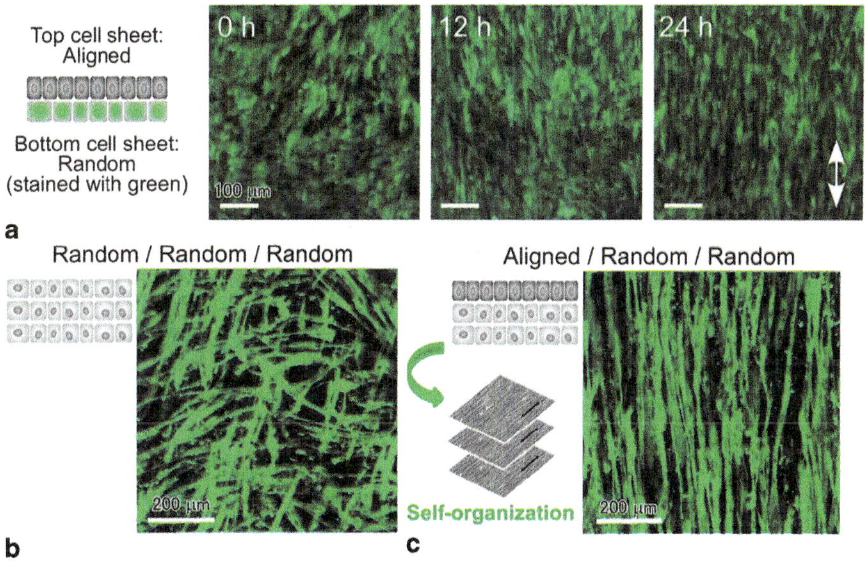

Fig. 22.15 Self-organization behavior induced by layering an anisotropic myoblast sheet in multilayered myoblast sheets. **a** A cell sheet composed of aligned myoblasts (Aligned sheet) is layered onto a fluorescently stained cell sheet composed of randomly oriented myoblasts (Random sheet). The bottom sheet is stained with CellTracker™ *Green*. **b** Three random sheets are layered together and then cultured in differentiation media for 5 days to form myotubes. **c** One aligned sheet was layered onto two Random sheet, and then both are incubated in growth media during the self-organization. After the all myoblasts were aligned, the triple-layered cell sheets were cultured in differentiation media. Myosin heavy chain-positive myotubes were stained fluorescently [52]

tion, layered anisotropic fibroblast sheets are unable to exhibit the self-organization behavior [51]. Therefore, this orientation change of myoblasts probably occurs through a biological process mimicking native myoblast behavior. Considering in term of tissue engineering technology, this self-organization behavior can be used to uniquely organize tissue anisotropy. Currently, a number of tissue engineering researches focus on potentials for preparing thick tissue construction. Whereas organizing complex tissue structures is still challenging in the research field, native tissues create their well-organized 3D structures by themselves. In the cell sheet-based tissue construction, myoblasts show their native property to organize their oriented structures. Since all myoblasts self-organize and can be aligned with the top anisotropic sheet, eventually the cells form a biomimetic bundle architecture by just one anisotropic myoblast sheet and adding multiple normal myoblast sheets. This is advantageous, because no adjusting the alignment of multiple cell sheets is necessary. Myoblasts are able to adjust their 3D orientation precisely by themselves in the tissue construct. As a result, after the self-organization has been completed, multilayered cell sheets form a aligned myotube structure with a single direction. This cell sheet-based bioassembler technology provides a new potential for quickly constructing complex tissues composed of natively-oriented cell assemblies, particularly for skeletal muscle tissue.

22.5 Conclusion

Cell sheet-based bioassembler technology has contributed to the establishment of functional 3D tissue reconstruction. In addition, novel bioassembler methods for automated microcontact printing to provide surface micropatterning, and self-organization of multilayered tissues using an oriented cell sheet have a potential to quickly engineer complicated and functional tissue structures. Eventually, the development and improvements of those research and technologies would realize the creation of functional tissues and organs, leading to illuminating “sociocytology”.

References

1. Vacanti JP, Langer R (1999) Tissue engineering: the design and fabrication of living replacement devices for surgical reconstruction and transplantation. Lancet 354:S32–S34
2. Shinoka T, Breuer CK, Tanel RE, Zund G, Miura T, Ma PX, Langer R, Vacanti JP, Mayer JE Jr (1995) Tissue engineering heart valves: valve leaflet replacement study in a lamb model. Ann Thorac Surg 60:S513–S516
3. Puelacher W, Vacanti J, Ferraro N, Schloo B, Vacanti C (1996) Femoral shaft reconstruction using tissue-engineered growth of bone. Int J Oral Maxillofac Surg 25(3):223–228
4. Vacanti CA, Langer R, Schloo B, Vacanti J (1991) Synthetic polymers seeded with chondrocytes provide a template for new cartilage formation. Plast Reconstr Surg 88(5):753–759
5. Cao Y, Vacanti JP, Paige KT, Upton J, Vacanti CA (1997) Transplantation of chondrocytes utilizing a polymer-cell construct to produce tissue-engineered cartilage in the shape of a human ear. Plast Reconstr Surg 100(2):297–302

6. Hannachi IE, Yamato M, Okano T (2009) Cell sheet technology and cell patterning for biofabrication. Biofabrication 1(2):022002
7. Yamada N, Okano T, Sakai H, Karikusa F, Sawasaki Y, Sakurai Y (1990) Thermo-responsive polymeric surfaces; control of attachment and detachment of cultured cells. Makromol Chem Rapid Commun 11(11):571–576
8. Okano T, Yamada N, Sakai H, Sakurai Y (1993) A novel recovery system for cultured cells using plasma-treated polystyrene dishes grafted with poly (N-isopropylacrylamide). J Biomed Mater Res 27(10):1243–1251
9. Heskins M, Guillet JE (1968) Solution properties of poly (N-isopropylacrylamide). J Macromol Sci Chem 2(8):1441–1455
10. Kushida A, Yamato M, Konno C, Kikuchi A, Sakurai Y, Okano T (1999) Decrease in culture temperature releases monolayer endothelial cell sheets together with deposited fibronectin matrix from temperature-responsive culture surfaces. J Biomed Mater Res 45(4):355–362
11. Elloumi-Hannachi I, Yamato M, Okano T (2010) Cell sheet engineering: a unique nanotechnology for scaffold-free tissue reconstruction with clinical applications in regenerative medicine. J Intern Med 267(1):54–70
12. Haraguchi Y, Shimizu T, Yamato M, Kikuchi A, Okano T (2006) Electrical coupling of cardiomyocyte sheets occurs rapidly via functional gap junction formation. Biomaterials 27(27):4765–4774
13. Sakaguchi K, Shimizu T, Horaguchi S, Sekine H, Yamato M, Umezu M, Okano T (2013) In vitro engineering of vascularized tissue surrogates. Sci Rep 3:1316
14. Yang J, Yamato M, Kohno C, Nishimoto A, Sekine H, Fukai F, Okano T (2005) Cell sheet engineering: recreating tissues without biodegradable scaffolds. Biomaterials 26(33):6415–6422
15. Ohashi K, Yokoyama T, Yamato M, Kuge H, Kanehiro H, Tsutsumi M, Amanuma T, Iwata H, Yang J, Okano T, Nakajima Y (2007) Engineering functional two- and three-dimensional liver systems in vivo using hepatic tissue sheets. Nat Med 13(7):880–885
16. Soldatow VY, Lecluyse EL, Griffith LG, Rusyn I (2013) Models for liver toxicity testing. Toxicol Res 2(1):23–39
17. Kim K, Ohashi K, Utoh R, Kano K, Okano T (2012) Preserved liver-specific functions of hepatocytes in 3D co-culture with endothelial cell sheets. Biomaterials 33(5):1406–1413
18. Semler EJ, Ranucci CS, Moghe PV (2000) Mechanochemical manipulation of hepatocyte aggregation can selectively induce or repress liver-specific function. Biotechnol Bioeng 69(4):359–369
19. Dunn J, Yarmush M, Koebe H, Tompkins R (1989) Hepatocyte function and extracellular matrix geometry: long-term culture in a sandwich configuration. FASEB J 3(2):174–177
20. Bhandari RN, Riccalton LA, Lewis AL, Fry JR, Hammond AH, Tendler SJ, Shakesheff KM (2001) Liver tissue engineering: a role for co-culture systems in modifying hepatocyte function and viability. Tissue Eng 7(3):345–357
21. Khetani SR, Szulgit G, Del Rio JA, Barlow C, Bhatia SN (2004) Exploring interactions between rat hepatocytes and nonparenchymal cells using gene expression profiling. Hepatology 40(3):545–554
22. Ohno M, Motojima K, Okano T, Taniguchi A (2009) Maturation of the extracellular matrix and cell adhesion molecules in layered co-cultures of HepG2 and endothelial cells. J Biochem 145(5):591–597
23. Nahmias Y, Schwartz RE, Hu WS, Verfaillie CM, Odde DJ (2006) Endothelium-mediated hepatocyte recruitment in the establishment of liver-like tissue in vitro. Tissue Eng 12(6):1627–1638
24. Bhatia SN, Balis UJ, Yarmush ML, Toner M (1999) Effect of cell-cell interactions in preservation of cellular phenotype: cocultivation of hepatocytes and nonparenchymal cells. FASEB J 13(14):1883–1900
25. Bhatia SN, Yarmush ML, Toner M (1997) Controlling cell interactions by micropatterning in co-cultures: hepatocytes and 3T3 fibroblasts. J Biomed Mater Res 34(2):189–199
26. Khetani SR, Bhatia SN (2008) Microscale culture of human liver cells for drug development. Nat Biotechnol 26(1):120–126

27. Tsang VL, Bhatia SN (2004) Three-dimensional tissue fabrication. Adv Drug Deliv Rev 56(11):1635–1647
28. Calvert P (2007) Materials science. Printing cells. Science 318(5848):208–209
29. Tsuda Y, Kikuchi A, Yamato M, Nakao A, Sakurai Y, Umezu M, Okano T (2005) The use of patterned dual thermoresponsive surfaces for the collective recovery as co-cultured cell sheets. Biomaterials 26(14):1885–1893
30. Tsuda Y, Kikuchi A, Yamato M, Chen G, Okano T (2006) Heterotypic cell interactions on a dually patterned surface. Biochem Biophys Res Commun 348(3):937–944
31. Harimoto M, Yamato M, Hirose M, Takahashi C, Isoi Y, Kikuchi A, Okano T (2002) Novel approach for achieving double-layered cell sheets co-culture: overlaying endothelial cell sheets onto monolayer hepatocytes utilizing temperature-responsive culture dishes. J Biomed Mater Res 62(3):464–470
32. Adams DH, Eksteen B (2006) Aberrant homing of mucosal T cells and extra-intestinal manifestations of inflammatory bowel disease. Nat Rev Immunol 6(3):244–251
33. Qin D, Xia Y, Whitesides GM (2010) Soft lithography for micro-and nanoscale patterning. Nat Protocol 5(3):491–502
34. Kane RS, Takayama S, Ostuni E, Ingber DE, Whitesides GM (1999) Patterning proteins and cells using soft lithography. Biomaterials 20(23):2363–2376
35. Williams C, Tsuda Y, Isenberg BC, Yamato M, Shimizu T, Okano T, Wong JY (2009) Aligned cell sheets grown on thermo-responsive substrates with microcontact printed protein patterns. Adv Mater 21(21):2161–2164
36. Elloumi Hannachi I, Itoga K, Kumashiro Y, Kobayashi J, Yamato M, Okano T (2009) Fabrication of transferable micropatterned-co-cultured cell sheets with microcontact printing. Biomaterials 30(29):5427–5432
37. Tanaka N, Ota H, Fukumori K, Yamato M, Okano T (2013) Stamp-stiffness calibrated micro contact printing. In: Robotics and Automation (ICRA). IEEE International Conference on, 2013. IEEE, pp 2582–2587
38. Tanaka N, Ota H, Fukumori K, Yamato M, Okano T (2013) Multiple micro-contact printing of extra cellular matrix with fine alignment. In: Micro-NanoMechatronics and Human Science (MHS). International Symposium on, 2013. IEEE, pp 1–2
39. Tanaka N, Ota H, Fukumori F, Miyake J, Yamato M, Okano T (2014) Micro-patterned cell-sheets fabricated with stamping-force-controlled micro-contact printing. Biomaterials 35(37):9802–9810
40. Costa KD, Lee EJ, Holmes JW (2003) Creating alignment and anisotropy in engineered heart tissue: role of boundary conditions in a model three-dimensional culture system. Tissue Eng 9(4):567–577
41. Foolen J, van Donkelaar C, Nowlan N, Murphy P, Huiskes R, Ito K (2008) Collagen orientation in periosteum and perichondrium is aligned with preferential directions of tissue growth. J Orthop Res 26(9):1263–1268
42. Vunjak-Novakovic G, Tandon N, Godier A, Maidhof R, Marsano A, Martens TP, Radisic M (2009) Challenges in cardiac tissue engineering. Tissue Eng Part B Rev 16(2):169–187
43. Hooks DA, Trew ML, Caldwell BJ, Sands GB, LeGrice IJ, Smaill BH (2007) Laminar arrangement of ventricular myocytes influences electrical behavior of the heart. Circ Res 101(10):e103–e112
44. Lam MT, Sim S, Zhu X, Takayama S (2006) The effect of continuous wavy micropatterns on silicone substrates on the alignment of skeletal muscle myoblasts and myotubes. Biomaterials 27(24):4340–4347
45. Choi JS, Lee SJ, Christ GJ, Atala A, Yoo JJ (2008) The influence of electrospun aligned poly (ε-caprolactone)/collagen nanofiber meshes on the formation of self-aligned skeletal muscle myotubes. Biomaterials 29(19):2899–2906
46. Zhao Y, Zeng H, Nam J, Agarwal S (2009) Fabrication of skeletal muscle constructs by topographic activation of cell alignment. Biotechnol Bioeng 102(2):624–631

47. Sasagawa T, Shimizu T, Sekiya S, Haraguchi Y, Yamato M, Sawa Y, Okano T (2010) Design of prevascularized three-dimensional cell-dense tissues using a cell sheet stacking manipulation technology. Biomaterials 31(7):1646–1654
48. Haraguchi Y, Shimizu T, Sasagawa T, Sekine H, Sakaguchi K, Kikuchi T, Sekine W, Sekiya S, Yamato M, Umezu M (2012) Fabrication of functional three-dimensional tissues by stacking cell sheets in vitro. Nat Protocol 7(5):850–858
49. Le Grand F, Rudnicki MA (2007) Skeletal muscle satellite cells and adult myogenesis. Curr Opin Cell Biol 19(6):628–633
50. Takahashi H, Nakayama M, Itoga K, Yamato M, Okano T (2011) Micropatterned thermoresponsive polymer brush surfaces for fabricating cell sheets with well-controlled orientational structures. Biomacromolecules 12(5):1414–1418
51. Takahashi H, Nakayama M, Shimizu T, Yamato M, Okano T (2011) Anisotropic cell sheets for constructing three-dimensional tissue with well-organized cell orientation. Biomaterials 32(34):8830–8838
52. Takahashi H, Shimizu T, Nakayama M, Yamato M, Okano T (2013) The use of anisotropic cell sheets to control orientation during the self-organization of 3D muscle tissue. Biomaterials 34(30):7372–7380

Index

T. Arai et al. (eds.), *Hyper Bio Assembler for 3D Cellular Systems*,
DOI 10.1007/978-4-431-55297-0

Printed by Printforce, the Netherlands